E. S. Tansey

Archimedes

Volume 7

Archimedes

NEW STUDIES IN THE HISTORY AND PHILOSOPHY OF SCIENCE AND TECHNOLOGY

VOLUME 7

Archimedes has three fundamental goals; to further the integration of the histories of science and technology with one another: to investigate the technical, social and practical histories of specific developments in science and technology; and finally, where possible and desirable, to bring the histories of science and technology into closer contact with the philosophy of science. To these ends, each volume will have its own theme and title and will be planned by one or more members of the Advisory Board in consultation with the editor. Although the volumes have specific themes, the series itself will not be limited to one or even to a few particular areas. Its subjects include any of the sciences, ranging from biology through physics, all aspects of technology, broadly construed, as well as historically-engaged philosophy of science or technology. Taken as a whole, *Archimedes* will be of interest to historians, philosophers, and scientists, as well as to those in business and industry who seek to understand how science and industry have come to be so strongly linked.

Archimedes

Volume 7

New Studies in the History and Philosophy of Science and Technology

---

Reworking the Bench

Research Notebooks in the History of Science

edited by

FREDERIC L. HOLMES

Yale University School of Medicine, New Haven, USA

JÜRGEN RENN

and

HANS-JÖRG RHEINBERGER

Max Planck Institute for the History of Science, Berlin, Germany

KLUWER ACADEMIC PUBLISHERS

DORDRECHT / BOSTON / LONDON

Library of Congress Cataloging-in-Publication Data is available.

ISBN 1-4020-1039-7

Published by Kluwer Academic Publishers,
P.O. Box 17, 3300 AA Dordrecht, The Netherlands

Sold and distributed in North, Central and South America
by Kluwer Academic Publishers,
101 Philip Drive, Norwell, MA 02061, U.S.A.

In all other countries, sold and distributed
by Kluwer Academic Publishers, Distribution Center,
P.O. Box 322, 3300 AH Dordrecht, The Netherlands

Printed on acid-free paper

Printed and bound in Great Britain by Antony Rowe Limited

TABLE OF CONTENTS

Introduction vii

JÜRGEN RENN AND PETER DAMEROW / The Hanging Chain: A Forgotten "Discovery" Buried in Galileo's Notes on Motion 1

WILLIAM R. NEWMAN AND LAWRENCE M. PRINCIPE / The Chymical Laboratory Notebooks of George Starkey 25

ALAN E. SHAPIRO / Newton's Optical Notebooks: Public Versus Private Data 43

MARCO BRESADOLA / At Play with Nature: Luigi Galvani's Experimental Approach to Muscular Physiology 67

FRIEDRICH STEINLE / The Practice of Studying Practice: Analyzing Research Records of Ampère and Faraday 93

OHAD PARNES / From Agents to Cells: Theodor Schwann's Research Notes of the Years 1835-1838 119

H. OTTO SIBUM / Narrating by Numbers: Keeping an Account of Early 19th Century Laboratory Experiences 141

ANDREA LOETTGERS / Exploring Contents and Boundaries of Experimental Practice in Laboratory Notebooks: Samuel Pierpont Langley and the Mapping of the Infra-red Region of the Solar Spectrum 159

CHRISTOPH HOFFMANN / The Pocket Schedule. Note-taking as a Research Technique: Ernst Mach's Ballistic-Photographic Experiments 183

DANIEL P. TODES / From Lone Investigator to Laboratory Chief: Ivan Pavlov's Research Notebooks as a Reflection of His Managerial and Interpretive Style 203

HANS-JÖRG RHEINBERGER / Carl Correns' Experiments with Pisum, 1896–1899 221

JÜRGEN RENN AND TILMAN SAUER / Errors and Insights: Reconstructing the Genesis of General Relativity from Einstein's Zurich Notebook 253

GERD GRAßHOFF AND MICHAEL MAY / Hans Krebs' and Kurt Henseleit's Laboratory Notebooks and Their Discovery of the Urea Cycle – Reconstructed with Computer Models 269

FREDERIC L. HOLMES / Laboratory Notebooks and Investigative Pathways 295

JED Z. BUCHWALD / The Scholar's Seeing Eye 309

INTRODUCTION

EPISTEMOLOGICAL PREMISES

The published papers in which scientists report the results of their investigations are hardly ever literal accounts of the historical processes through which their authors have reached the conclusions they present. Once an investigation, or a publishable phase of a longer research project, has been completed, the actual pathway it followed becomes largely irrelevant to the investigator, who is expected to marshal the best arguments and evidence available to support the claims she wishes to make. Sometimes the case to be presented sufficiently resembles the process of discovery so that the order of presentation may recapitulate the order of investigation; but temporal rearrangements, omissions of false or aborted trails, and other retrospectively unessential steps, are made routinely, with no intention to falsify a record of discovery. Consequently, historians aiming to reconstruct the historical routes to landmark discoveries have long sought other forms of documentation to fill the gaps left open by the published reports of the completed work.

Research records composed of notes and protocols have long played a role in these efforts to understand the origins of what have come to be seen as the established milestones in the development of modern science. Their subordinate role, however, is symbolized by their presence as mere footnotes in many editions of the classical works of science, and the rarity of the publication of research notes relating even to very prominent milestones in the advancement of science.

The use of research records to probe the nature of scientific investigation itself is a recent development in the history of science. With Eduard Dijksterhuis, we could address them as a veritable "epistemological laboratory" (Dijksterhuis in Marshall Clagett, ed., *Critical Problems in the History of Science*, 1962). The purpose of a workshop entitled *Reworking the Bench: Research Notebooks in the History of Science*, held at the Max Planck Institute for the History of Science in Berlin in November, 1998, was to bring together historians who have been exploiting such resources, to share their experiences; to compare the similarities and differences in the materials they had used and the problems for which they had used them; and to measure the potential and scope for future explorations of "science in the making" based on such forms of documentation. The contributions which form this volume are based on papers presented at this workshop or written afterward by participants in the discussions.

Besides their character as traces of a past scientific activity, there is, however, another aspect of research notes, that has yet to receive proper attention. Research notes – or scribblings, as some forms of them might aptly be called – represent a special genre of scientific writing. They are literary activities in their own right, circumscribing a space that lies between the materialities of experimental arrangement, or the unexplored potentials of theoretical formalisms, and the structured formats of printed communication that are released eventually to the scientific community. This intermediate space is shaped, on the one hand, by individual idiosyncrasies, such as are amply illustrated in the case studies of this volume, and, on the other hand, by local, national, or temporal styles – yet to be studied by historians – that have governed the manner in which scientific activity was to be recorded. This space thus belongs to a particular arena for the formation of discourse, but it also escapes those confines, and it has paradoxical features. Research scribblings are, in one respect, very near to, are even quasi parts of, the instruments and objects of research, such as the written display of an arrangement of controls in a biological experiment; at the same time, in their usually elliptic character, they bear an element of subjectivity, unruliness, and privacy from which they must be freed if they are to become elements of a scientific text. In this intermediate space the objects of research have been set to paper, but have not yet become prose. The paper, the protocol, and the notes – whether casually or systematically recorded – are still integral to a materially mediated environment where the subjectivity of the scientist gives free play to its innovative potential. Here much of the individual style of scientific discovery is expressed and consequently can be captured there. By the publication of this volume we hope to promote and encourage the exploration of this uniquely intimate space, to which documents such as those discussed by the participants in the workshop give us salient points of entry.

A HISTORIOGRAPHICAL APERÇUE

The historians of science of the wartime and first post-war generations who established the field as a professional scholarly discipline focused on the great canonical achievements of the past. Their attention was fixed, in particular, on the landmark changes in scientific thought from which the foundations of the modern branches of science were believed to have emerged. Only gradually, as they moved beyond the recapitulation of these accomplishments to undertake reconstructions of the practices behind the achievements, to follow their genesis as well as to interpret the meaning of the finished products, did they seek persistently in archival repositories the types of materials that would document such developments. In this process, many new archival sources were unearthed, but sometimes also scholarly work based on unpublished sources that had already been done at the end of the nineteenth century was ignored.

This point can be illustrated particularly in the long history of scholarly study of the critical role in the emergence of modern science played by Galileo. The substantial collection of theoretical and experimental research notes kept by Galileo and his circle of collaborators and students had attracted considerable interest among historians of science by the turn of the century, but their work was then largely forgotten until Stillman Drake renewed interest in Galileo's manuscripts. Drake (1978) exploited these materials selectively to show that, contrary to the earlier claims of Alexandre Koyré, Galileo had actually performed the experiments he described in his published works. Around the same time, a few of the other great "founders" of the modern sciences, who had left substantial repositories of their work in progress began to draw intense scholarly scrutiny. The rich records of his successive efforts to solve the problems of terrestrial mechanics and planetary motion left by Isaac Newton have provided material for several decades of historical interpretation, and continue to sustain scholarly activity. The retrieval and publication of the notebooks of Charles Darwin on geology, the transmutation of species, and metaphysical enquiries, resulted from a growing interest in the life sciences in the wake of the evolutionary synthesis, and has given rise to an active cluster of scholarly reconstructions of the succession of ideas and evidence from which Darwin's concept of natural selection emerged. Howard Gruber's *Darwin on Man* (1973; 1981), is an early, and still rare example of the combination of such reconstruction with a theoretical study of the nature of scientific creativity.

For a long time laboratory notebooks attracted even less attention than other types of research records, such as correspondence, for example. Philosophers of science in the first part of the twentieth century had given priority of place to the construction of theory, and did not show great interest in the particulars of experimentation. Symbolic of the relative neglect of sources that document experimental activity itself are the laboratory notebooks of Antoine Lavoisier. Twelve bound volumes, recording most of the experiments Lavoisier carried out from the start of his investigation of the "processes that fix and release airs" in 1773, to his studies in fermentation and the composition of organic substances in 1789, have been readily accessible since the end of the nineteenth century. In *La revolution chimique*, published in 1895, Marcellin Berthelot provided an abstract of the contents of these volumes, including a chronology of the topics of the experiments, and selected comments made by Lavoisier in the notebooks, but without giving the substantive descriptions of his procedures or results. Subsequent historians have cited portions of Berthelot's abstracts to summarize phases in Lavoisier's experimental activity, but until recently no one gave more than passing attention to the original notebooks. Berthelot's abstract satisfied them, so long as historians remained more interested in the development of Lavoisier's ideas than in his experimental practice. The challenges to analytical philosophy of science brought about by the writings of Thomas Kuhn and others, and the increasing interest in the nature of scientific practice itself have, during the last three decades, changed these priorities sufficiently to greatly enhance the value currently attributed to documents such as Lavoisier's laboratory notebooks.

A pioneer in the serious investigation of the contents of laboratory notebooks and the illumination they can provide concerning the process of discovery has been the late Mirko Grmek. Given the task of ordering and cataloguing the huge collection of manuscripts left by Claude Bernard, Grmek saw that the many volumes of laboratory notebooks included among those documents provided a means to reexamine the well-known retrospective narratives of his experimental discoveries that Bernard himself had given in *An Introduction to the Study of Experimental Medicine* (1865) and in other published writings. Grmek showed that for several of his major discoveries, Bernard's later narratives were incompatible in significant ways with the experimental record.

Since the 1980's there has been a small but growing stream of historical work based on, or at least involving laboratory records kept in a broad range of centuries and disciplines. The editors of the present volume have wanted to draw attention to this work, which remains overshadowed by other more highly visible fashions in the field; to raise a number of overarching issues arising from the nature of this work, and to encourage the further application of the methods illustrated in the individual contributions to frame historical narratives, and to provide deeper epistemological perspectives on the research process.

A NEW BRANCH OF THE HISTORY OF SCIENCE?

The contributions to this volume bear witness to the spread of interest in "science in the making" that is leading historians, philosophers, and sociologists increasingly to base their accounts on information that research notebooks can provide. Notebooks represent a peculiar source that can demand more ingenuity, more technical effort and experience, and more background knowledge for the reconstruction of the activity they document than does the typical interpretation of published sources. Those who have undertaken the study of such notebooks have, therefore, had to become specialized in such techniques as deciphering difficult handwriting, reading between the lines of statements expressed in compressed, elliptical form, and inferring the intentions behind pages that record only operations. In some cases a feeling for paper qualities, ink colors, and watermarks, or the ability to transform intuitively small hints from the size of a piece of paper or the smell of a box has been essential to transform scattered records into a chronological order. The individual papers represented here illustrate how diverse the problems posed by particular records can be. Are there, nevertheless, common perspectives that emerge from these singular pursuits? Those who participated in the workshop on which this volume is based recognized that beneath the many superficial dissimilarities in their specific endeavors lay a deeper commonality of commitment, and a shared craft. The variety of approaches adapted to particular circumstances were connected in many ways to a broad and increasing array of techniques and methods available to historians who want to study notebooks. They raise the possibility that some of the tacit knowledge of individual scholars can be transformed into the precepts of a methodological canon.

With a few exceptions, those scholars whose work is included in this volume have had to journey to the archives in which research records are preserved, or to have obtained them from the scientists who kept them, and to work with the original documents themselves, or with xerox copies prepared for them personally. The publication of such records has remained a problem. Only when records take the form of more or less coherent texts, as in the case of Darwin's notebooks, have they been accessible to traditional editorial methods. In other cases, as in the publication of Galileo's archival materials, the records of experimental activity have been left out because of their fragmentary, non-discursive nature. The opportunity to make high-quality reproductions of research records electronically available on the internet together with work on them, has now radically changed this situation. It entails the option to make such materials much more widely available to the community of historians of science thus creating new challenges for scholarly work on these sources.

Rapidly advancing technologies are also providing an arsenal of analytical tools to disentangle some of the deeper mysteries of research notebooks. Traditional sidelight photography to reveal traced lines in manuscripts can now be supplemented by the analysis of inks through the methods of nuclear physics. The results of such examinations, together with information about paper sizes, watermarks, and handwriting styles can be incorporated into databases freely accessible to the community of scholars. The study of research notebooks is thus taking part in a rapidly evolving methodology of manuscript research, which has become a field comparable to archaeology or palaeontology, combining the traditional investigative hermeneutics of the humanities with the analytical acumen of the natural sciences. The solution of notebook puzzles that have hitherto occupied single scholars alone can now become collective enterprises. As far as we can foresee, however, many of those historians who choose to reconstruct science in the making will still experience the toil and the pleasure of tracking down original documents that may have landed somewhere in an archive, or that may still be in the possession of a living scientist or her heirs.

The problems encountered in reconstructing research processes documented by notebooks are, of course, not limited to technical considerations concerning their physical condition. Most research notes are taken for the benefit of a very restricted audience, sometimes only for the individual researcher himself. Because the recorded information serves in this way only as a supplement to memory or to the constant informal communication that takes place among the members of a research group, it may be stated in highly abbreviated, and to outsiders cryptic form. For a successful analysis, the modern interpreter must bridge the gulf between the notes that survive and their original context which has disappeared, reconstructing the knowledge, the skills, the objects, and the instruments that were taken for granted by the author of the notebook. To gain access to this implicit background knowledge, it is necessary to go beyond the notebook itself, to exploit other sources such as published papers, contemporary textbooks, institutional records, biographical information, and material artefacts.

It is in the combination of information drawn from notebook records and from other sources to reconstruct the processes of investigation that the historian must especially exercise critical judgement to avoid arbitrary choices. In some cases, a continuous notebook record can provide a solid core, around which the reconstruction can be built. In other cases, where the records themselves or their surviving parts are more fragmentary, the structure of an investigation must be retrieved from and based on contextual information, and the notebooks used to fill in meaning at critical junctures. Cogent examples of both extremes, as well as intermediate situations, appear in the individual contributions to this volume. In general, the historian must work in both directions – from inside out, and from outside in – interpreting notebook details within the broader framework, and understanding larger outcomes as integrations of the smaller steps revealed only in the notebook record.

The study of research notebooks opens particularly powerful avenues of access to the microhistorical examination of the processes of scientific change. Such an approach comes as close as is practically possible to witness the production of scientific novelty at the very point of its emergence. It can reveal, not only the interplay between thought and action in the day-by-day movement of the investigator toward outcomes that he could not anticipate beforehand, but that become in retrospect logical progressions; but also the role played by material minutiae and contextual conditions, that are not visible in the published papers representing the conclusions eventually reached. Such microhistorical reconstructions can challenge accepted epistemological notions of scientific progress, and can even lead us to rethink what a scientific discovery means. Collectively, such studies may set new standards for historical interpretation, calling into question conventional modes of narration, conquering new historical terrain, and offering new explorations of fundamental epistemological questions.

Microhistory alone, however, can provide only isolated fragments of broader insight. The thoughts and actions of individual scientists or research groups make sense only within the larger contexts within which they work. Thus microhistorians must orient their studies within constellations of persistent problems that link generations of scientists, and whole fields of knowledge in a given historical period. They must be attentive to the constant flow of information, both formal and informal, to personal relations, competition and cooperation, that link a given investigator with contemporary scientists pursuing similar problems, with the traditions that effect both the style and substance of the work, and with many other factors of the culture at large.

Because the microhistory of scientific investigation deals with the processes of individual thinking, acting, and representation, historians who rely on research notebooks to gain access to these processes must also be attentive to other fields that provide insights into their general nature. The cognitive sciences, psychology, literary history, the philosophy of science, and the sociology of science, are among the fields that contribute methods of great potential value for the intense scrutiny of science in the making. Only by bringing these larger perspectives to bear can we make the study of scientific research notebooks a field of potentially general interest, rather than an esoteric playground for parochial specialists.

In summary, if microstudies of the scientific enterprise based on research notebooks are intimately interwoven with other ways to approach history of science and with knowledge and methods drawn from other disciplines, they have the potential to become not merely a new subspecialty within the history of science, but rather a point of crystallization for a truly interdisciplinary historical epistemology.

* * *

Each of the contributors to this volume has become involved in the study of research notebooks through his or her own personal scholarly trajectory, and most have learned by trial and error the techniques they have applied to the use and interpretation of such sources. Of necessity their individual choices, rather than a systematic overview of the subfields of the history of science by the editors, has dictated the topics and periods included here, as well as the particular historiographical problems discussed. The individual papers intersect one another in a variety of ways. Rather than to group them according to any one of a number of possible systematic criteria, we have placed them in a simple chronological order.

For the seventeenth century, Jürgen Renn and Peter Damerow report on an analysis which extracts, from sheets of paper on which he constructed parabolas and other curves, new insights about how Galileo discovered the fundamental law of free fall and at the same time question the notion of discovery as it is traditionally used. Bill Newman and Lawrence Principe show from the laboratory notebooks of George Starkey, an American-born alchemist who influenced Robert Boyle, that alchemy was not a mere projection of psychic processes, as has often been maintained, but that alchemists performed experiments with considerable skill. From the relatively few surviving direct records of Isaac Newton's optical experiments, Alan Shapiro concludes that in his experiments on the "colored circles" now known as Newton's rings, Newton went to great lengths to achieve a quantitative accuracy that is not revealed in his published accounts of the experiments.

The eighteenth century is represented primarily by a reconstruction of the early experiments of Luigi Galvani on animal electricity, in which Marco Bresadola is able to elucidate the "interplay between problems, instruments, experiments, and interpretations that characterized his laboratory practice." One of the several cases discussed in Frederic Holmes' paper concerns the well-known experiment on fermentation performed at the end of that century by Antoine Lavoisier.

Fittingly for an era in which experimentation became a robust, large-scale activity carried on by professional scientists in well-demarcated disciplines, seven of our cases come from the nineteenth century. Friedrich Steinle discusses the early experiments on electro-magnetism by André-Marie Ampère, and later ones by Michael Faraday. Steinle demonstrates particularly how much more difficult the task of the historian is when confronted by the sparse, chaotic traces of experimental activity left by Ampère, than it is when he can work with such well ordered notebook records as those kept by Faraday. Ohad Parnes finds that the research notebooks of Theodor Schwann are unusual in the fact that they probably served at once as records of experimental activity and as drafts for the publications that Schwann drew from that activity. Otto Sibum subjects to deep analysis selected pages from the laboratory notebooks recording James Prescott Joule's experiments on the mechanical equivalent of heat, and shows that their format derived more from his experience in the practice of brewing than from any formal scientific style of record-keeping. Dealing with thirty laboratory notebooks recording the experiments carried out over 18 years by Samuel Pierpont Langley with an instrument Langley invented to observe the infra-red region of the solar spectrum, Andrea Loettgers examines this immense record selectively to reveal how the laboratory was organized, and how a new and fragile scientific instrument was gradually stabilized. The notebooks of Ernst Mach, Christoph Hoffmann finds, are incoherent, neither systematic nor closely dated; yet, in conjunction with other published and unpublished evidence, the sketches and cryptic comments they contain illuminate significantly the intimate creative mental processes through which Mach came to interpret his important experiments on the motion of projectiles through the air. Daniel Todes employs the scattered remaining laboratory records of Ivan Pavlov to highlight Pavlov's transition from an individual investigator to the manager of a large laboratory organized like a factory. Hans-Jörg Rheinberger retrieves, from laboratory protocols left by Carl Correns, evidence for a gradual shift in the meaning for Correns of his experimental crosses with maize plants and peas from reproductive physiology to transmission genetics. In addition, he sheds new light on the question to what extent Correns' reading of the famous paper of Gregor Mendel shaped his own experimental program.

The twentieth century is auspiciously represented by Jürgen Renn's and Tilman Sauer's report on the genesis of Albert Einstein's theory of general relativity, based on a notebook in which Einstein recorded his theoretical explorations in search of the gravitational field equations. Gerd Graßhoff and Michel May demonstrate how computer reconstructions can aid in the interpretation of research pathways documented in laboratory notebooks. The case they present, the discovery of the ornithine cycle by Hans Krebs in 1932, is also one of the cases discussed in Holmes' paper. For the final case discussed by Holmes, the origin of the Meselson-Stahl experiment that provided experimental confirmation of the mode of replication of DNA predicted by the Watson-Crick model, no conventional notebook record exists, but an equivalent chronological trace can be followed through the log of the analytical centrifuge used in the experiments, and the surviving ultra-violet absorption films that constitute the immediate results of the experiments. This case may point toward the different sorts of records to which historians may have recourse to reconstruct still more recent scientific research, in which other forms of inscription have increasingly displaced the traditional notebooks or sheets of paper on which the investigator wrote down in his own hand what he wished to preserve from the events of the day in the laboratory or the study.

The cases discussed in this volume span a sufficient range of problems, styles of records, sciences, and centuries, to provide the rough outlines of an array of techniques and problems on which future scholars can build. It is the hope of the editors and the participants in the workshop from which the volume emerged, that their examples will stimulate more historians than have hitherto taken part in such studies to examine the many unexploited opportunities for studies of surviving research notebooks and of the traces of historical investigations hidden in them. If so, our provisional repository of samples of their use can eventually be replaced by a more systematic survey of both the challenges such documents pose and the insights they can yield.

Frederic L. Holmes
Jürgen Renn
Hans-Jörg Rheinberger

JÜRGEN RENN AND PETER DAMEROW*

THE HANGING CHAIN: A FORGOTTEN "DISCOVERY" BURIED IN GALILEO'S NOTES ON MOTION

One of the main motives for looking into sources such as notebooks is to date the discoveries of the heroes of science that are documented by their publications. In the case of Galileo's major discoveries it has thus been established that he found his law of fall around the year 1604 and that, several years later only, he discovered the parabolic shape of the trajectory of projectile motion. We claim that for both discoveries a much earlier dating becomes unavoidable if such sources are interpreted under a different perspective, not under the limited perspective of determining precise dates of isolated pioneering discoveries, but under the perspective of a historical epistemology that does not adhere to such a simple concept of discovery.[1] The dating of Galileo's achievements is, however, not our primary concern here but rather the context in which his discoveries were made. It will be shown that for Galileo the projectile trajectory was closely related to another curve which he erroneously thought to be much easier to understand. We claim that the alleged identity between the two different curves provided him with a conceptual framework for the interpretation of his famous discoveries on projectile motion which differs from that of classical mechanics, thus undermining the common understanding of these discoveries.

This needs some explanation. Contrary to the agreement about the dating, there is no unanimity among recent historians of science concerning the sources from which Galileo derived his major discoveries. The assumptions about his sources range in fact from pure empirical evidence achieved exclusively by means of careful experimentation, on the one hand, to predominantly theoretical speculation in direct continuation of scholastic traditions, on the other hand. In spite of the wide range of different reconstructions of the discovery process, however, a simple fact has nearly been completely neglected both by the older and the more recent literature: for Galileo, a close connection exists between the parabolic trajectory and, as mentioned above, another curve, and this curve is the catenary, the curve of a hanging chain.

This neglect is all the more astonishing as the connection is explicitly made a subject of discussion in his final word on the matter, the *Discorsi*. In the course of

---

* Max Planck Institute for the History of Science, Berlin

Frederic L. Holmes, Jürgen Renn and Hans-Jörg Rheinberger (eds.), Reworking the Bench: Research Notebooks in the History of Science, 1–24 © *2003 Kluwer Academic Publishers. Printed in Great Britain*

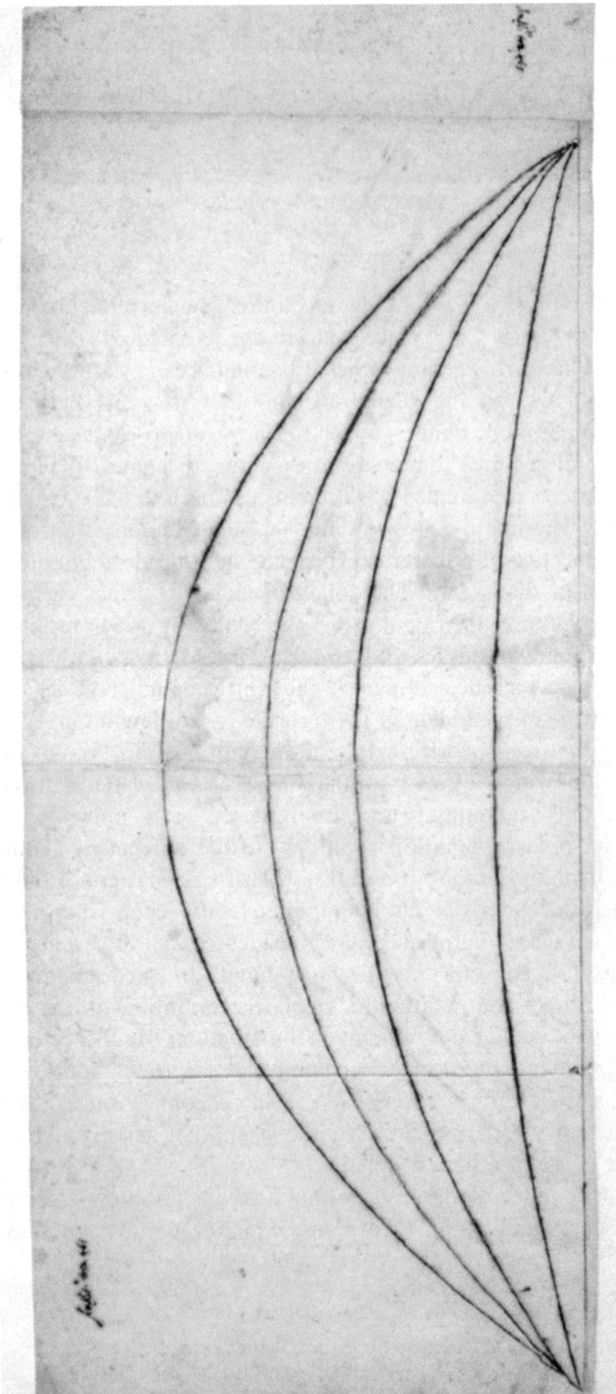

Figure 1. Folio 41/42 of Galileo's notes on motion with curves produced by means of a hanging chain.

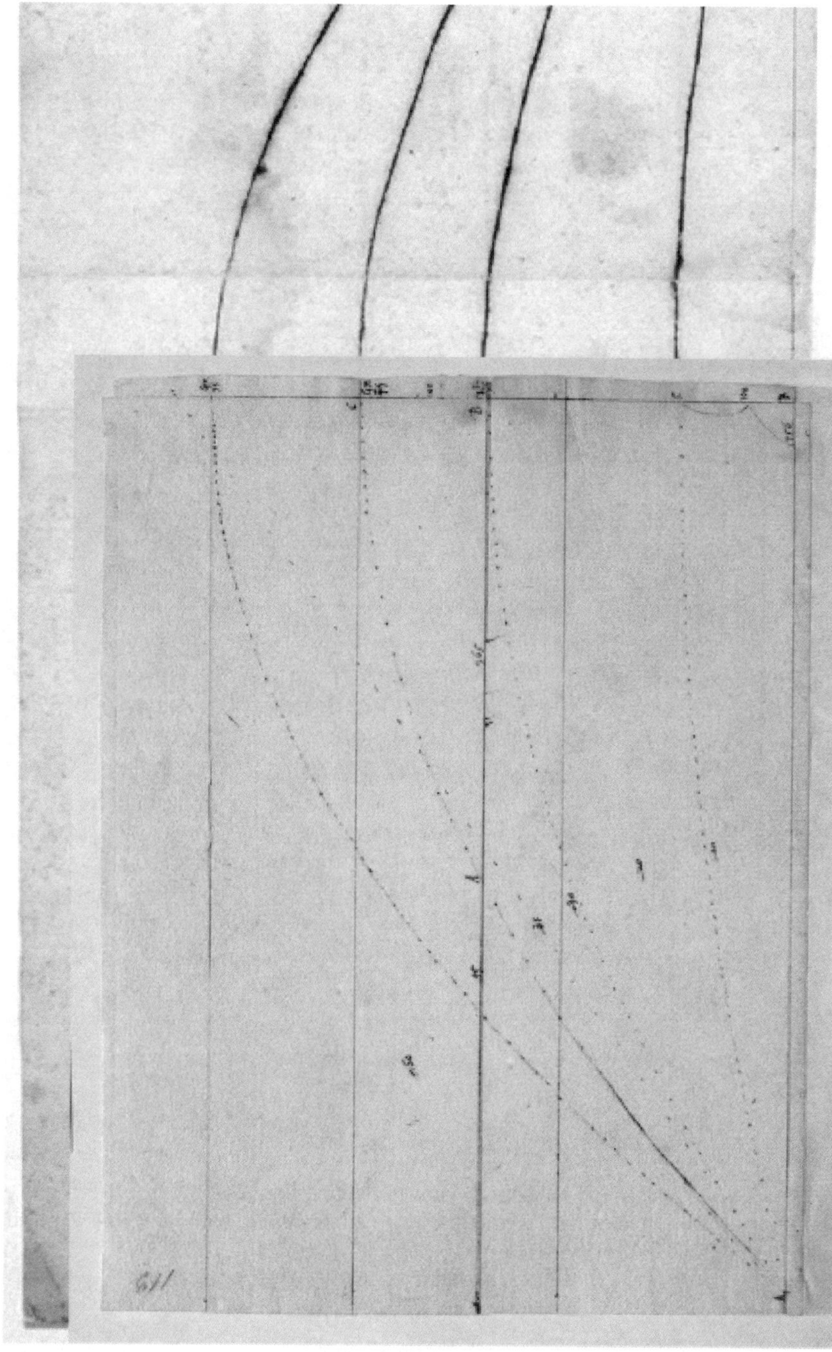

Figure 2. Folio 41/42 of Galileo's notes on motion (at the back) used as a template for drawing projectile trajectories on folio 113 (in front).

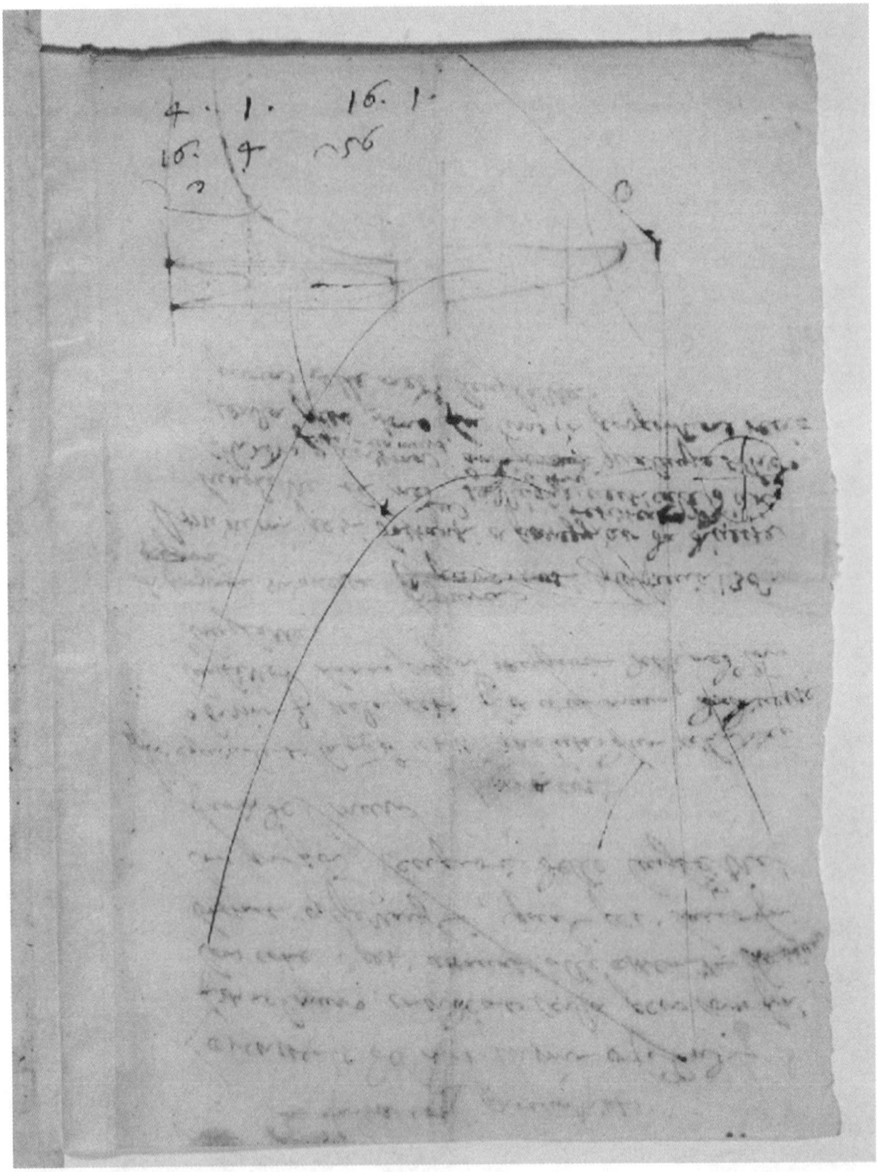

Figure 3. Sheet of paper found in Galileo's copy of the first edition of his *Discorsi* with two parabolic curves generated by an inked ball thrown along an inclined plane, inserted near the place where this method is described.

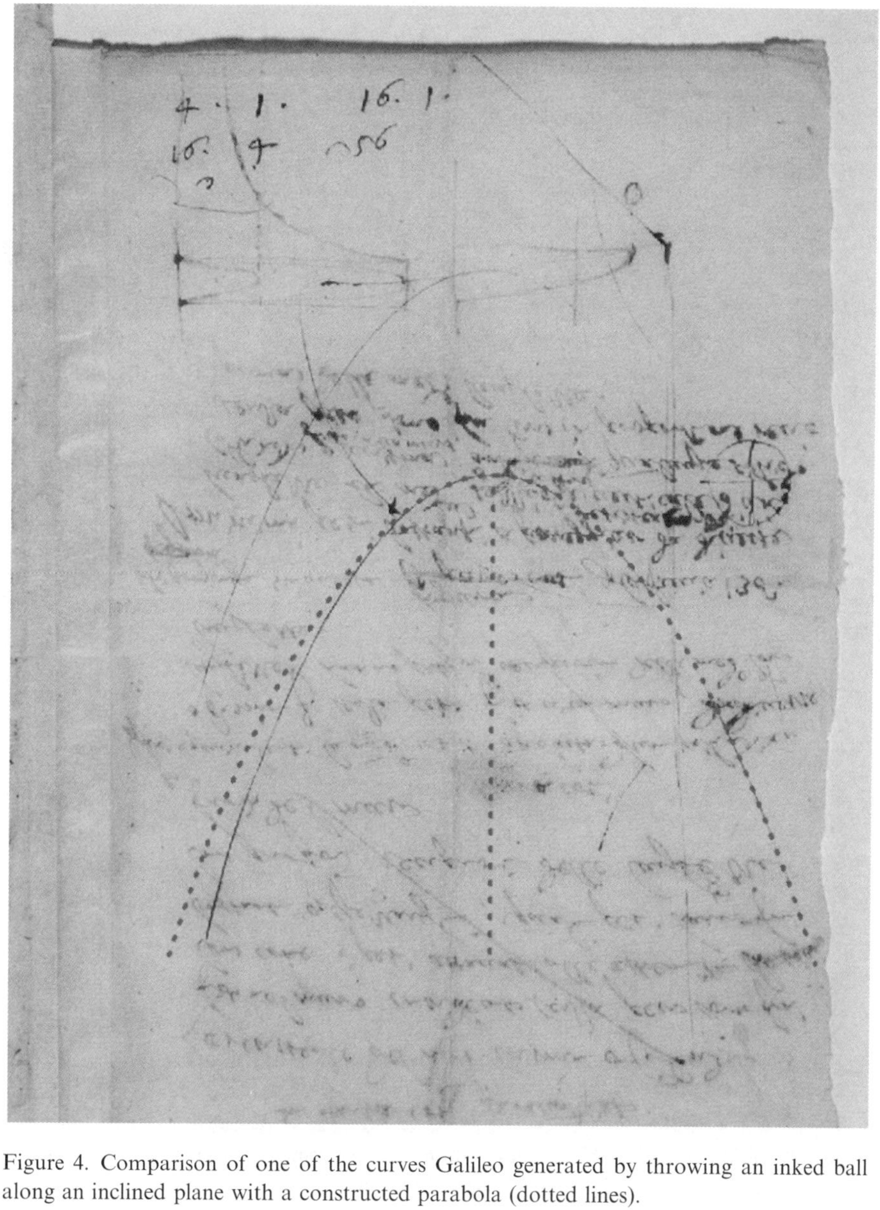

Figure 4. Comparison of one of the curves Galileo generated by throwing an inked ball along an inclined plane with a constructed parabola (dotted lines).

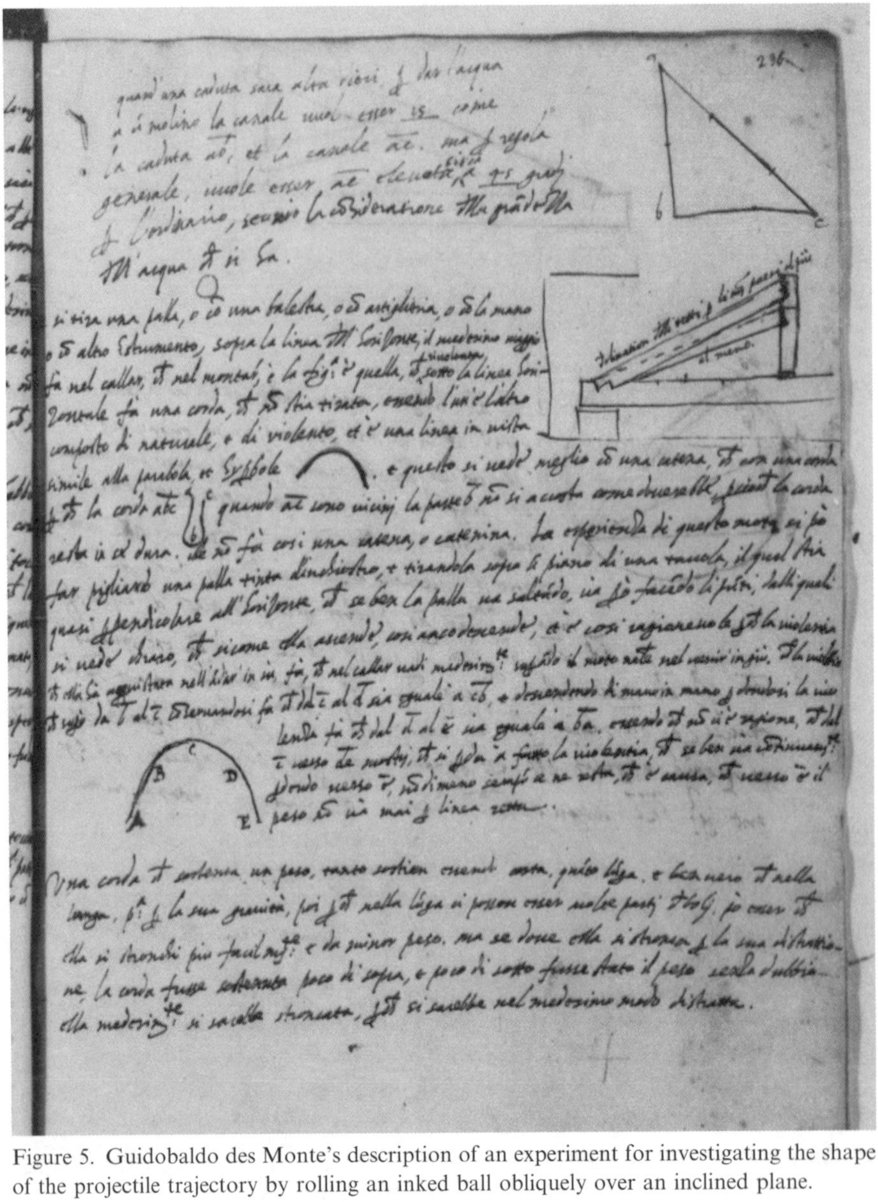

Figure 5. Guidobaldo des Monte's description of an experiment for investigating the shape of the projectile trajectory by rolling an inked ball obliquely over an inclined plane.

the discussions of the *Second Day*, Galileo's spokesman Salviati describes two methods of drawing a parabola, one of them involving the trajectory of a projected body, the other one using a hanging chain:[2]

> I use an exquisitely round bronze ball, no larger than a nut; this is rolled on a metal mirror held not vertically but somewhat tilted, so that the ball in motion runs over it and presses it lightly. In moving, it leaves a parabolic line, very thin, and smoothly traced. ...
>
> The other way to draw ... the line we seek is to fix two nails in a wall in a horizontal line .... From these two nails hang a fine chain .... This chain curves in a parabolic shape, so that if we mark points on the wall along the path of the chain, we shall have drawn a full parabola.

Galileo's claim that the curve of a hanging rope or chain is a parabola is obviously wrong. It leaves us therefore with a puzzle which can be resolved only by studying the role of this claim in the real discovery process materialized in his notes.

Did Galileo really use these methods? For the hanging chain this question can easily be answered. Among Galileo's notes on mechanics, the famous codex 72,[3] there is a folded sheet of rough paper designated as folio 41/42 that has obviously been used for drawing catenaries just as Galileo described in the *Discorsi* (fig. 1). Chains of different length were fixed with two nails, their shapes were copied to the paper, and by allowing ink to seep through little holes pierced with a needle in the paper, the resulting curves could be copied to as many other pages as desired. Such a page is folio 113 recto (fig. 2). It shows a drawing which contains curves produced by using the actual folio 41/42 as a template. In the drawing of this page the curves represent projectile trajectories of oblique gun shots projected at various angles.

For the other method of drawing parabolas, the one using a ball rolling over an inclined plane, the evidence of the truth of Galileo's claim is somewhat more indirect. Galileo's personal copy of the first edition of the *Discorsi* contains numerous corrections, notes, and additions mostly by the hand of his disciple Vincenzio Viviani.[4] On a sheet of paper inserted close to Galileo's description of the latter method for drawing parabolas (fig. 3) one finds two curves which show the typical characteristics of such a method: the indications of the bouncing of the ball at the beginning and the slight deformation of the parabola at the end due to friction.

We have verified the nature of the curve in Galileo's copy of the *Discorsi* by comparing it with a constructed parabola and with the results of a controlled repetition of the experiment with modern equipment (fig. 4).

In addition to the description of the experiment in the *Discorsi*, there is also evidence from a much earlier period. At the end of a notebook of Guidobaldo del Monte, a correspondent, benefactor, and close associate of Galileo in his early research on mechanics, there is a protocol with a description of the projectile experiment (fig. 5).[5] Contrary to Tartaglia's generally accepted view that the curve

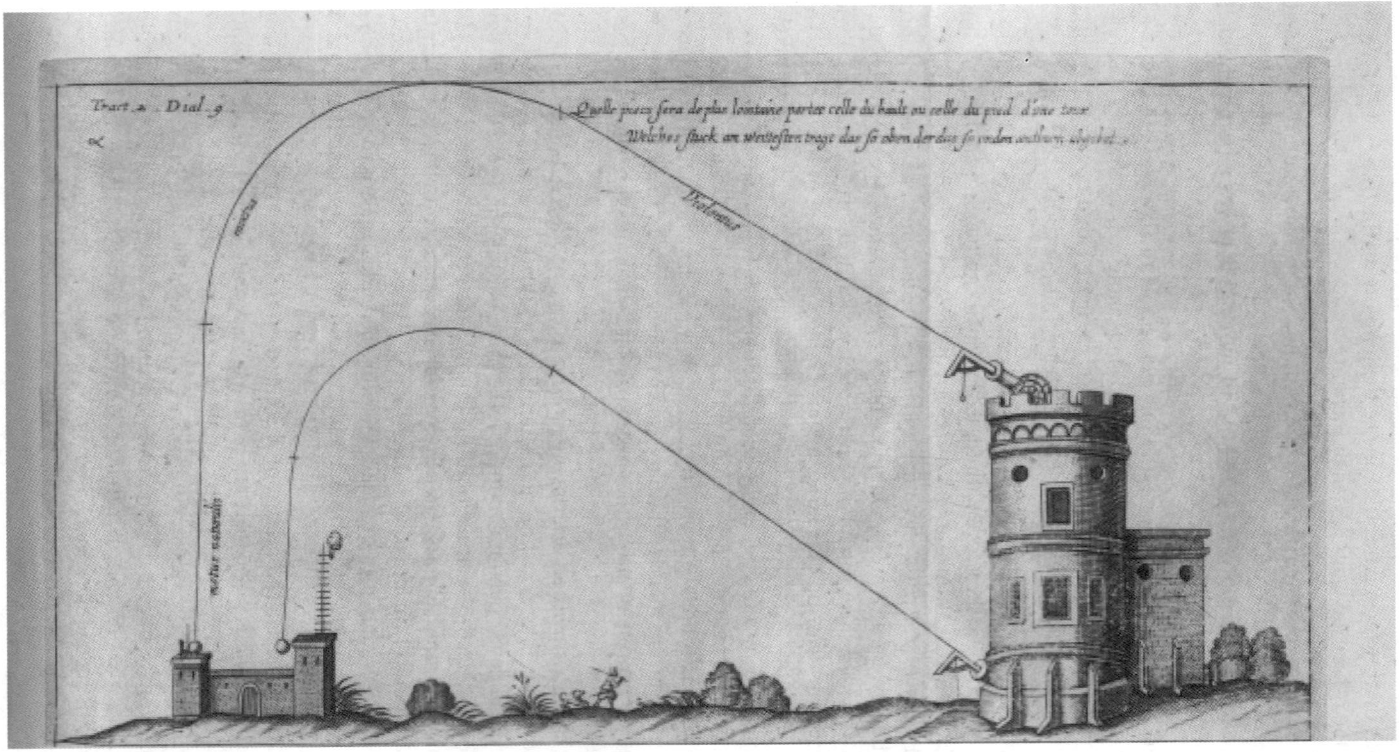

Figure 6. Ballistic curves drawn according to the Aristotelian understanding of the combined effect of forced and natural motion, reproduced from the 1628 French edition of Diego Ufano's widespread treatise on artillery, first published in 1612.

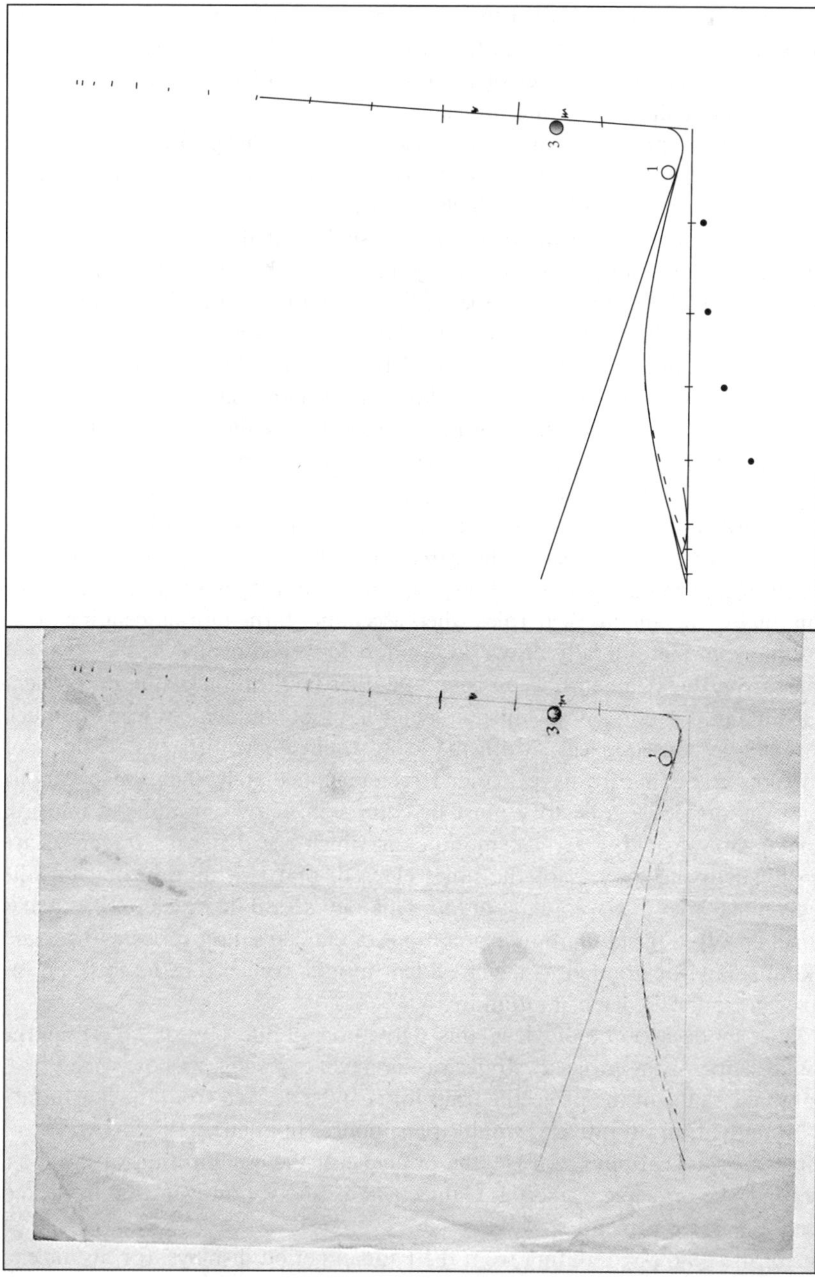

Figure 7. Folio page 175 recto of Galileo’s notes on motion showing a drawing with an erroneous construction of a projectile trajectory which can be inferred from compass marks (black dots in the figure on the right).

must be asymmetrical for principal reasons, this protocol claims the symmetry of the trajectory. The reason for the general acceptance of Tartaglia's view up to that time was its perfect agreement with the traditional Aristotelian understanding of the essentially different character of the forced motion of the ascending and the natural motion of the descending projectile (fig. 6).

We were able to provide strong evidence showing that this protocol was written in August 1592 when Galileo visited Guidobaldo on his way to Padua, that is more than 10 years before the period to which the discoveries of the law of fall and the parabolic shape of the projectile trajectory is usually dated.[6]

Even more surprising than this early dating is, however, the theoretical interpretation of the symmetric shape of the trajectory offered in the protocol. Following the prevailing interpretations of Galileo's discoveries one would have to assume that such an experiment must have definitely refuted the traditional theory that the trajectory results from an interplay of forced and natural motion. In the protocol, however, precisely this interplay is used to *explain* the symmetry. And now comes the point: This surprising explanation is based on a comparison of the trajectory with a hanging chain where a similar interplay of force and the natural tendency downwards is assumed to take place, but here self-evidently producing a symmetric curve. To be precise, the explanation for the unexpected symmetry suggested by this comparison is that the two tendencies act jointly and in the same way, mutually exchanging their roles when ascending turns into descending. This conclusion is, in fact, literally drawn in Guidobaldo's protocol.

Apart from the symmetry, there is still another implication of the experiment recorded in Guidobaldo's note which is somewhat more hidden but leads to much more *dramatic* consequences. Mathematically trained scientists like Galileo or Guidobaldo could hardly have avoided seeing immediately that the parabolic curve of the projectile trajectory must be composed of two motions, a uniform horizontal motion and a vertical motion for which the distances traversed are proportional to the squares of the times elapsed, that is a motion obeying the allegedly much later discovered *law of fall*. This solves a riddle which until now has remained unsolved in standard interpretations of Galileo's discoveries: How could he take this law for granted as a self-evident insight even before he took up his serious work on a theory of motion in 1604?

From a modern point of view, this derivation of the law of fall from the parabolic shape of the projectile trajectory immediately suggests how, vice versa, the parabolic shape of the projectile trajectory can be derived from the law of fall. Such a derivation, however, would presuppose familiarity with two basic principles of classical physics, first, the principle of the composition of motions, and second, the principle of inertia. Galileo's notes show, however, that he lacked precisely these two principles.

One of the folio pages dating from the Paduan period displays, for instance, a construction aimed at determining the shape of the trajectory of oblique projection (fig. 7).[7] Compass marks still visible on the paper make evident that Galileo constructed the projectile trajectory by compounding a decelerated motion along

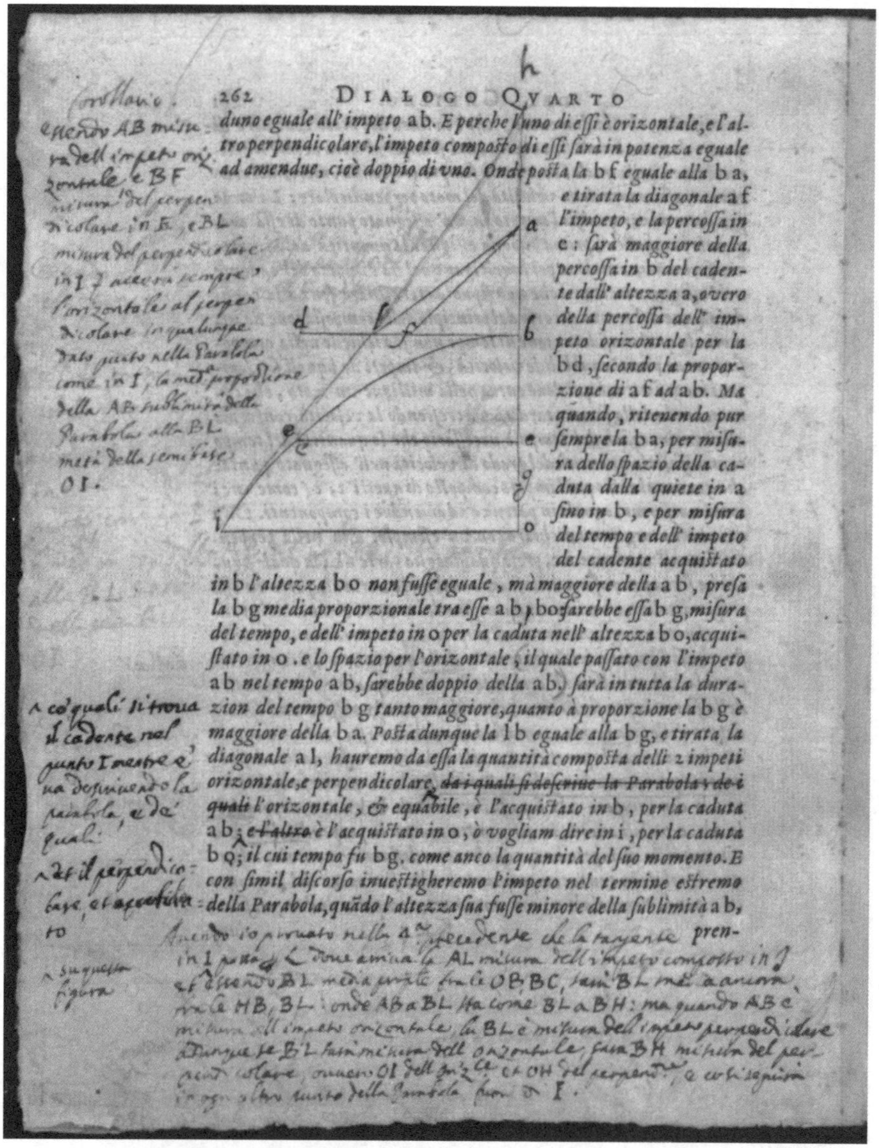
262 DIALOGO QVARTO

duno eguale all'impeto ab. E perche l'uno di essi è orizontale, e l'altro perpendicolare, l'impeto composto di essi sarà in potenza eguale ad amendue, cioè doppio di vno. Onde posta la bf eguale alla ba, e tirata la diagonale af l'impeto, e la percossa in e: sarà maggiore della percossa in b del cadente dall'altezza a, overo della percossa dell'impeto orizontale per la bd, secondo la proporzione di af ad ab. Ma quando, ritenendo pur sempre la ba, per misura dello spazio della caduta dalla quiete in a sino in b, e per misura del tempo e dell'impeto del cadente acquistato in b l'altezza bo non fusse eguale, mà maggiore della ab, presa la bg media proporzionale tra esse ab, bo sarebbe essa bg, misura del tempo, e dell'impeto in o per la caduta nell'altezza bo, acquistato in o. e lo spazio per l'orizontale, il quale passato con l'impeto ab nel tempo ab, sarebbe doppio della ab, sarà in tutta la durazion del tempo bg tanto maggiore, quanto à proporzione la bg è maggiore della ba. Posta dunque la lb eguale alla bg, e tirata la diagonale al, hauremo da essa la quantità composta delli 2 impeti orizontale, e perpendicolare, da i quali si descriue la Parabola, e de i quali l'orizontale, & equabile, è l'acquistato in b, per la caduta ab; e l'altro è l'acquistato in o, ò vogliam dire in i, per la caduta bo; il cui tempo fu bg. come anco la quantità del suo momento. E con simil discorso inuestigheremo l'impeto nel termine estremo della Parabola, quãdo l'altezza sua fusse minore della sublimità ab, pren-

Figure 8. Emended diagram in Galileo's hand copy of his *Discorsi*.

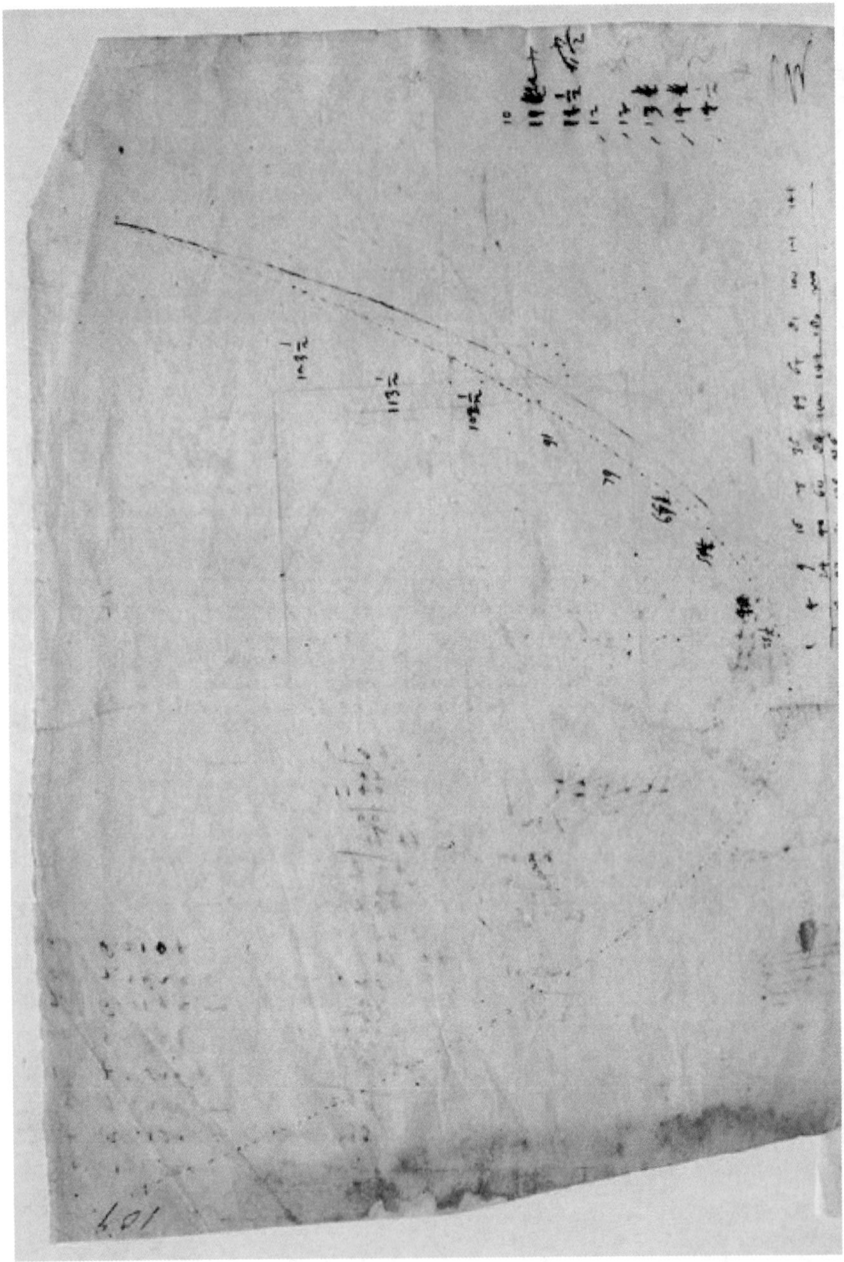

Figure 9. Folio page 107 recto of Galileo's notes on motion showing a comparison of an empirically produced catenary with a mathematically constructed parabola.

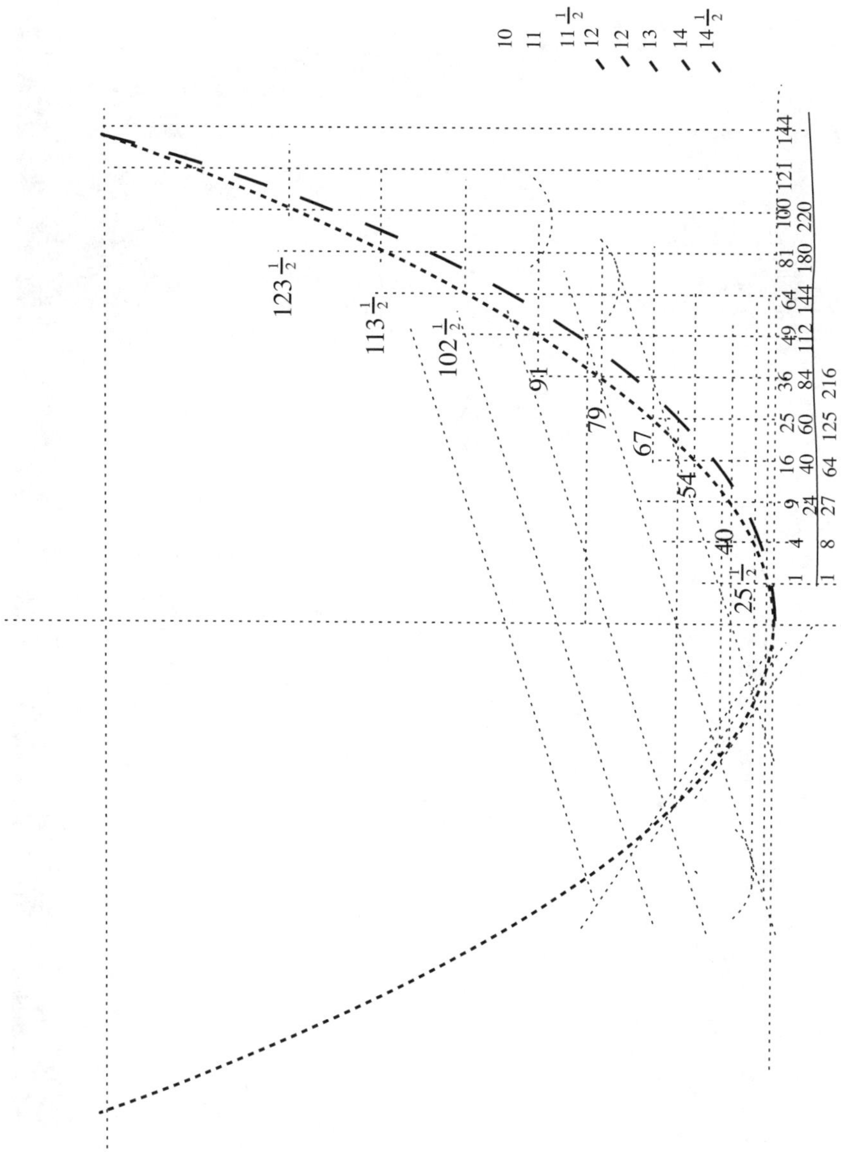

Figure 10. Construction lines drawn without ink (thin, dotted lines) on folio page 107 recto of Galileo's notes on motion.

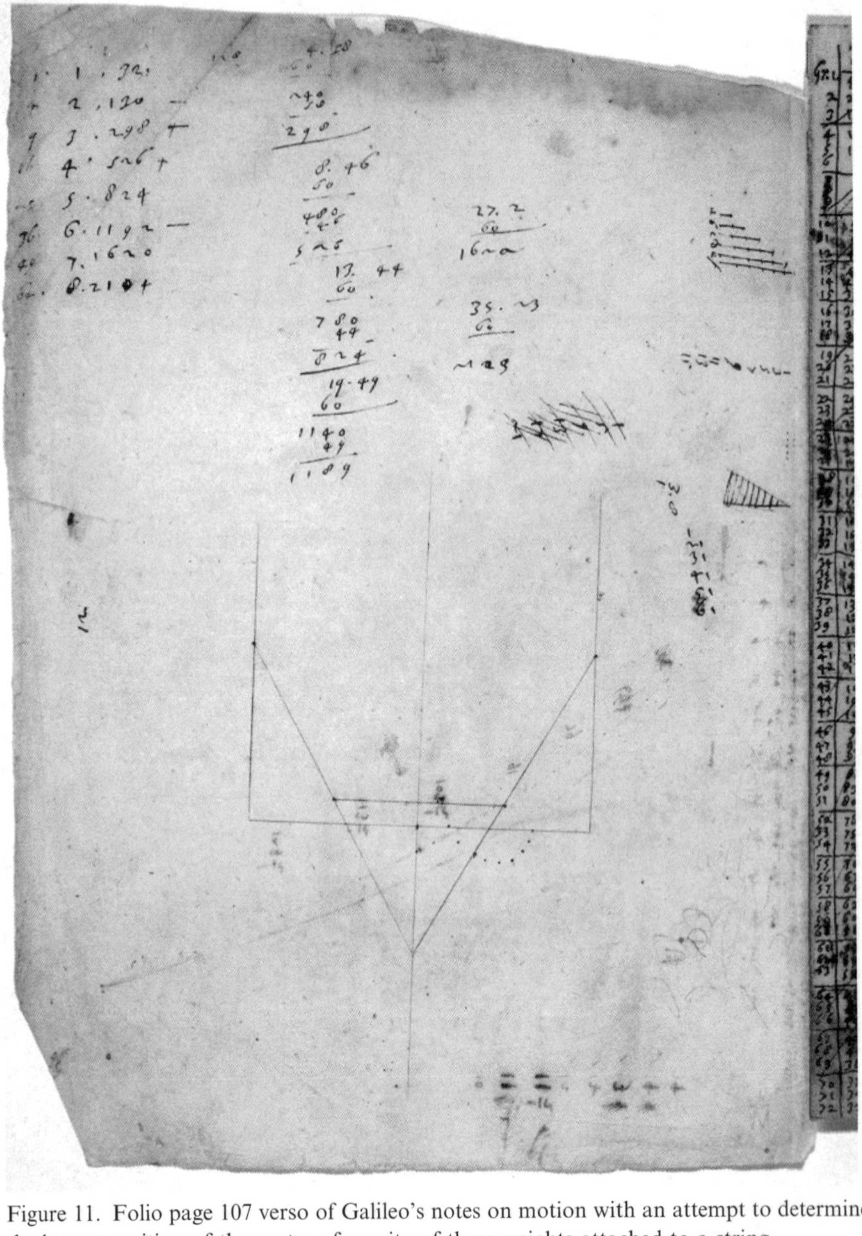

Figure 11. Folio page 107 verso of Galileo's notes on motion with an attempt to determine the lowest position of the center of gravity of three weights attached to a string.

an inclined plane representing the direction of the projection and an accelerated downward motion according to the law of fall. This seemingly plausible composition is, of course, fallacious. Galileo would have had to assume a uniform motion in the direction of the projection. His construction shows, however, that he obviously was, at least at that time, not familiar with the principle of inertia.

We have shown elsewhere that even Galileo's final publication does not yet contain the general principle of inertia.[8] The correct superposition of an oblique inertial motion and free fall is only found in annotations to Galileo's personal copy of the *Discorsi*, inserted by him or by one of his disciples (fig. 8).

It fits into this picture that we found striking evidence in Galileo's notes that he never gave up the idea of an equal shape of the trajectory and the catenary and that in various ways he tried to find a confirmation for this fallacious assumption. One example is given by folio page 107r, again dating from the Paduan period (fig. 9).[9] The page contains two curves with a common upper endpoint and a common zero point at the bottom, distinguished by a different curvature.

Close inspection of the original folio has furthermore revealed a considerable number of compass marks and construction lines drawn without ink (fig. 10). Although, with the exception of a couple of numbers, there is no text on the folio which could explain the function of the curves, a meticulous investigation of the folio including an analysis of the composition of the different inks on the page by means of Particle Induced X-ray Emission made it possible to establish beyond any doubt that one of the curves is a geometrically constructed parabola whereas the other curve is a catenary produced, as described in the *Discorsi*, by means of a hanging chain.[10] The numbers on the page represent a failed attempt to find a rule for the obvious deviation of the two curves.

This shows that Galileo did not draw the consequences from the projectile experiment that would have been drawn in the context of classical physics. And, moreover, he stuck to his identification of catenary and parabola despite the fact that he had indubitable empirical evidence of the difference of both curves. This is not as suspect as it appears but reflects merely the fact that experiments do not prove or disprove a theoretical assumption per se. In fact, a possible explanation for the deviation has already been given in the protocol of the projectile experiment in Guidobaldo's notebook, which states that due to the stiffness of the material a hanging rope may fail to fit precisely to a parabola.

There are further folio pages among Galileo's notes on motion which help to understand why he did not dismiss the idea that the curve of the projectile trajectory and the catenary are identical.

One of the most surprising discoveries made in our work on Galileo's notes on motion concerns the reverse of the very folio that documents his comparison of an empirically produced catenary with the parabola (fig. 11).[11] This page has until now been interpreted erroneously as being the record of his famous inclined plane experiment. In fact, it has turned out to contain an essential part of an ingenious but failed attempt to find a proof of the parabolic shape of the catenary. The purpose of the drawing in the center of this page would have remained

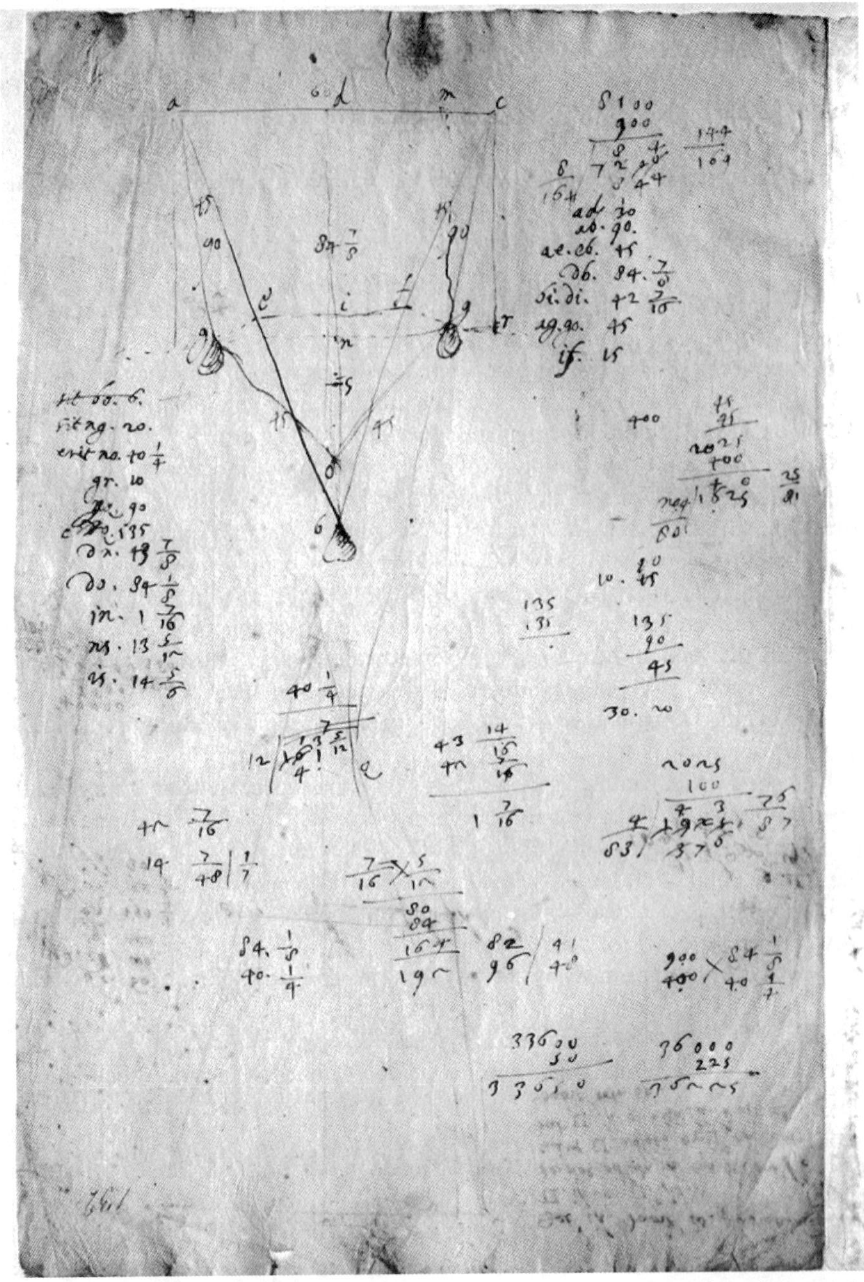

Figure 12. Folio page 132 recto of Galileo's notes on motion concerning three weights attached to a string.

undiscovered if another folio dating from the Paduan period, namely folio 132, had not been preserved (fig. 12).[12]

On the obverse of this folio, Galileo worked more explicitly with two different geometrical constellations of a hanging string fixed with its ends to two points *a* and *c* of a horizontal line, the middle of the line between these two suspension points being designated by the point *d*. In the first position the hanging string is pulled down by a neatly drawn weight, fixed in its middle at point *b*, so that the string forms a triangle *abc*. In the second constellation two further weights have been fixed precisely in the middle of each half of the string at points *e* and *f* pulling the string outwards on circles around the suspension points *a* and *c* towards two points which are both designated by the letter *g*. These additional weights raise the first weight from point *b* to point *o*.

The meaning of this drawing can be unambiguously reconstructed from the accompanying calculations. We were able to interpret them consistently as a check of whether the three weights attached to the string are lying on a parabola with the result that the values he compared were different by about ten percent.

How could Galileo know how far the two attached weights would pull the string away from its initial position? An answer to this question is indicated by a final calculation on the page after the parabola check. Galileo calculated the height of the center of gravity of the assumed final position, obviously in order to compare the result with the position of the center of gravity of the original constellation which he had calculated previously. This calculation suggests that he was already aware of the fact that the center of gravity reaches its lowest point when the three weights attached to the string are in equilibrium.

That this was in fact the idea Galileo had in mind is indubitably confirmed by the seemingly obscure drawing on the reverse of the folio 107 containing his empirical check of the parabolic form of the trajectory. A close inspection of the precise drawing revealed a great number of invisible construction lines which make the purpose of the drawing evident (fig. 13). For a sequence of different positions of a string with three weights Galileo constructed and marked by ink spots the positions of the center of gravity of the right half of the symmetric constellation, thus using this geometrical construction to empirically find its lowest position (fig. 14).

Why did Galileo try to fit the constellation of a string with three weights to a parabola? The answer can be inferred only from the epistemological context. Given Galileo's conviction that the catenary is parabolic, he would have established, if he had been successful, an important step towards a proof of this assumption within the conceptual framework of his preclassical theoretical thinking. If he could have shown that for any number of weights attached to a string they would tend to take their final position on a parabola, this would have been true also for an infinite number of weights, that is for a chain or rope. But Galileo necessarily failed and thus the whole idea did not appear in the publication of his results in the *Discorsi*.

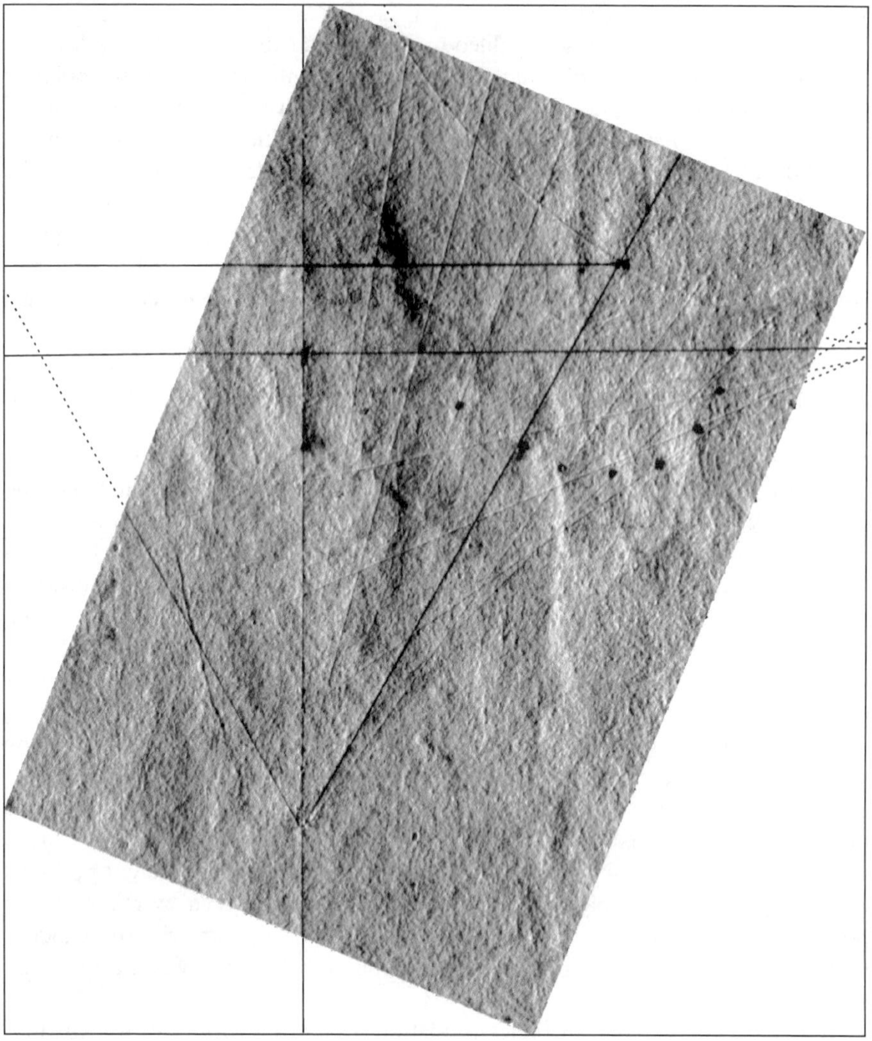

Figure 13. Detail of folio page 107 verso of Galileo's notes on motion showing parts of the figure in the center of this page together with construction lines drawn without ink.

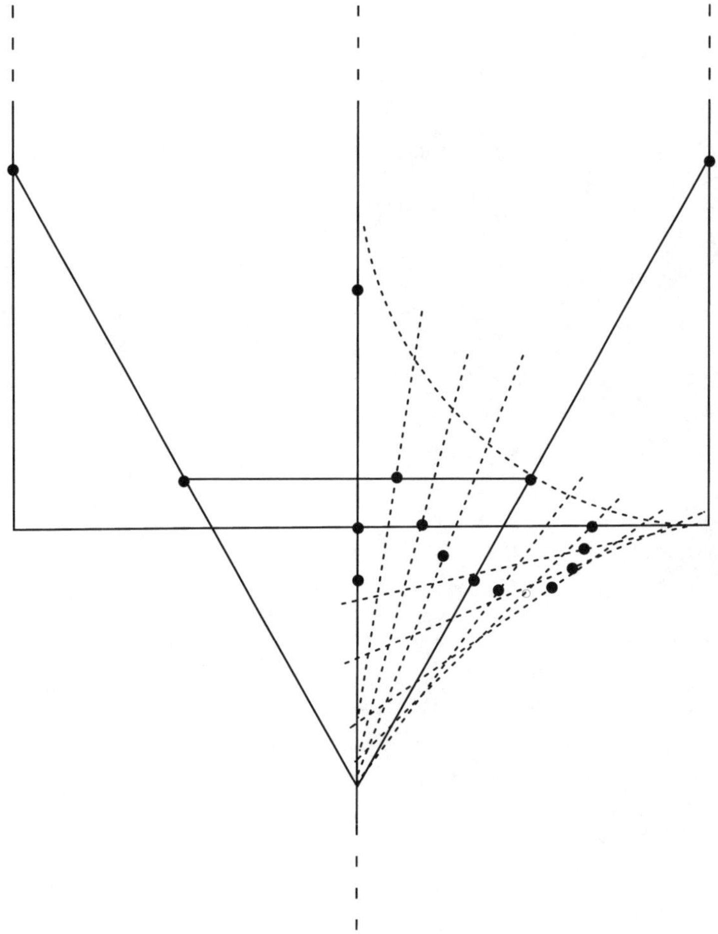

Figure 14. Galileo's construction for finding the lowest position of the center of gravity of three weights attached to a string.

Galileo's failed attempts to empirically validate and theoretically prove the parabolic shape of the catenary are, however, not the last ones documented by his notes. There is overwhelming evidence that the futile search for a satisfactory proof of the parabolic shape of the projectile trajectory directed his attention again to the alleged close relation between the latter and the curve of a hanging chain.

Galileo's *Discorsi*, composed near the end of his life, ended with the *Fourth Day* which is the last part he managed to bring into a satisfying form. He actually planned to complete the *Discorsi* with a *Fifth Day* which, among other topics, was intended to comprise a proof of the alleged parabolic shape of the catenary and an explanation of the practical utility of chains for determining projectile trajectories in artillery.

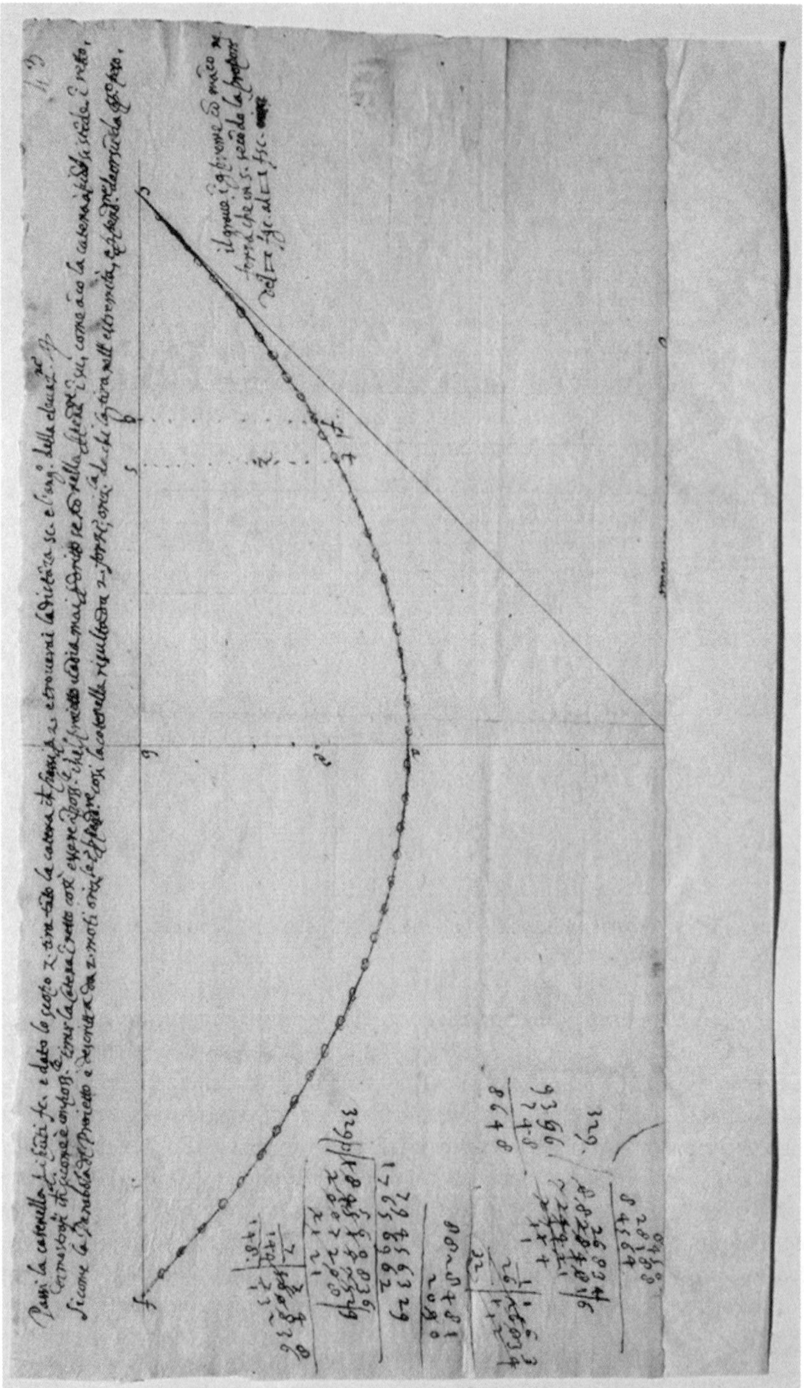

Figure 15. Folio page 43 recto of Galileo's notes on motion with a sketch of a proof for his erroneous assumption that the shape of the catenary is parabolic.

An important source for the reconstruction of the proof Galileo intended to publish is provided by later recollections of his disciple, Vincenzio Viviani, who was living with Galileo from late 1638 until the latter's death in 1642. Viviani writes:[13]

> Now all I have left to say is how much I know about the use of chains, promised by Galileo at the end of the *Fourth Day*, referring to it as he intimated when, he being present, I was studying his science of projectiles. It seemed to me then that he intended to make use of some kind of very thin chains hanging from their extremities over a plane surface, to deduce from their diverse tensions the law and the practice of shooting with artillery to a given objective.

Viviani continues with a sketch of Galileo's proof of the parabolic shape of the catenary, a proof that is based on Galileo's theory of the strength of materials.

The reliability of Viviani's report is confirmed by a folio page in Galileo's own hand, folio page 43r, which we were able to identify among his notes on motion (fig. 15).[14]

The page contains the drawing of a chain line (marked by little rings) representing a projectile trajectory. The text along the line connecting the suspension points concerns the practical utility of the chain for determining the angle of artillery shots at a given target:

> Let the little chain pass through the points *f* and *c*, and, given the target *z*, stretch the chain so much that it passes through *z*, and you will find the distance *sc* and the angle of elevation etc.

The short note written next to the chain line is particularly important in our context because it perfectly resembles Viviani's later recollection:

> the heavy body in *g* presses with less force than in s according to the proportion of the rectangle *fgc* to the rectangle *fsc*

Just as in the longer explanation by Viviani, this short text identifies the downward force in the points of a chain with the limit resistances of a beam supported at both ends as is determined in one of Galileo's theorems. The application of his proposition on the stability of beams to the hanging chain is essentially based on the idea made explicit in Viviani's later recollection that a chain can be conceived of as a beam which is cut in small pieces and linked in a way that makes them move down in proportion to the moments or forces acting on them as if they were weights suspended from the corresponding points on the beam.

From the viewpoint of classical mechanics this argument leads only to an approximation of the catenary because it is presupposed that the matter of the chain is equally distributed along the horizontal whereas it is, in fact, equally distributed along the hanging chain itself. Galileo, of course, did not have the mathematical means necessary to take the length of the hanging chain within a given horizontal interval into account.

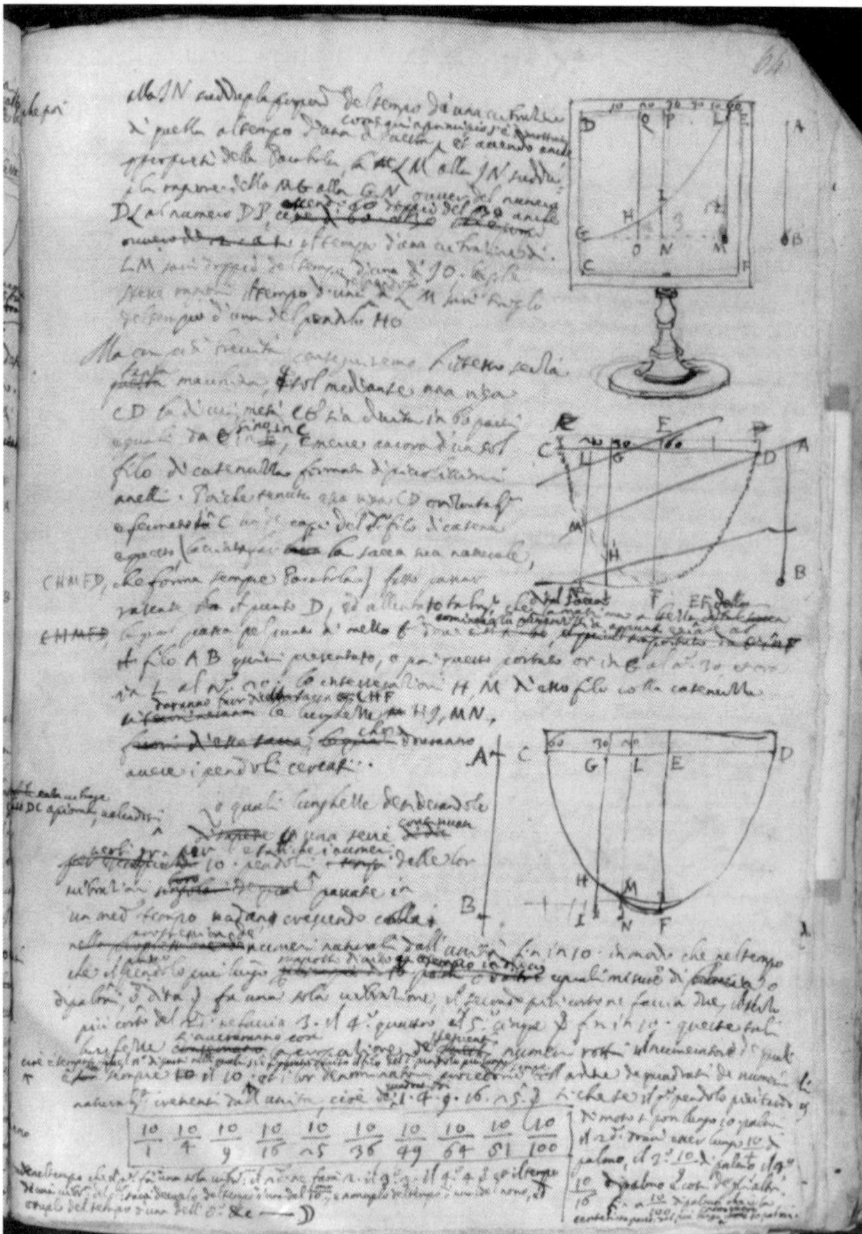

Figure 16. Vincenzio Viviani's application of a hanging chain for determining the length of a pendulum so that it swings with a given period of time.

The misfortune of the *Fifth Day* of the *Discorsi* did not also entirely seal the fate of the chain as a subject of the theory of motion. It saw a striking revival in a context we had to neglect here, that of Galileo's study of the motion of the pendulum. His faithful disciple Viviani not only made sure that Galileo's proof of the parabolic shape of the catenary was preserved for posterity but also thought of giving practical significance to Galileo's discovery. He designed an instrument consisting of a horizontal rod with a chain hanging underneath and used the supposedly parabolic shape of the catenary for determining the lengths of pendulums swinging with a given period of time (fig. 16).[15]

In conclusion, let us briefly review the evidence we have discussed. When and how was the law of fall discovered? According to standard criteria in the history of science it must have been the moment when Galileo and Guidobaldo del Monte performed their projectile trajectory experiment together. But the interpretation they presented for the puzzling symmetry of the trajectory was different from the one provided later by classical physics. Instead of interpreting the trajectory as compounded of an inertial motion in the direction of the shot and free fall, they assumed that the symmetry must result from a symmetric relation between force and natural downward tendency that they conceived as being the same as in the case of a hanging chain.

Given that neither Guidobaldo nor Galileo initially recognized the theoretical consequences of the outcome of their experiment, would it not be better to date their discovery not to the year when the experiment was performed but rather to a later time when its implications in classical physics were derived? Should not the fallacious original interpretation be considered a mere event in the context of discovery, which is irrelevant for the theoretical justification of the parabolic shape of the projectile trajectory? But this makes things even worse. First, as we have seen, Galileo never gave up his idea of a common interpretation of the trajectory and the catenary. Second, what from a later perspective appears as an error was at the time of Galileo the first solution to a problem never before tackled, the theoretical derivation of the catenary, an ingenious achievement unrecognized by historians of science. And last but not least, what Guidobaldo and Galileo applied was not the cunning reasoning of exceptional minds but the commonly accepted conceptual framework of Aristotelian physics. It therefore makes no sense to search for the moment of discovery outside the ordinary work documented in the working papers of people like Galileo. Of course, it may well be that the personal notes of the heroes of science sometimes document highlights or turning points in the development of human knowledge. But what we normally gain from the study of such personal notes is, as we have seen in the case of Galileo's notes on the hanging chain, an understanding of productive intellectual work structured by categories which are difficult to understand because they are not the scientific concepts of our times.

NOTES

[1] This paper gives a short summary of arguments extensively presented elsewhere, see Renn, Damerow, and Rieger 2002.
[2] Galilei 1974, p.142f.
[3] See Galilei ca. 1602–1637. An electronic representation of Galileo's notes on mechanics has been made available by a joint research project of the Max Planck Institute for the History of Science (MPIWG) in Berlin, the Biblioteca Nazionale Centrale and the Istituto e Museo di Storia della Scienza (IMSS), both in Florence; this electronic representation is freely accessible from the websites of the IMSS, www.imss.fi.it and the MPIWG, www.mpiwg-berlin.mpg.de, see also Damerow and Renn 1999 (1998).
[4] See Galilei after 1638 folio page 90v.
[5] See del Monte ca. 1587–1592.
[6] For an extensive discussion see Renn, Damerow, and Rieger 2002.
[7] See folio 175v in Galilei ca. 1602–1637.
[8] See Damerow, Freudenthal, McLaughlin, and Renn 1992.
[9] See folio 107r in Galilei ca. 1602–1637.
[10] See Working-Group 1996.
[11] See folio 107v in Galilei ca. 1602–1637.
[12] See folio 132 in Galilei ca. 1602–1637.
[13] See Viviani 1674, p. 105f.
[14] See folio 43r in Galilei ca. 1602–1637.
[15] See folio page 164r of Viviani after 1638.

REFERENCES

Damerow, P., Freudenthal, G., McLaughlin, P., and Renn, J. (1992), *Exploring the Limits of Preclassical Mechanics* (New York: Springer).

Damerow, P., and Renn, J. (1999 (1998)), "Galileo at Work: His Complete Notes on Motion in an Electronic Representation," *Nuncius* 13 (2): 781–789.

del Monte, G. (ca. 1587-1592), *Meditantiunculae Guidi Ubaldi e marchionibus Montis Santae Mariae de rebus mathematicis* (Paris: Bibliothèque Nationale de Paris), manuscript, Lat. 10246.

Galilei, G. (1974), *Two New Sciences*, ed. S. Drake (Madison: The University of Wisconsin Press).

Galilei, G. (after 1638), *Discorsi (annotated copy of Galileo)* (Florence: Biblioteca Nazionale Centrale, Florence), Manuscript, ms. Gal. 79.

Galilei, G. (ca. 1602-1637), *Notes on Motion* (Florence: Biblioteca Nazionale Centrale, Florence), Manuscript, ms. Gal. 72.

Renn, J., Damerow, P., and Rieger, S. (2002), "Hunting the White Elephant: When and How did Galileo Discover the Law of Fall?," *Galileo in Context*, ed. J. Renn (Cambridge: Cambridge University Press). First published in *Science in Context* 13 (3-4).

Viviani, V. (1674), *Quinto libro degli Elementi di Euclide, ovvero scienza universale delle proporzioni, spiegata colla dottrina del Galileo* (Florence: Condotta).

Viviani, V. (after 1638), *Notes on Mechanical Problems* (Florence: Biblioteca Nazionale Centrale, Florence), Manuscript, ms. Gal. 227.

Working Group (1996), *Pilot Study for a Systematic PIXE Analysis of the Ink Types in Galileo's Ms. 72: Project Report No. 1*, Preprint 54 (Berlin: Max Planck Institute for the History of Science).

WILLIAM R. NEWMAN* AND LAWRENCE M. PRINCIPE**

THE CHYMICAL LABORATORY NOTEBOOKS OF GEORGE STARKEY†

INTRODUCTION

Few areas in the history of science stand to gain more by examination of laboratory notebooks than alchemy. The luxuriant imagery permeating early modern alchemical texts that were intended for manuscript circulation or publication is often absent from the dry recipes and processes described in the working notebooks kept by alchemists for their own personal use. In practical terms, this means that the notebooks, when they can be found, provide important tools of interpretation that allow us to penetrate the complex maze of symbolism often found in the "public" texts prepared for the eyes of others. More than this, the very existence of such notebooks belies the common opinion that alchemy was concerned more with visionary experiences and otherworldly speculation than it was with the facts of the laboratory. In the present paper, we shall compare the printed texts and laboratory notebooks written by one of the most famous alchemical authors of the seventeenth century in order to illustrate these points.

In 1667, a curious text filled with the extended conceits typical of early modern alchemy was published by the Dutch printing firm Janson and Weyerstraet of Amsterdam. The work, called *Introitus apertus ad occlusum regis palatium (An Open Entrance to the Closed Palace of the King)*, and attributed to one "Eirenaeus Philalethes" (A Peaceful Lover of Truth) went on to become one of the most celebrated texts in the history of early modern alchemy: it was extensively commented upon by Isaac Newton, favorably received by John Locke, and diligently read by Robert Boyle.[1] Its author employs the full panoply of traditional alchemical cover-names – *Decknamen* – to describe the veiled processes that he employs. The author tells us that in order to make the Philosophers' Stone, the agent of metallic transmutation, it is necessary to begin with a "chaos," a primordial matter rather like Aristotle's *protē hylē* from which one must make a special "sophic mercury." The latter, when sealed up with gold and heated, is supposed to mature, eventually, into the Philosophers' Stone. In order to capture the peculiar flavor of this text, let us consider the following passage:

---

* Indiana University, Bloomington, IN
** Johns Hopkins University, Baltimore, MD

Frederic L. Holmes, Jürgen Renn and Hans-Jörg Rheinberger (eds.), Reworking the Bench: Research Notebooks in the History of Science, 25–41 © 2003 Kluwer Academic Publishers. Printed in Great Britain

> Let four parts of our fiery dragon be taken, which hides the magical Chalybs [steel] in its belly, and nine parts of our magnet. Mix [them] by means of torrid Vulcan, in the form of a mineral water, on which a scum will float which must be rejected. Throw out the shell and retain the kernel, purge thrice with fire and sun, which will be easy if Saturn has seen his own form in the mirror of Mars. Thus our chameleon or chaos will come to be, in which all secrets lie *in potentia*. This is the infant hermaphrodite, who was infected in his cradle by the rabid Corascene dog, whence he raves with perpetual hydrophobia, although water lies closer to him than any other natural thing. But he fears and flees it, Oh [horrid] fate! But there are in Diana's woods two doves, which pacify his insane rabies.[2]

Philalethes goes on to say that the rabid hermaphrodite must be assuaged with Diana's doves and then drowned in water. He will then re-emerge as a "blackening dog." After having turned into an eagle, the former dog must finally fly away from the dead doves seven times. The result will be a brilliant, solvent substance, the sophic mercury.

Some of these themes were in turn illustrated in the 1695 collection of Philalethes' *Opera omnia*, published in Modena (Fig. 1).

Beneath the coiled snake to the left, one can make out two dogs, one of them picking at a reclining figure. This is probably a reference to the above passage, where the rabid Corascene dog bites the hermaphrodite, who is then drowned by the doves of Diana, only to resurface as a blackening dog.

The rococo imagery of the *Introitus apertus,* with its language of venomous monsters, murderous gods, and ravening hermaphrodites was enough to make

Figure 1. From [Eirenaeus] Philalethes *Opera omnia* (Modena: Fortunianus Rosatus, 1695).

Richard Westfall – attempting to explain Philalethes' influence on Isaac Newton – turn despairingly to religious and extra-rational factors.[3] Nor was he alone in invoking the irrational as a means of making sense out of such alchemical imagery. Betty Jo Teeter Dobbs, in her famous *Foundations of Newton's Alchemy*, explicitly used the analytical psychology of Carl Jung to decipher the bizarre symbolism of alchemy, and her example has been followed more recently by Marco Beretta, Allison Coudert, and a host of other historians.[4] Indeed, Jung himself commented on the very Philalethes passage that we just cited: Jung interprets the passage as a description of the healing of the unconscious and conscious minds by their unification in what he calls "the self." Thus Jung says – "It is clear that [the rabid dog] refers to a psychic disturbance ... ."[5] The drowning of the dog and its subsequent apotheosis are symbols of the disturbed psyche transforming itself.

However tortured this interpretation may seem to the psychological agnostic, it perfectly reflects the widespread Jungian view that alchemy was not fundamentally concerned with the facts of chemistry, but was, rather, a sort of proto-psychology. Jung expressed this view programmatically as follows:

> [In alchemy] we are called upon to deal, not with chemical experimentations as such, but with something resembling psychic processes expressed in pseudo-chemical language.[6]

According to the Jungian interpretation, alchemists were concerned less with chemical reactions than with psychic states taking place within themselves. The practice of alchemy involved the use of "active imagination" on the part of the would-be adeptus, which led to a hallucinatory state in which he "projected" the contents of his psyche onto the matter within his alembic.[7] The Jungian alchemist literally "saw" his own unconscious expressing itself in the form of bizarre archetypal images, which were "irruptions" of the collective unconscious into his conscious mind. Because he views the primary role of alchemy in the light of the unconscious, Jung and his followers today pointedly devalue the chemical content of alchemical texts. The alchemist's "experience had nothing to do with matter in itself," and consequently, the attempt to decipher alchemical texts from a chemical point of view is quite "hopeless."[8]

This position is rendered utterly untenable by the fact that the author of the *Introitus apertus* himself gave a chemical interpretation to the passage cited above. It is now an established fact that "Eirenaeus Philalethes" was actually George Starkey, a native of the New World and graduate of Harvard College who immigrated to London in 1650.[9] In the early 1650's Starkey was closely allied with Robert Boyle; the two prepared iatrochemical medicaments together, and it is clear from Starkey's letters to Boyle that the latter subsidized Starkey's research.[10] The earliest of Starkey's letters to Boyle, written in spring 1651, contains a section called "A Key into Antimony," which Betty Jo Dobbs published from a Latin transcript in the hand of Newton, mistakenly thinking that it was a Newtonian composition.[11] In this "Key" Starkey reveals that the sophic mercury must be made from "star regulus of antimony," a crystalline form of metallic antimony that is produced by heating nine parts of stibnite, or antimony sulfide, with four parts of iron. The iron reduces the metallic antimony by combining with the sulfur in the stibnite [$Sb_2S_3 + 3Fe \rightarrow 2Sb + 3FeS$]. The "Chalybs" of the *Introitus*, then, is the iron, the "magnes" the stibnite, and the "chaos" is the refined antimony. The term "chaos" refers in part to the volatility of antimony, in the same sense that J. B. Van Helmont used when he derived the term "gas" from "Chaos." As Starkey also tells Boyle, the two doves of Diana refer to two parts of refined silver, which must be fused with the regulus in order to make the metallic antimony capable of amalgamating with quicksilver, the water in which the dog is drowned. The product of this recipe is the "sophic mercury," which must be sealed up with gold and heated, to produce a metallic "vegetation" that Starkey thought would ultimately mature into the philosophers' stone. Modern laboratory replication has shown that Starkey's process does indeed yield striking dendritic growths that require no hallucination to appear in the shape of trees.[12]

An analysis of Starkey's letters makes it clear that he was far from the Jungian stereotype of alchemist as self-absorbed mystic. Yet we must not forget that Boyle was Starkey's patron, and that it was therefore incumbent on the impoverished American to impress his Maecenas at every opportunity. For this and other reasons, Starkey invented the fictive character of Eirenaeus Philalethes, whom he claimed to be an associate of his, still living in New England. Starkey asserted that the anonymous adept had given him manuscripts to circulate, among them the *Introitus apertus,* a bit of the Philosophers' Stone (actually the argyropoetic stone) and some sophic mercury. He was attempting, or so he told Boyle, to replicate Philalethes' recipe for the latter.[13] From this intentional obfuscation it is quite clear that we can expect to learn little of Starkey's real sources for the sophic mercury from his letters to Boyle.

Indeed, Starkey went to great lengths to make Boyle and his associates think that he had learned his alchemical secrets from dreams and revelations made especially to him, by God.[14] One such dream describes Starkey's visit from his own "Eugenius" or tutelary spirit, who revealed the secret of Van Helmont's marvelous alchahest, or universal solvent, to Starkey. There was only one problem – the revelation, as Starkey informed Boyle, was "veiled, as it were, by fog," and "was more obscure than Paracelsus himself." Starkey then gives a chymical explanation of his dream, but the elaborate oneiric framing, largely absent from his laboratory notebooks, contains a further revelation for the modern reader: while Starkey's letters to Boyle are extremely helpful in explaining both the theory and the practice underlying the Philalethes texts, they do not tell us the pathways by which Starkey arrived at that theory and practice.

It is therefore fortunate that a number of Starkey's laboratory notebooks have survived from the 1650's, the very time when he was composing his major Philalethan writings: British Library mss. Sloane 3711 and Sloane 3750, and Royal Society ms. 179 are the most important of these.[15] An abundance of citations in these notebooks reveal that Starkey's major source for his experimentation with antimony was the Prussian iatrochemist Alexander von Suchten, who wrote an influential *Antimonii mysteria gemina*, published in the early seventeenth century. The second part of this was translated into English, probably by a member of the Hartlib Circle, Starkey's friend Dr. Robert Child.[16] Suchten openly describes the very process that Starkey reveals in coded language in the *Introitus*, complete with the two parts of purged silver that are necessary in order to make the antimony and mercury amalgamate.[17] Starkey's repeated references to Suchten in the notebooks – especially in Sloane 3750 – make it absolutely clear that he was using the German alchemist as his fundamental source, and that the story of the mysterious American adept was an embellishment.

But the notebooks are significant for far more than their revelation of Starkey's major source. Their straightforward and orderly character is the very antithesis of the menagerie-gone-berserk that populates Starkey's Philalethan writings. Starkey's notebooks are perhaps unique in the history of alchemy in providing the plaintext that allows us to decode the author's veiled public works into operational processes. Indeed, in the following, we will argue that Starkey's notebooks are models of scientific clarity, in marked contradistinction to the image of alchemy propounded by the Jungians and found in all-too-many historical treatments of the subject.

GENERAL CHARACTERISTICS OF STARKEY'S NOTEBOOKS

The first thing that strikes one's eye when viewing Starkey's notebooks is the presence of standard, recurring, marginal tags. These are not the usual entries such as "Recipe," and "NB" that are typical of alchemical codices and recipe-books from the Middle Ages onward. Rather, Starkey has tags such as *Processus conjecturalis, Conclusio probabilis, Quaere, Proba, Observatio, Animadversio,* and even *igne refutata.* These formulaic entries strongly recall the standard divisions of a scholastic *quaestio disputata* into elements such as *quaestio, responsio, oppositum, dubitatio, argumentum quod sic, argumentum quod non, refutatio*, and *vera solutio.* One of Starkey's entries in Sloane 3750 even divides his argument into numbered affirmative and negative instances – he calls these *conclusiones probabiles negativé* and *affirmativé*.[18] We know that arguments of this general form provided the stock-in-trade of Harvard education during the 1640's, when Starkey was an undergraduate. One such disputation survives in the notebooks of Jonathan Mitchell, a Harvard AB of 1647: the argument was disputed on April 3, 1646.[19] Mitchell's disputation concerns a metaphysical issue, namely whether a cause remains present in its effect. What concerns us is the form of the argument, which is divided into *Quaestio, Neg[atio], Oppositum*, and *Responsio.* This form of argumentation at Harvard was frequently accompanied by numbered *Theses* as well. A subsequent folio of the Mitchell ms. contains numbered rhetorical theses drawn from the work of Peter Ramus, the famous Calvinist dialectician who was murdered in the St. Bartholomew's Day massacre in 1572. These notes in turn are reminiscent of Starkey's numbered journal entries: the practice at Harvard was to extract pithy nuggets from primary sources and then to arrange them into a

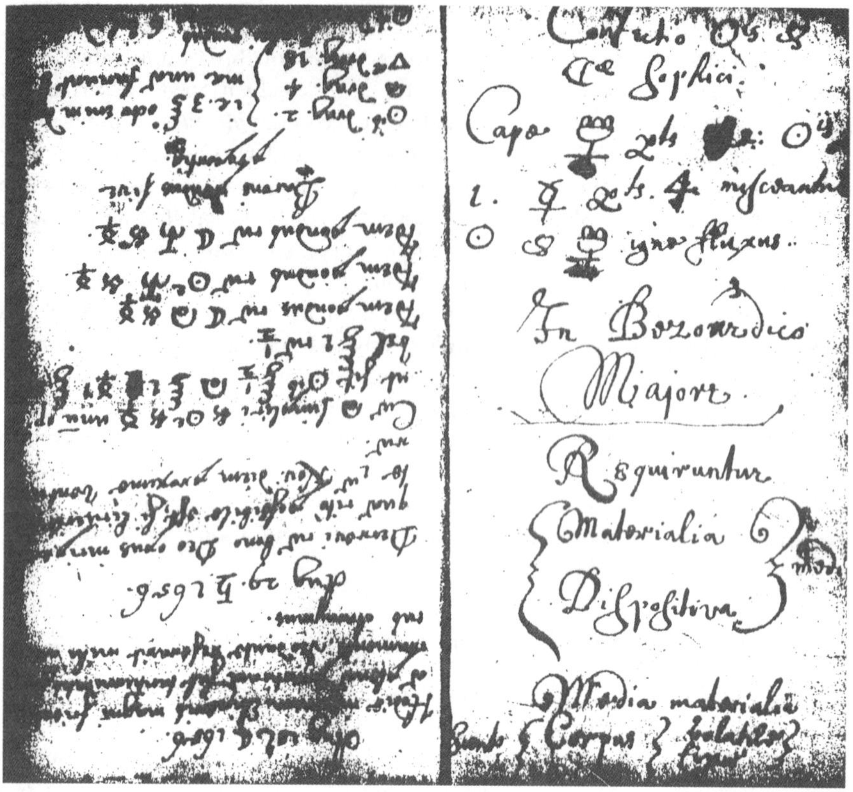

Figure 2a. From Royal Society ms. 179.

numbered sequence. This was sometimes referred to as "epitomizing" a work: candidates for the AM were supposed to prepare a "Synopsis, or Compendium" of some art, also called a "System," based on these methods.[20]

The formal similarities between Starkey's notebooks and those of other seventeenth-century Harvard graduates extend into the area of dichotomy-charts as well. Popularized by Peter Ramus in the mid-sixteenth century, these outlines were organized primarily around bifurcations indicated by swung brackets. An example of such a chart appears in Royal Society ms. 179, a Starkey notebook from the mid–1650's (Fig. 2a and 2b).

Starkey was unable to expand his chart in the usual fashion due to the small format of this notebook: if he had done so, it would look like this (Fig. 3).

The term "Greater Bezoar" here is a trope for the Philosophers' Stone, which, like the quasi-legendary bezoar stone, was supposed to be a universal antidote. The chart lays out the various materials and methods that are required in order to produce the sophic mercury and to convert it, with an addition of gold, into the Philosophers' Stone. If we compare Starkey's chart to that found in a so-called *Logic System* prepared or copied by William Partridge, a Harvard A.B. of 1689, the same method of organization is also presented (Fig. 4).

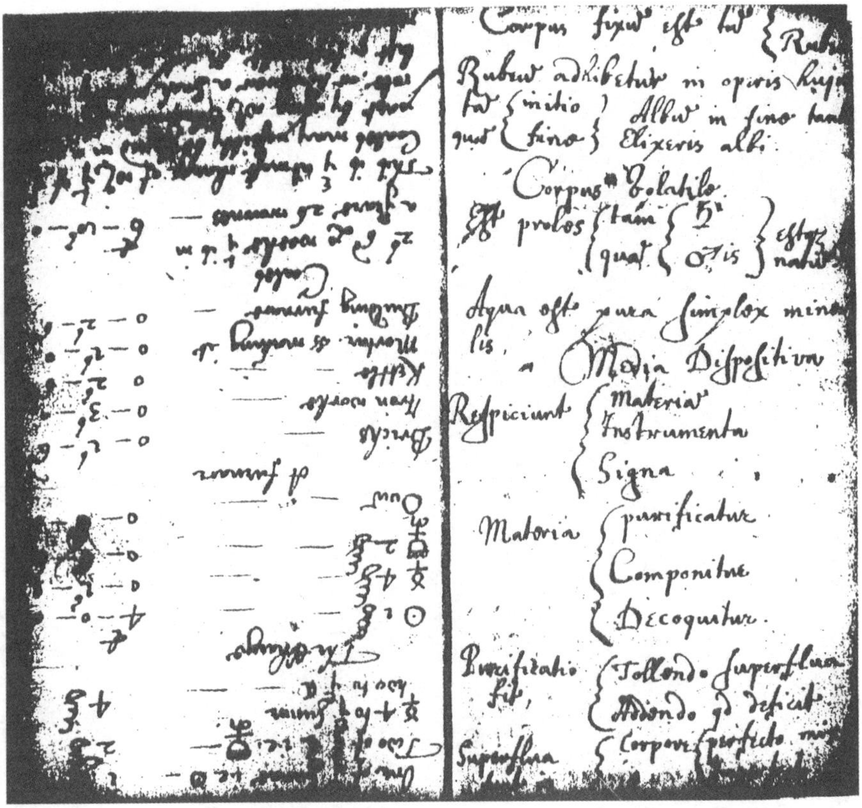

Figure 2b. From Royal Society ms. 179.

The existence of these dichotomy charts in both Starkey's and Partridge's notebooks is consistent with the pervasive influence of Peter Ramus at early Harvard. A famous letter from Leonard Hoar, a Harvard A.B. of 1650 who went on to become President of the College between 1672–1675, expressly advises his nephew Josiah Flynt to study the note-taking "method of the incomparable P. Ramus," as a means of navigating his Freshman year.[21]

We can see, then, that Starkey's laboratory notebooks retain traces of the note-taking and study techniques that he was taught at Harvard College. Indeed, Starkey even tells us in his exposition of Helmontian medicine, *Natures Explication*, that while a student at Harvard he engaged in a disputation concerning potable gold, and that he "thought that the Logical heads of invention, especially according to *Ramus,* would not fail to unfold ... this whole mysterie."[22] As a result of this conviction he wrote a "Congest of methodical Arguments which might unfold how [to make potable gold]," no doubt a "system" of the type that we have described.[23] There are very strong indications, then, that the form of Starkey's notebooks owes a considerable debt to the scholastic methods that he acquired at Harvard College. At the same time, we should point out that such

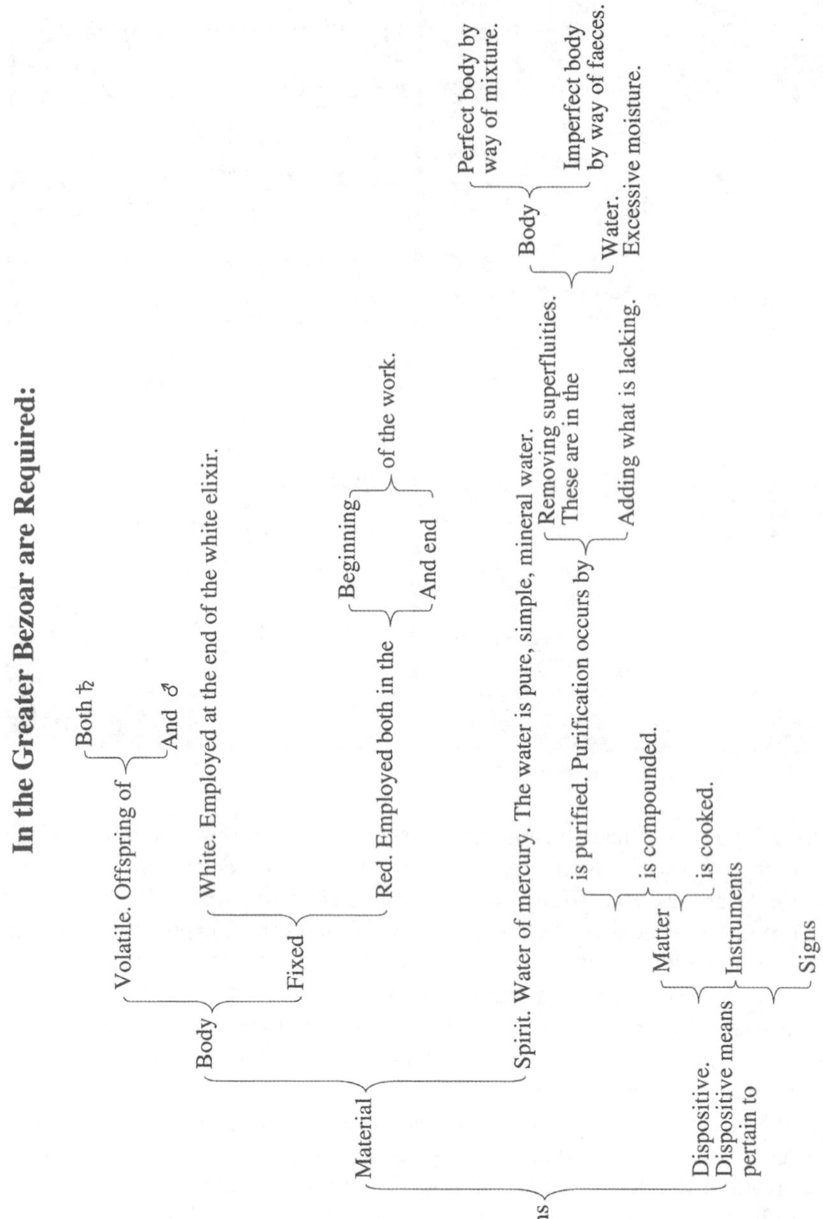

Figure 3. Translated expansion of Starkey's dichotomy chart found in Royal Society ms. 179.

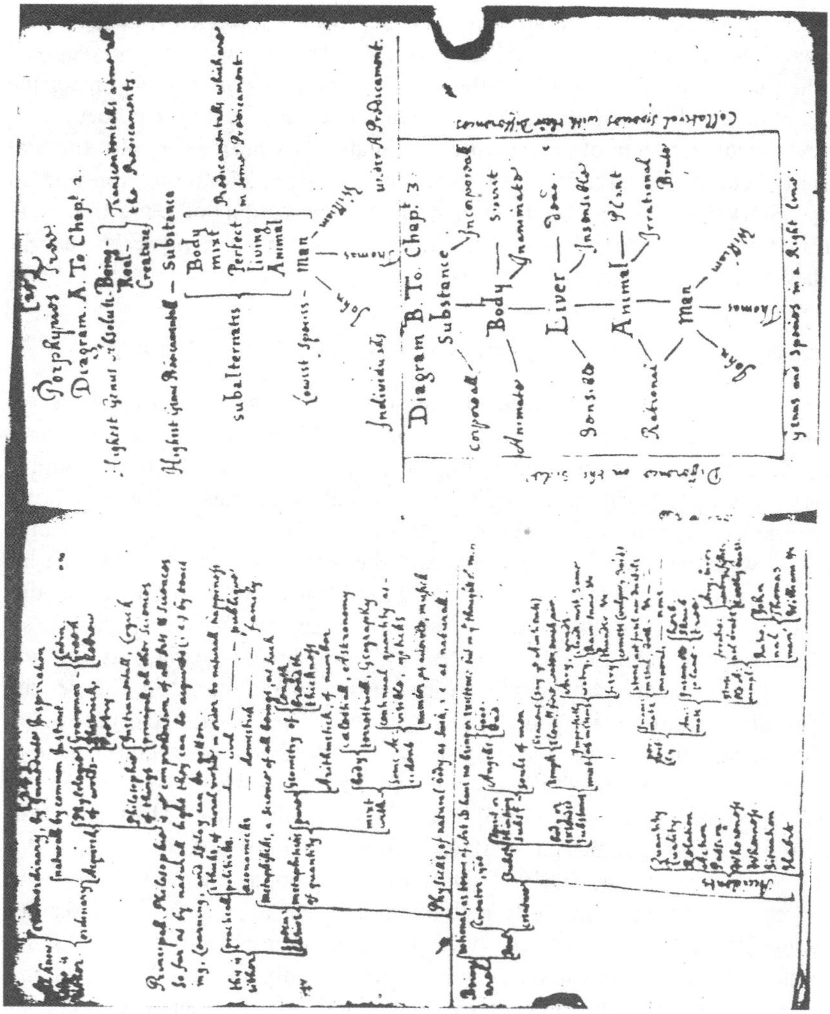

Figure 4. Chart from William Partridge, "Logic System," Yale University, Vault ms. Partridge.

features as Ramist dichotomies and numbered observations were not lacking in the chymical textbooks that Starkey had at his disposal. The famous *Alchemia* of Andreas Libavius, published in 1597, contains dichotomy charts, and both J. B. Van Helmont and Angelus Sala provide numbered lists of observations in their iatrochemical works.[24] What is striking, however, is that Starkey for the most part avoids such features in his own published works, both in the works of chrysopoeia published under the name of Philalethes, and in his works on chymical medicine published under his own name.[25] This lends credence to the argument that Starkey's peculiar system of note-taking was in large part a transfer from the oral and scribal culture of early Harvard rather than a mere adaptation from printed sources. Starkey seems to have been respecting demarcations of genre that were inculcated into his tender brain while he was still a *dominus imberbis*, a "Sir Beardless" at Harvard College.

SOME SPECIFIC EXAMPLES FROM STARKEY'S NOTEBOOKS

Now that we have a general idea of the surprisingly methodical character of Starkey's notebooks, let us pass to some examples. Sloane 3750 has a number of entries devoted to antimony, including the production and uses of the sophic mercury. A passage beginning on folio 23v, and dated "August 29 [1653]," shows that Starkey's experimental program operated on two levels. He was deeply concerned with the cost of the reagents and the labor involved in producing his sophic mercury: he therefore strove to maximize efficiency of production in this area. At the same time, he thought that the sophic mercury was capable of dissolving metals and minerals into their first principles – namely sulfur and mercury. The sophic mercury was therefore a powerful analytical tool for arriving at experimental knowledge of metallic substances. These two goals, productive and analytical, emerge quite clearly from the passage under consideration, especially when compared with Starkey's 1651 letter to Boyle.

The entry for "August 29 [1653]" bears the marginal tag "Most expedient preparation of the antimonial mercury." What Starkey describes under this heading is a process for heating 23 troy ounces of his sophic mercury, mixed with 11 ounces of additional antimony regulus. The object is to make more regulus amalgamate with the sophic mercury without the addition of further silver. After a description of the experiment, Starkey provides a numbered list of *probabilitates* which led him to devise this process. First, he points out, the regulus is a "chaos" from which all the other metals can be drawn. This is a reference not only to the volatility of the chaotic antimony, but to Starkey's belief that antimony was a prime-matter-like substance out of which the metals could be extracted. Dozens of attempts are found in his notebooks to produce the metals in this fashion, which he ultimately owed to Suchten. The implication here seems to be that the chaotic regulus should therefore contain silver *in potentia,* so that the sophic mercury should be able to combine with additional regulus without the need for more silver.

He then points out in the second "Probability" that the melting point of antimony regulus when combined with silver is lower than that of either silver or antimony fused independently, and that the two substances could then be kept in flux together at a temperature slightly above that of molten lead (327° C). This is a correct observation, since the metal and semi-metal form a eutectic alloy when combined, which melts at a lower temperature than either taken separately. Finally, Starkey notes that liquid mercury remains at the bottom of the digestion apparatus at a temperature above that of molten lead, and therefore it should combine directly with the regulus, which will be molten at that temperature, as he noted in the second "Probability." The combination of these observations suggested to Starkey that he should try the experiment of combining sophic mercury and regulus in a systematic fashion.

After this section, Starkey discusses the economic advantages of this new technique, which we will call the "digestion-process." He states that he can now reduce his cost by 16 and one-half shillings per pound of mercury, since he can avoid buying three or four ounces of silver. In addition, the laborious pounding, washing, and grinding that were formerly involved in making additional sophic mercury can now be replaced by simple heating in a flask. Finally, Starkey adds that this process of extended heating will also make it easier to amalgamate copper with the sophic mercury. The process in this case is to begin by fusing antimony regulus with copper to produce an alloy called "the net," a substance employed by Newton as well, and made famous by the work of Betty Jo Dobbs.[26] Starkey quaintly observes that this is the net "in which the poets say Mars and Venus were trapped and captured by Vulcan," in reference to the roles of copper (Venus), iron (Mars), and fire (Vulcan) in its production.[27] Starkey's idea is that this "net" should then be mixed with the sophic mercury and digested with it just as the normal regulus was.

After having described his process for digesting the sophic mercury with regulus or the "net," and giving the reasons that led him to it, Starkey then introduces a new problem. In order to understand this, we must recall that Starkey views his sophic mercury as an analytical agent, which separates metallic substances – including regulus of antimony – into their constituent sulfur and mercury. He views the digestions described above as working precisely by means of such analysis, followed by a unification of the mercurial principle in the antimony with the sophic mercury. With this background, we can understand his further ruminations. Starkey points out that he has repeated the experience of making the sophic mercury itself at least a hundred times. This is a reference to the famous process described in Starkey's Key, the section of his 1651 letter to Boyle that was later transcribed by Newton. In that document, Starkey described the careful grindings of quicksilver with the antimony-silver alloy that had to be followed by washings and further grindings. Each repetition of the grinding and washing carried off some blackness, so that the amalgam was eventually left "as bright as burnished Silver."

In a part of Starkey's letter only recently discovered in a notebook of John

Locke's, Starkey suggests that this black water that was washed from the mercury be allowed to settle, and the black powder then dried out.[28] If this powder is burned in a crucible to become a sort of ash, it can then be reduced to regulus by simple melting. Starkey says that if this new regulus is weighed, it will be found to be less by only ¼ to ⅓ relative to the original weight of antimony regulus amalgamated with the mercury. He then points out to Boyle that this demonstrates that the sophic mercury has not gained any significant weight from the regulus, but rather a "spiritual impregnation" or "acuation." The point of the experiment as presented in the 1651 letter is precisely one of demonstration: Starkey wants to show Boyle that the sophic mercury "grows not in pondus but in fermentall vigour."[29] If we now return to Sloane 3750, we shall see a very similar experiment – not carried out for demonstrative purposes, however, but for the sake of production. Here Starkey points out again that most of the regulus that was added in the production of the sophic mercury can be recovered from the black washings. As in his letter to Boyle, he says that the dried washings must have their "external combustible sulfur" burned off before their reduction. Here, however, Starkey adds that a glass-like scorious matter separates itself from the regulus during its reduction. This scorious matter, along with the combustible sulfur, is now the object of Starkey's anxieties about his digestion-process. Since in Starkey's digestion-process, the sophic mercury works on the regulus with which it is digested by separating the pre-existent sulfur and mercury of the regulus, Starkey fears, reasonably enough, that the separated "sulfur" will eventually impose a mechanical barrier between the antimony and the sophic mercury, thus halting the absorption of the "mercury" of antimony into sophic mercury. He therefore adds another step to the digestion process: at some point it will be necessary to grind and wash the mixture of sophic mercury and regulus of antimony, in order to eliminate the excess "sulfur" that has been liberated.

We see, then, that Starkey musters the same experiment in different contexts for both demonstrative and productive goals. This is fully in accord with his notion of alchemy as an applied science, having both a theoretical and practical component. It is fair to say, nonetheless, that his notebooks focus on the productive rather than the demonstrative. Nonetheless, Starkey's notebooks are filled with explicit examples of laboratory experience used to test or illustrate theories, as the following examples will demonstrate. One of his notebooks contains a so-called *Disquisitio philosophica* that states the following general claim:

> Alcalies are true sulfur fixed with salt, and therefore far more agreeable for preparing sulfurs, which, once reduced into the nature of salt, evince a much greater energy than when elevated into a fatty substance. The immense sulfurous stink which is perceived when sulfur is precipitated back from alcali by vinegar or wine argues for this . . . .[30]

Here the laboratory experience of sulfurous fumes provides direct evidence of the "energy" released during the precipitation of "sulfur" from a lixivium by means of a weak acid. The experiment also provides evidence for the claim (drawn originally from Van Helmont) that alkalies are a form of sulfur rendered non-volatile by a salt, since Starkey assumes that the mutual attraction of the alkali and the "sulfur" results from a kinship of nature.

Starkey's notebooks are generally more tightly linked to specific marginal tags than in the case of his *Disquisitio philosophica*, however. In another example, he begins an entry in the early 1650's with the heading "Observations," followed by some thoughts that have occurred to him while reading Van Helmont. Van Helmont accepted the traditional alchemical theory that minerals and metals are composed of the two principles mercury and sulfur, but deployed a corpuscular interpretation of this, asserting that metallic corpuscles are composed of alternating shells of the principles, and that these shells could be loosened or made to exchange places within a given particle.[31] As Starkey points out, Van Helmont also believed that the mineral acids acted on minerals and metals because of the sulfurous component in the latter. Thus, Starkey asserts, "it is conformable to reason that the sulfur in minerals (in which there is no strong union with the mercury) can be turned outward by corrosives ... ."[32] Therefore, Starkey proceeds, "wouldn't this work better by how much the acid is stronger?"[33] If so, he asks, couldn't one facilitate the process by strengthening his aquafortis (nitric acid) with salt, or niter and ammonium chloride? Starkey then outlines a procedure for dissolving stibnite – antimony sulfide – in such fortified aquafortis and then extracting its so-called sulfur with a lixivium of salt of tartar. He suggests that the "sulfur" may then be isolated either by subliming it or by precipitating with spirit of vinegar.

What is interesting about this entry is that Starkey, after outlining the means of testing his hypothesis, then actually did test it. In this instance the evidence lies in subsequent marginal addenda to his comments, where he says that the sublimation from salt of tartar outlined above was "hardly possible" *(vix possibile)*, but that the precipitation with spirit of vinegar worked "better" *(melior via)*. Thus it is clear that Starkey did test his inference that the Helmontian theory of extraversion implies that a stronger acid will better separate the "sulfur" from crude antimony. But he did not test the Helmontian theory of extraversion itself: this he assumed to be evident. Starkey's primary interest lay in drawing out the implications of a theory for the sake of production, and then trying out the resulting process. He was quite willing to accept the evidence of his senses when an experiment failed, as the the occasional process crossed out and followed by "not possible – opinion refuted by the fire" makes clear.[34]

Such experiments – and we use the term advisedly – make up the bulk of Starkey's notebooks. He remarks, for example, that crude antimony, tartar, and saltpeter flash and thunder like gunpowder when exposed to flame, and that the product of this combustion may be reduced by augmenting the flame.[35] His description of this process is followed by five "Observables" *(Observabilia)*, as he calls them. First, he states the obvious, saying that a separation of the metallic part of the antimony from a "burning, inflammable sulfur" occurred. Second, it is clear that the metallic antimony still contains sulfur, since fresh saltpeter makes it deflagrate again. Third, he notes that the alkalized salts made from the burnt up tartar, which absorb the sulfur when the fire is augmented, are corrosive. Fourth, that the slag quickly turns green when exposed to air, and turns one's fingers yellow, whence it is called "golden sulfur." And fifth, that the separation of the salts from their sulfur by precipitation with vinegar releases a stench and leaves the solution in which the sulfur was dissolved clear.

These observations are followed by eight ruminations called by Starkey "Probable Conclusions" *(Conclusiones probabiles)*. These consist primarily of further queries regarding the separation of the sulfur from metallic antimony, and we need not go into them here. In passing, however, Starkey notes that this method of separating the sulfur of antimony "is of less cost and easier to carry out" than others.[36] This important comment underscores the fact that Starkey's chief goal for these processes was in fact practical. Indeed, he hoped to use his antimonial product as an internal purgative, a gentle diaphoretic and emetic.[37] Such practical goals underlie virtually all of Starkey's laboratory work: he was not intent on devising tests to disprove the Paracelsian theory of matter, as was his mentor Van Helmont, or his student, Robert Boyle. Starkey's alchemy was more closely related to the industrial and pharmaceutical chemistry of today, focusing on the discovery of new compounds by means of the most efficient and cost-effective methods. Laboratory research directed at the discovery of new chemical products and the manufacture of existing ones by more economical means is obviously no less "experimental" than the testing of grand hypotheses. All the same, this enterprise did undeniably engage Starkey in careful testing of individual hypotheses, even when the theories tested did not address themselves to determining the ultimate foundations of matter. Within the numerous examples of "conjectural processes" and "probable conclusions" that populate his work, one finds assumptions about the activity of matter subjected to rigorous scrutiny, and even "refuted by the fire." How can this be brought into conformity with the widespread perception that alchemy was primarily a matter of "psychic processes expressed in pseudo-chemical language"? The answer, quite simply, is that it cannot be and should not be. The laboratory notebooks of George Starkey demonstrate that even in his own private musings the "last great philosophical alchemist" deviated strikingly from the picture that the psychologically influenced historiography has led us to expect.[38] Instead of giving evidence of "psychic disturbance" or even cloaking themselves in the vibrant imagery of the *Introitus apertus ad occlusum regis palatium*, Starkey's notebooks display a remarkable integration of scholastic discipline combined with experimental skill. In the forge of his chymical furnace Starkey managed to harden the tools of seventeenth-century pedagogy into an instrument of scientific research that would not be unfamiliar even to his modern descendants.

NOTES

† We have prepared an annotated edition and translation of Starkey's correspondence and laboratory notebooks, to be published in due course. In addition, we are composing a study of Starkey's collaboration with Robert Boyle, which will focus extensively on the notebooks of the two men and the role of Helmontian chymistry in their work.

[1] William R. Newman, *Gehennical Fire: The Lives of George Starkey, an American Alchemist in the Scientific Revolution* (Cambridge, MA: Harvard University Press, 1994), pp. 2, 54–83, 228–239. For the connection with Locke, see Lawrence M. Principe, *The Aspiring Adept* (Princeton: Princeton University Press, 1998), pp. 175–179. Locke owned three works of Eirenaeus Philalethes: the *Enarratio methodica* of 1678, the *Introitus apertus* of 1667, and *Secrets Reveal'd* of 1669: see John Harrison and Peter Laslett, *The Library of John Locke* (Oxford: Oxford Bibliographical Society, 1965).

[2] Eirenaeus Philalethes, "Introitus apertus ad occlusum regis palatium," in J. J. Manget, *Bibliotheca chemica curiosa* (Geneva, 1702), vol. II, 663–664: "Sumantur Draconis nostri ignei, qui in ventre suo Chalybem occultat magicum, partes quatuor, Magnetis nostri partes novem, misce simul per Vulcanum torridum, in forma mineralis Aquae, cui supernatabit spuma rejicienda. Testam repudia, nucleumque selige, purga tertia vice, per ignem et Solem, quod facile fiet, si Saturnus in speculo Martis suam formam aspexerit. Fiet inde Chamaeleon sive Chaos nostrum, in quo latent omnia arcana virtute non actu. Hic est infans Hermaphroditus, qui a primis suis incunabilis per Canem Corascenum rabidum morsu infectus est, unde perpetua hydrophobia stultescit, insanitque, imo licet aqua sic sibi quavis re naturali propinquior, tamen illam horret ac fugit, o fata! Sunt tamen in sylva Dianae binae Columbae, quae rabiem suam insanam mulcent."

[3] Richard S. Westfall, *Never at Rest* (Cambridge: Cambridge University Press, 1980), pp. 20–22, 299–301. See also Westfall, "The Role of Alchemy in Newton's Career," in M. L. Righini-Bonelli and W. R. Shea, *Reason, Experiment, and Mysticism in the Scientific Revolution* (New York: Science History Publications, 1975), pp. 198–199, 213–214.

[4] Betty Jo Teeter Dobbs, *The Foundations of Newton's Alchemy* (Cambridge: Cambridge University Press, 1975), pp. 26–35. Marco Beretta, *The Enlightenment of Matter* (Canton, MA: Science History Publications, 1993), p. 77, n. 6 and p. 331. Allison Coudert, *Alchemy: The Philosopher's Stone* (London: Wildwood House, 1980), pp. 148–160. See also Gareth Roberts, *The Mirror of Alchemy* (London: The British Library, 1994), p. 7, and p. 66; William H. Brock, *The Norton History of Chemistry* (New York: Norton, 1993), p. 17 and p. 678; E. J. Holmyard, *Alchemy* (New York: Dover, 1990; first edition, 1957), pp. 163–164, 176.

[5] Carl Jung, *Mysterium conjunctionis* (Princeton, 1970, second edition), p. 155.

[6] Carl Jung, "The Idea of Redemption in Alchemy," in Stanley Dell, ed., *The Integration of the Personality* (New York: Farrar & Rinehart, 1939), pp. 205–280; p. 210 for quotation (a modified translation of "Die Erlösungsvorstellungen in der Alchemie," *Eranos-Jahrbuch 1936* (Zürich: Rhein-Verlag, 1937), pp. 13–111; a retranslated and much expanded version of the original Eranos lecture appears in *Psychology and Alchemy*, pp. 227–471).

[7] Ibid., p. 215. The German text is quite unequivocal in its reference to "hallucinations" (Jung, "Erlösungsvorstellungen," pp. 23–4): "Wie die beiden vorhergehenden Texte, so beweisen auch Hoghelande's Ausfuehrungen, dass waehrend der praktischen Arbeit halluzinatorische oder visionaere Wahrnehmungen erfolgten, die nichts anderes sein koennen als Projektionen unbewusster Inhalte."

[8] Jung, "Redemption in Alchemy," p. 213, p. 206.

[9] William R. Newman, "The Authorship of the *Introitus apertus ad occlusum regis palatium*," *Alchemy Revisited: Proceedings of the International Conference on the History of Alchemy at the University of Groningen 17–19 April 1989*, ed. Z.R.W.M. von Martels (Leiden, 1990), pp. 139–144.

[10] See our forthcoming study of the Starkey-Boyle collaboration. See also William R. Newman, *Gehennical Fire: The Lives of George Starkey, An American Alchemist in the Scientific Revolution* (Cambridge, MA: Harvard University Press, 1994), pp. 54–83.

[11] William R. Newman, "Newton's *Clavis* as Starkey's *Key*," *Isis* 78 (1987), pp. 564–574.
[12] For the chemical replication of this process and "growth" of gold into a "philosophical tree" by means of the sophic mercury, see Principe, "Apparatus and Reproducibility in Alchemy," in *Instruments and Experimentation in the History of Chemistry*, ed. F. L. Holmes and Trevor Levere (Cambridge, MA: MIT Press, 2000); and for a view of the history of the process before Starkey, see Principe, "Chacun à son Gout: Experimental Continuity and Theoretical Diversity in Sixteenth to Eighteenth-Century Chymistry," in *In Search of An Identity: Chymists and Chymistry in the Seventeenth Century*, ed. Brigitte van Tiggelen and Lawrence M. Principe, forthcoming in *Sudhoffs Archiv*, Beihefte, 2000.
[13] "Starkey to Boyle," [spring 1651], in *Oxford University, Bodleian Library*, ms. Locke c44, p. 149.
[14] Newman, *Gehennical Fire*, pp. 64–67.
[15] All of the surviving notebooks, along with Starkey's correspondence, have been edited, translated, and annotated by us and will appear in due course.
[16] Alexander von Suchten, *Antimonii mysteria gemina* (Leipzig: Jacob Apel, 1604). The attribution to Child is made in University of Glasgow, ms. Ferguson 163, p. 97: "Alexander van Suchten of the Secrets of [antimony] translated out of the high Dutch by D[r] Child."
[17] Alexander von Suchten, *Second Treatise of Antimony Vulgar*, University of Sheffield, Hartlib Papers 16/1/48–63. The text of HP 16/1/48–63 being corrupt at this point, we have had to rely on the printed version – Alex. von Suchten, *Of the Secrets of Antimony* (London: Moses Pitt, 1670), pp. 76–79.
[18] British Library, ms. Sloane 3750, 6r.
[19] Samuel Eliot Morison, *Harvard College in the Seventeenth Century* (Cambridge, MA: Harvard University Press, 1936), vol. 1, pp. 143–144.
[20] Morison, vol. 1, pp. 149, 155–157. For Starkey's own comments on these methods, see his *Natures Explication and Helmont's Vindication* (London: Thomas Alsop, 1657), p. 20.
[21] Morison, vol. 2, pp. 639–644.
[22] For Starkey's education at Harvard College, see Newman, *Gehennical Fire*, pp. 18–39, 46–50.
[23] Starkey, *Natures Explication*, pp. 35–36.
[24] Angelus Sala, *Opera Medico-Chymica quae extant omnia* (Rothomagi: Ioannis Berthelin, 1650), pp. 212–213 (numbered *porismata* from *Chrysologia*); pp. 345–346 (*Anatomia vitrioli: dubia* and *responsiones*). J. B. Van Helmont, *Opuscula medica inaudita* (Leiden: Elsevier, 1648), p. 37a (numbered *notabiles* from *De lithiasi*; Van Helmont, *Ortus medicinae* (Leiden: Elsevier, 1648), pp. 500–501 (numbered *notabilia* following *mechanica*, from *Ignotus hospes morbus*).
[25] There are exceptions, however, as in Starkey's *Natures Explication*, pp. 314–317, where one finds numbered points very similar in style to those of Starkey's notebooks.
[26] Dobbs, *Foundations*, pp. 161–164, 167.
[27] British Library, ms. Sloane 3750, fol. 25v.
[28] Oxford University, Bodleian Library, ms. Locke c44, p. 149.
[29] Starkey has another demonstrative goal for this experiment as well, as appears in his *Sir George Riplye's Epistle to King Edward unfolded*, pp. 29–31. Here too he is intent to show that in making the sophic mercury, one must "extract not the Pondus, but the celestial virtue," yet he is also eager to demonstrate that the vulgar quicksilver is purged by the regulus in the sophic mercury's manufacture, just as the regulus is purged by the sophic mercury in Sloane 3750, 26r: "So is it in our body, the fermental spirit that is in it, is scarce a third part of the whole, the rest is of no value, yet all is joyned in the composition, and the feculent corporeous, part of the body comes away with the dreggs of the Mercury. But beyond the example given of a grain, it may be observed that the hidden and spiritual vertue of this our body, doth purge and purifie its matrix of water, in which it is sowen, that is, it makes it cast forth a great quantity of filthy earth, and a great deal of Hydropical saline moisture. For instance make thy washings (for a tryall) with pure and clean fountain-water, weigh first a pint of the same water, and take the exact weight of it, then wash thy compound eight or ten times, save all the faeces, weigh thy body and Mercury exactly, weigh thy faeces being very dry, then distill or sublime all that will sublime a very little quick Mercury will ascend; then put the

Residue of the faeces in a crucible, set them on the coals, and all the faeculency of the Mercury will burn like a coal, yet without fume; when that is all consumed, weigh the remaining faeces, and thou shalt find them to be two thirds of thy body, the others being in the Mercury, weigh the Mercury which thou sublimest, and the Mercury prepared by itself, and the weight of both will not recompence thy Mercury weight by farre. So then boile up thy water to a skin, in which thou madest the Lotions, for that is a thick water: and in a cool place thou shalt have Christals, which is the salt of Mercury Crude, and in no way fit for Medicines; yet it is a content for the Artists to see how the Heterogeneyties of Mercury are discovered . . . ." This passage is inspired by Alexander von Suchten's analysis of antimony in his *Second Treatise of Antimony vulgar*, as in Hartlib Papers 16/1/56a-b, but Starkey has added additional steps so that the procedure becomes an analysis of mercury as well.

30 British Library, Sloane ms. 3711, f. 6v: "Alcalia sunt verum [sulphur] cum sale fixatum . . . . Hoc arguit immensus [sulphur]eus faetor qui percipitur, dum ex alcali rursum per Acetum . . . praecipitatur [sulphur] . . . ."

31 Newman, "The Corpuscular Theory of J. B. Van Helmont and its Medieval Sources," *Vivarium* 31 (1993): 161–191.

32 British Library, Sloane ms. 3750, f. 1v: "Ergo rationi consentaneum est [sulfur] in mineralibus (in quibus nulla est fortis unio illius cum [mercurio] per corrosiva extraverti . . . ."

33 British Library, Sloane ms. 3750, f. 1v: "Ergo quo 'fortior' [melior deleted] AF, quidni melior?"

34 British Library, Sloane ms. 3750, f.7r: "Non possibile. Opinio Igne refutata."

35 British Library, Sloane ms. 3750, f. 3r.

36 British Library, Sloane ms. 3750, f. 3v: "Saltem haec via minoris est impensi, faeliciorisque tractationis."

37 British Library, Sloane ms. 3750, f. 3r.

38 The expression "last great philosophical alchemist of the seventeenth century, the anonymous Eirenaeus Philalethes," occurs in Dobbs, *Foundations*, p. 52. Dobbs, of course, did not realize that Starkey and Philalethes were one.

ALAN E. SHAPIRO*

NEWTON'S OPTICAL NOTEBOOKS: PUBLIC VERSUS PRIVATE DATA

Newton's optical investigations, especially as gathered in his *Opticks* (1704), are widely regarded as representing a decisive, historical change in experimental practice. Recently I have been attempting to characterize the nature of his experimental research with more precision, and his laboratory "notebooks" have provided a revealing view of him at work in his laboratory. Newton's "laboratory notebooks," however, do not conform to our current conception of that term. In the first place, he did not carry out his experiments in a laboratory, but rather in his private rooms. In the second place, his notes are undated and, after the early years, unbound.[1] Newton did date some of his al/chemical experimental records, but as William Newman has shown in his paper, a tradition of dating alchemical records was apparently already established by Newton's day. The lack of dated entries in his notebooks has not been a serious impediment. Modern scholars can generally determine accurately the place of a document within a sequence of related ones, and establish its date to within a few months to a year or so.

Newton's records differ in a more fundamental way from laboratory notebooks, in that they were not intended as a permanent record of his experiments. He evidently discarded his raw experimental notes after an investigation was concluded and written up. The only sheet of raw data that survives is from his experiments on diffraction.[2] It most probably survived because it was part of his last, incomplete experimental investigation, and he expected to take it up at a later date. Theorizing, experimenting, and writing were part of a single process for Newton.[3] In the early 1690's in composing those parts of the *Opticks* that represented new experimental research – those on the colors of thick plates and diffraction – he proceeded successively from draft to experiment to the deduction of new laws, and back to new drafts, and so on. For later reference he saved drafts, summaries, and preliminary papers as a record of his investigatory path. Raw data was apparently just that for Newton, raw. Data assumed significance for him only within a sequence of experiments and measurements and a broader context that explained the phenomenon and its properties.

Despite repeatedly denouncing hypotheses, Newton's experimental, optical research was in part theory driven, or to use his term, hypothesis driven. In each of

* University of Minnesota, Minneapolis, MN

Frederic L. Holmes, Jürgen Renn and Hans-Jörg Rheinberger (eds.), Reworking the Bench: Research Notebooks in the History of Science, 43–65 © *2003 Kluwer Academic Publishers. Printed in Great Britain*

his optical investigations – the colors thin films, diffraction, and the colors of thick plates – he formulated a physical model that was mathematizable and used it to devise experiments, deduce laws, control his measurements, and generally to conceive of the phenomenon. In the principal record that I will examine, "Of ye coloured circles twixt two contiguous glasses," most probably from 1671, Newton began with a sequence of propositions that he then succeeded in confirming experimentally.[4] Many of these propositions were deduced from his physical model based on hypothetical entities, light corpuscles and vibrations of the aether, and some proved to be erroneous. That Newton found them to be erroneous by further experimentation shows that he would readily abandon a physical model if the experimental evidence was sufficiently convincing.

His notebooks show Newton trying to integrate physical explanation, mathematical description, and experimental results, and an unrelenting quest to eliminate sources of error. As we shall see, however, he suppressed much of the most exciting parts of his quantitative research and left few clues for subsequent generations on how to carry out mathematical and quantitative experimental research. His accounts of his qualitative experimental investigations – most notably, the theory of light and color that ultimately formed Book I of the *Opticks* – were, on the contrary, sufficiently detailed to serve as an exemplar for later generations. My concern will be both with the experimental methods by which Newton achieved his results and the dilemmas he subsequently confronted in presenting them publicly.

"Coloured Circles" is a working paper intended for his own use. There is some evidence that the data were not directly entered here until Newton convinced himself of their meaningfulness. On a few occasions he tells us they are only a sampling of his measurements. For example. he wrote "Suspecting y^{t} ye pressure might much vary ye glasses figures I tryed w^{th}out pressing them together, & y^{t} many times ... *Two or three of the observatio[ns] follow.*"[5]

"COLOURED CIRCLES"

It is unnecessary to retrace the history of the colors of thin films here, but it should be noted that Newton learned about them from Hooke's account in the *Micrographia* of the colors seen in sheets of mica.[6] Hooke had conjectured that the appearance of the colors was periodic, though he confessed he was unable to measure the thickness of such thin films to demonstrate this property. Newton's key breakthrough was his insight that if he put a lens (which is really just a segment of a circle) on a glass plane, then by a principle from Euclidean geometry on the tangents to circles (Euclid's *Elements,* III, 36) he could readily determine the thickness of the thin film of air formed in the gap between them simply by measuring the circles' diameter. If in Figure 1 a convex lens *ABC* is placed on a glass plate *FBG* and illuminated and viewed from above, a set of concentric colored circles – now aptly known as Newton's rings – produced by the thin film of air *ABCGBF* will be seen through the upper surface of the lens.

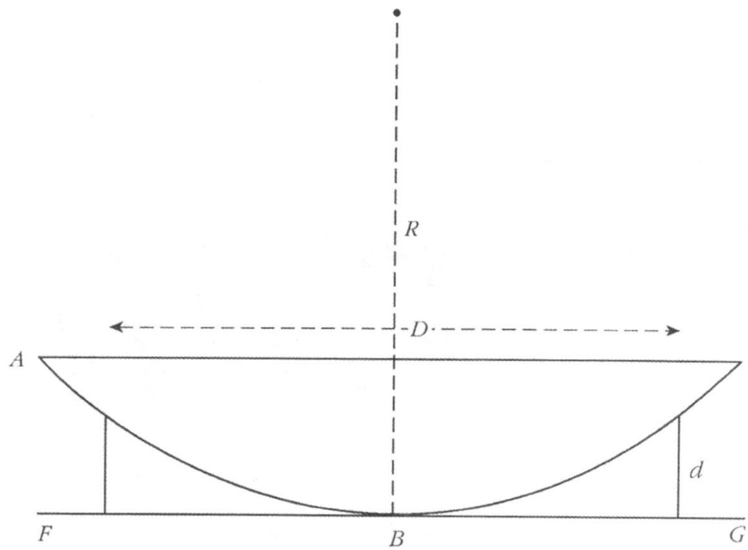

Figure 1. Newton's method for determining the thickness d of a thin film of air between a spherical lens and a plane.

The circles in Figure 2 will form an alternating sequence of bright and dark rings and their common center, the point of contact A, will be surrounded by a dark spot.

Let the diameter of a colored circle in Figure 1 be denoted by D, the thickness of the air film producing that circle by d, and the radius of the lens by R. Then by Euclid's theorem $d = D^2/8R$. To establish that the circles do appear at integral multiples of some definite thickness, or that the appearance of the colors is a periodic phenomenon, Newton simply had to measure the diameter of successive circles and see if their squares increased as the integers. He found that the thickness of the film for successive circles does increase in arithmetic progression, or expressed in the modern form of an equation,

$$d = \frac{D^2}{8R} = \frac{mI}{2}, \tag{1}$$

where I is an interval such that for m odd the ring is a bright one and for m even a dark one. As I shall explain later, in Newton's physical model I represents the pulse length of a vibration of the aether, although publicly he would interpret this length as the experimentally observed value of the thickness of the film producing a particular color.

In his numerous descriptions of this experiment, Newton says surprisingly little about the experimental arrangement that he used. We know little more than that he built some kind of a frame for the glasses, which were tied together, that he measured the diameters of the colored circles with a compass, and that he believed

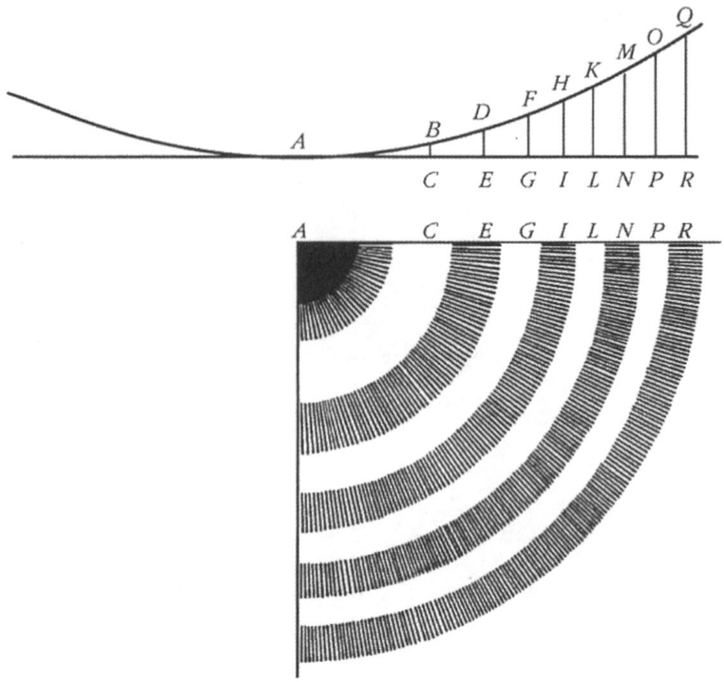

Figure 2. Newton's explanation of the periodicity of the circles formed with light of a single color in Newton's rings. He has drawn one quadrant of alternating dark and bright rings, and the corresponding thickness of the air film between the glasses that produces those circles. (After Newton's drawing in the "Hypothesis," CUL, MS Add. 3970, f. 545r.)

that he could determine them to fractions of 1/100th of an inch. By this method Newton could measure accurately thicknesses of two-hundred thousandth of an inch. Newton seems to have had this insight while reading the *Micrographia,* quickly carried out a rough and ready test in 1666, and recorded it in his early notebook essay "Of Colours."[7] His results, though quantitatively very crude, were sufficient to demonstrate – at least to his satisfaction – that the appearance of the colors was a periodic phenomenon and to provide a measure of the periodicity: that is, that thickness of the film at integral multiples of which the colors reappeared.

Satisfied with this fundamental result and that his method worked, Newton set it aside and pursued other research until 1671, when he undertook a serious investigation of the colors of thin films, which he recorded in "Of ye coloured circles twixt two contiguous glasses." He began with a series of propositions on colored rings that he intended to prove experimentally. These show that Newton had done a substantial amount of experimental investigation before preparing "Coloured Circles." Some of the propositions, such as the first, report results that he had already deduced in 1666:

> Prop 1. That their areas are in arithmeticall proportion, & soe ye thicknesse of ye interjected [film.] Or y^{t} the spaces wch ye rays pass through twixt circle & circ[l]e are in arithm prop[ortion]

Others, however, present new results that could only be derived experimentally. The sixth proposition, for example, states "That if the glasse bee illuminated by coloured light, that wch is most refrangible makes the least circles. & the *thicknes of a pulse* for ye extreame red, to that for ye extreame purple is greater y^{n} 3 to 2 & scarce greater y^{n} 5 to 3. Viz about 9 to 14 or 13 to 20."

Still other propositions imply a deduction from a mechanical model based on the motion of light corpuscles and their interaction with the vibrations that they stir up in the aether. Evidently Newton had worked out the derivations elsewhere, for none are presented here. The following propositions are stated explicitly in terms of the "motion," "force," and "percussion" of the corpuscles:

> Prop 2 That they swell by ye obliquity of the eye: soe y^{t} the [di]ameter of ye same circle is as ye secants of ye rays obliquity [in] ye interjected filme of aire, or reciprocally as ye sines of its obliqui[ty]; that is, reciprocally as ye part of the *motion of ye ray* in ye said filme of aire wch is perpendicular to it, or reciprocally as y^{e} *force* it strikes ye refracting surface wth all.
>
> Prop 3 And hence ye spaces wch ye rays passe through twixt [y^{e}] circles in one position to the said spaces in another position are as ye squares of ye said secants or reciprocally as ye quares of ye sines, *motion, or percussion.*
>
> Prop 4 That if ye medium twixt ye glasses bee changed ye bignes of ye circles are also changed. Namely to an eye held perpendicularly over them, the difference of their areas (or ye thicknesses of ye interjected medium belonging to each circle) are reciprocally as ye subtilty of ye interjected medium or as y^{e} *motions* of ye rays in that medium.[8]

"Motion" is a technical term for Newton that usually means momentum but may also be velocity. In his published scientific works Newton would never adopt such an approach for his presentation, since it mixes the hypothetical – terms that refer to properties of light corpuscles and vibrations of the aether that I have put in italics – with experimentally confirmed principles. Newton insisted that we "not mingle conjectures with certainties."[9] Nonetheless, these propositions do reveal the intimate relation of physical modeling, mathematical derivation, and experiment in his private work.

Newton immediately sets out to demonstrate these propositions experimentally (see Fig. 3).[10] In Table 1 he presents in column two his measurements of the diameters of the first six dark rings at various angles of incidence, which are indicated in the first column by the sine of the angle of obliquity; the second set of measurements were made with "The Glasses being not altogether so hard pressed together."[11] The rings are in fact not well defined, and Newton had to overcome the obstacle of working with white light in which the rings rapidly overlap and

	Circulorum diametri	Quadrata diametrorum	Quadrat diff	Med Diff	
	1. 2. 3. 4. 5. 6	1. 2. 3. 4. 5. 6	1,2. 2,3. 3,4. 4,5. 5,6.		
1000	$26\frac{1}{3}$ .37 .45 $.51\frac{2}{3}$ $.57\frac{1}{3}$ $.62\frac{2}{3}$	693.1369.2025.2635.3287.3927	676. 656. 610. 652. 640	647	46
667	32 $.44\frac{2}{3}$ $.54\frac{2}{3}$ $.62\frac{3}{4}$ .70. *	1024.1995.2988.3938.4900. *	971. 993. 950. 962. *	969	55
435	$39\frac{1}{2}$ .55 .67 $.77\frac{1}{2}$ $.86\frac{2}{3}$.*	1561.3025.4489.6006.7511. *	1464.1464.1517.1505. *	1488	73
320	45 .62 .77 .89 .99 .*	2025.3844.5929.7921.9801. *	1819.2085.1992.1880. *	1944	81
206	57 .80 .98 . * . * *	3249.6400.9604. * . * . *	3151.3204. * . * . *	3177	72

	Circulorum diametri	Diametrorum quadrata	□orum differentiae	Diff: Med:	
	1. 2. 3. 4. 5. 6	1. 2. 3. 4. 5. 6	1,2. 2,3. 3,4. 4,5. 5,6.		
1000	$23\frac{1}{2}$ $.34\frac{1}{3}$ $.42\frac{1}{3}$ .49 $.54\frac{3}{4}$ 60	552.1179.1792.2401.2997.3600	617. 613. 609. 596. 603	610	58
667	29 $.41\frac{3}{4}$ $.51\frac{1}{2}$ $.59\frac{3}{4}$ .67 .*	841.1742.2652.3570.4489. *	901. 910. 918. 919. *	910	69
435	$36\frac{1}{2}$. 52 .64 $.74\frac{1}{4}$ $.83\frac{1}{4}$.*	1332.2704.4096.5513.6930. *	1372.1392.1417.1417. *	1400	67
206	$53\frac{1}{2}$. $76\frac{1}{3}$ $.93\frac{1}{2}$ .108 . * *	2862.5826.8742.11664. * *	2964.2916.2921. * . *	2934	72

Table 1.

mix to form a uniform white. Although he could observe more than twenty circles by isolating a simple spectral color from a spectrum, the intensity was then too low to make any measurements. In column three he formed the squares of the diameters, because it is these that should according to equation (1) increase by a constant quantity if the appearance of the colors is periodic. In the fourth column, he calculated the differences of these squares – which should be a constant equal to the first value – and in the next column he calculated the simple average or mean. In Figure 3 we can see how much he measured and remeasured to get what he thought were the best possible measurements. In the last column, which he inserted after the table was initially composed, he entered the difference between the average of the difference of the squares of the diameters and the square of the diameter of the first circle.[12] According to Newton's explanation of the formations of the rings these two quantities should be equal, but they were clearly not so. Throughout "Coloured Circles" Newton worked with mean and not singular values.

Newton's use of averages was almost unprecedented in the 17th century. Averages were occasionally used since the last part of the sixteenth century, but primarily for observations of singular entities, such as the position of a star or the angle of magnetic declination, which could not possibly be theoretically predicted. All of the documented uses of the arithmetic mean that I am aware of before Newton were in astronomy, navigation, or terrestrial magnetism, and not in laboratory physics, and certainly not in observations made to verify a law.[13] Indeed, quantitative experimental science scarcely existed before Newton. One of the few earlier examples of quantitative experiment, the work of Boyle, Hooke, Power, and Towneley on Boyle's law a decade before Newton's work on the colors

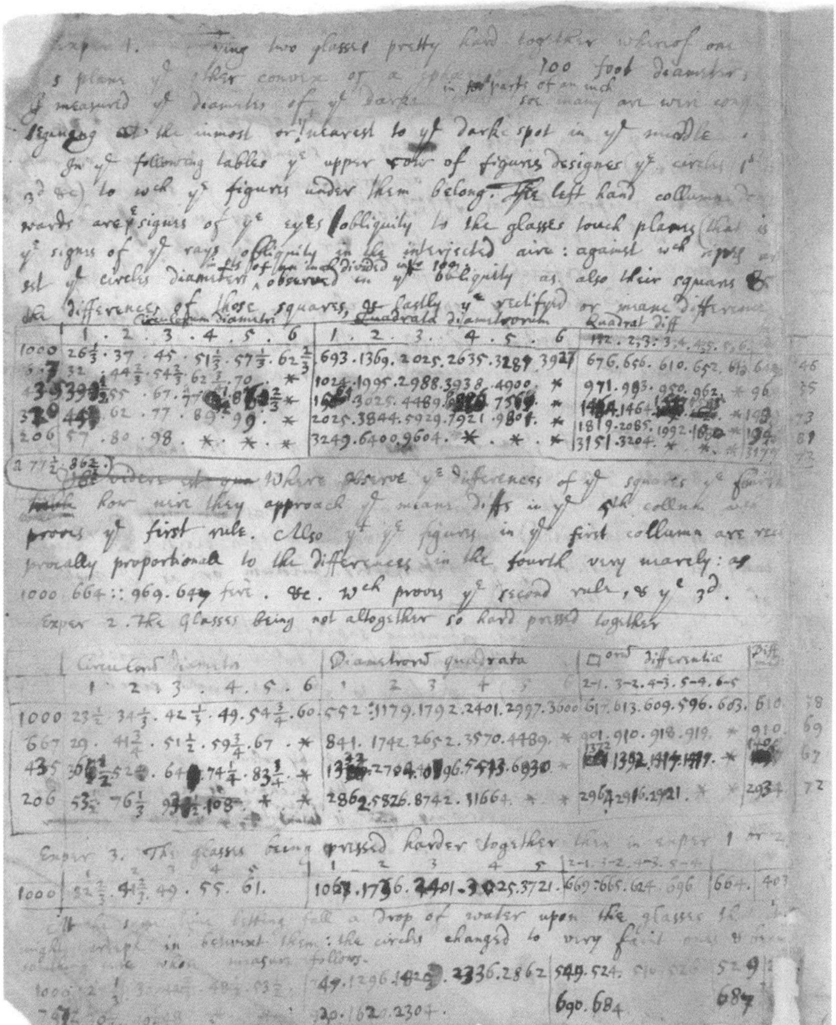

Figure 3. The second page, ms. Add. 3970, f. 350v, of "Of y^{e} coloured circles twixt two contiguous glasses," on which Newton begins his experimental investigation of Newton's rings. Part of f. 353r is shown because it includes the last column of the tables, which was added after the tables were initially composed. The two folios are actually one folded sheet of paper. (By permission of the Syndics of Cambridge University Library.)

of thin films, does not exhibit any sophisticated analysis of the data.[14] Westfall has described how Newton raised "quantitative science to a wholly new level of precision.... He boldly transported the precision of the heavens into mundane physics...."[15] Newton was applying new mathematical techniques to analyze his data and seek out sources of error. Newton, however, would suppress all use of averages in his published work

Newton wrote: "observe ye differences of ye squares ye fourth [column] how nere they approach ye meane diffs in ye 5th collum wch proves ye first rule," that is, that the rings increase in arithmetic progression.[16] This is the key result that eluded Hooke, for it establishes that the recurrence of the rings is periodic. He also observed that the data proved the second and third propositions on oblique incidence. Both of these laws are erroneous and, despite Newton's claim, his data actually prove (to within half a per cent) what we know to be the correct law.[17] It is important to note though that Newton subsequently caught his blunder and that he always cast a critical eye on both his theoretical and experimental results. This episode does show how theory was in some respects driving, though certainly not dominating, his investigation. The proposition was derived by his corpuscular theory of light, and he was eager – too eager – to test and confirm that law.

In his third experiment Newton put a drop of water between the lenses in order to prove his fourth proposition on the variation of the thickness of the film with the index of refraction, which is a simple extension of the law in equation (1),

$$d = \frac{D^2}{8R} = \frac{mI}{2n}, \tag{2}$$

where n is the index of refraction. He found that

$$\frac{d_a}{d_w} = \frac{n_w}{n_a} = \frac{624}{529} = \frac{4}{3.187} = [1.254].^{18}$$

This yielded $n = 1.254$, which is 6% smaller than the canonical value for the index of refraction of water, namely, 4/3, which Newton had a little earlier experimentally confirmed.[19] He therefore repeated the experiment, and this time he got

$$\frac{d_a}{d_w} = \frac{615}{448} = \frac{4}{2.9132} = [1.373]$$

"W^{ch} proportion," he observed, "is here greater then 4 to 3 as before it was lesse," specifically, it was 3% larger. He conjectured that this variation was caused by the pressure he had to apply to the glasses to get them in contact, for the pressure would flatten the lens and make the first diameter too large. He also suspected that "y^e endeavour of y^e aire from betwixt y^e glasses" might also contribute to the discrepancies.[20] Once again he repeated the experiment, and he now found:

$$\frac{d_a}{d_w} = \frac{731}{531} = \frac{4}{3.015} = [1.327].$$

Newton was content with this result (which scarcely differed from 4/3), though he was still clearly troubled by the divergence of his results because he devoted the next two pages of his notes to evaluating the effect of pressing the glasses together.

It was at this point that he went back to his tables and added the last column with the difference between the square of the diameter of the first circle and the mean value of the differences of the squares of the diameters. He noted that, "it may bee observed y^{t} ye difference twixt the first squares & ye meane difference of the squares in the last collumns are not much unequall unlesse where it may bee suspected that there is a disturbance made by pressing of the glasses."[21] Newton was now systematically applying his mathematics to an investigation of the experimental results themselves and not solely to natural phenomena. In his determination of the index of refraction, he calculated the ratios with the average values of his measurements, not from singular measurements.

Let me extend my analysis a little more widely to try to locate the origin of Newton's evident unease with his data. The data in his experiments were fairly good, but not great. In Table 1 the difference between the square of the diameter of the first circle (column 2) and the average value of the differences of the squares of the diameters (column 5), which Newton calculated in the last column of the table, varies from nearly 9% (line 1 in the second set of measurements) to 2%, and most of them are around 5 or 6%. Newton was very sensitive to experimental error and always had an intuitive awareness of how good his experimental results could be. From my work on Newton's optical paper over many years, I would estimate that he judged that in optics his errors should be less than 5% and ideally of the order of 2% to 3%. Clearly, by this criterion his problems with the first circle would have made him uneasy. If we calculate the largest difference between the mean value (column 5) and the individual values of the differences of the squares (column 4), we find that it varies from about 7% (the first value in line 4 in the first set of measurements) to 1%, but most range from 1 to 2%. Moreover, the difference between the average values (column 5) in the first two sets of measurements ranges from 3% to 8% (the last line in each set). Newton suspected that something was wrong and worked hard to resolve the discrepancies. His first guess, as we saw, was that pressing the lens made the first circle too big. From various experimental tests he convinced himself that the pressure on the lens was responsible for some of the error, but not all. He suspected that the glasses were not precisely shaped but could not demonstrate it, and he plaintively wrote,

> I am apt to suspect that there was some imperfection in the glasses t[o]o wch cheifly caused these varietys namely that the one glasse was not exactly plane nor the other exactly sphaerical & so ye experiment must needs bee various w^{n} made at severall places of the glasses. [Now ye least imperfection of this nature is sufficient to make considerable variation in ye experiment ...] And this I rather suspect because many times I could not imagin how the cause should bee from various pressures. At least these two causes together were sufficient to produce the sd disagreements.[22]

At a later date Newton drew a line in the manuscript and his symbol for something to note well, a fist with a pointing finger. He declared that he finally resolved the problem:

> ☞I find at last that one side of the greate glasse is more convex then the other though both were ground on the same toole. . . . soe that one makes ye circles about 45 parts greater then the other if applyed to ye same part of the plane of the other smaller glasse, but that is not an exact plane neither & thence it happened that in some cases one side of the greate glasse made ye difference of ye squares about 580 when ye other made y^m about 620 & againe ye same side of the g^t glase if applyed to another part of the small one made y^m about 600 wch ye other made them about 640. Soe that in generall they may bee esteemed about 610 if the glasses had beene regular & also w^{th}out pressure for that noe dout had some effect upon them. And besides all this there were severall little furrows or waves (not scratches) wch made some irregularity but they were but few & I was carefull to choose that part of the glasse where there were the fewest. And one glasse was w^{th}out any at all. Yet many times they imposed upon mee.[23]

Newton had begun his investigation confidently assuming that the lens was truly biconvex, since he had ground it himself, with a radius of 50 feet. He now concluded that the two sides of the lens had different radii, neither equal to 50 feet, although he did not calculate the corrected radii here.[24] The determination of the absolute value of the radius directly affected only the determination of the vibration length, though the asymmetry of the lens rendered the other results unstable and less reliable. Newton was assiduous in tracking down the cause(s) of his inconsistent results. After determining that the deformation of the lens by pressing on it could not account for the anomalous variations, he found two more sources of error, the shape of the lens and the "little furrows," both of which would change the effective thickness of the air film.

Newton's approach to locating his source of error assumed that that the law of arithmetic progression was valid. He then compared his measured values with those predicted by the law and relentlessly sought to eliminate all but the smallest deviations that he recognized to occur inevitably in any measurement. While this is a natural assumption and initially valuable in seeking out sources of error, it can be perilous. It can lead one to miss other physical effects that are in part responsible for the deviations. Indeed, in the early 1690s, when Newton was compelled by his investigation of the colors of thick plates (described in the next section) to repeat this experiment, he discovered two physical effects that must be introduced to establish the law of arithmetical progression more rigorously.[25]

Before showing how Newton presented the results of his experiments to the public, I want to recount one more, clever experiment in which he determined the ratio of the extreme vibration lengths. First I must briefly describe the physical model that he used to devise and interpret his experiments. Newton did not explain his model in the "Observations," because he believed that hypotheses and certain demonstrations must be kept separate. Rather, he composed a separate piece, "An

Hypothesis explaining ye properties of Light discoursed of in my several papers," that he sent to the Royal Society together with his "Observations," adding his standard caveat "that no man may confound this wth my other discourses, or measure ye certainty of one by th'other."[26] He assumed that light consists of particles emitted by luminous sources, and that when these particles encounter the aether that suffuses all bodies, they excite vibrations in it, "as stones do in water when thrown into it."[27] These vibrations, which move faster than the light particles, interfere with their motion by reflecting them at a condensation where the aether is too dense to allow the particles to pass through, and transmitting them at a rarefaction. Depending on the length of a vibration, for a given thickness of the film the vibration will be at or near a condensation or a rarefaction at the lower surface of the film and cause the alternating bright and dark rings. Thus in Figure 2, at the point of contact *A* of the two lenses, where the glasses are continuous, the light will pass through and form the central dark spot *A*; at the first condensation *C* they will be reflected and produce the first bright ring *C*; at the first rarefaction *E* they will be transmitted again and produce a dark ring *E*; at the next condensation *G* they will be reflected once more, and so on. With this physical mechanism Newton was able to introduce periodicity into an emission theory of light. Newton emphasizes that he considers the particles to be light and not the vibrations, "I suppose light is neither ye aether nor its vibrating motion," which is simply an effect of light.[28]

As we saw, Newton initially tried to incorporate both the motion of the corpuscles and the vibrations in his account of Newton's rings and mathematically described the properties of the circles in terms of the "motion," "force," and "percussion" of the corpuscles and the "thickness of a pulse." He soon came to recognize that this approach yielded contradictory and erroneous results, and afterwards he abandoned his attempt to mathematize the relation of the motion of the corpuscles to that of the vibrations and focused on the vibrations alone, which could be readily mathematized. The vibration model proved to be a powerful one. It helped him to think fruitfully about what was happening physically and to mathematize the phenomena. He was able to deduce new results, such as the size of the rings when the film has a different index of refraction than air, or to tackle new phenomena, such as the colors of thick plates. To be sure, his methodology required that he purge such hypothetical entities from his public accounts of his investigations. The vibration length, which in the *Opticks* became the interval of fits, corresponds to half the wave length in the modern (that is, 19th century) wave theory and is thus a fundamental optical quantity.

From his theory of color Newton knew that sunlight was a mixture of different colors and thus that the different colors must have different vibration lengths. "After many vaine attempts" he was unable to determine the vibration lengths for particular colors because light of simple colors was too weak too measure, but he cleverly used his vibration model to deduce the ratio of the length of the red to the violet vibrations.[29] When there was a very small gap between the lens and plate, Newton gradually cast the spectral colors on them; and by counting the succession of light and dark rings as the colors passed between the two extremes, he was able to determine how many more violet than red pulses fit into this small space. Then, in red light, he gradually pressed the lens until it was in contact with the plate; by counting the succession of light and dark rings, he now found how many red pulses fit in this same space. Adding the two numbers therefore yielded how many violet fit pulses fit in the space. Once again Newton repeated the experiment a number of times with his eye in different positions and reported the results in tabular format:[30]

Experiments	1	2	3	4	5
Differ of half pulses twixt red & purple	3 or more	7	5	$3\frac{1}{2}$	$3\frac{1}{2}$
half pulses made by y^e^ redd light	6	13	9	$6\frac{1}{2}$	6
halfe pulses made by y^e^ purple rays.	9 or more	20	14	10	$9\frac{1}{2}$

Table 2

The third row in Table 2 is determined, as was just noted, by adding the results in the first and second rows. The numbers in the top row indicate the position of his eye, 1 and 5 being "almost" perpendicular to the lens, 2 and 4 at "about" 30° or 40° from the perpendicular, and 3 "very oblique" or at "about" 60° or 65° from the perpendicular. He determined the ratio of "the thickness of a pulse" of blue to that of red was about 13 to 20 or 9 to 14, but chose the latter since – in a nice bit of numerology – it was the same as the index of refraction from air to glass. He also concluded "y^t^ ye obliquity alters not ye proportion."[31] This experiment established the last of the opening propositions, the sixth, and concluded this stage of his investigation of the colors of thin films. In a notebook Newton later summarized and extended some of the its results. "The thicknesse of a vibration," an important result that he did not calculate in "Coloured Circles" because of the problem of the radius of the lens, was now set at "1/81000 or 1/80000 part of an inch."[32]

"OBSERVATIONS"

Some time after Newton completed "Coloured Circles" (probably less than a year later), he wrote a more extended piece on the colors of thin films, his "Observations," that he intended to include with his reply to Hooke's criticism of his theory of color in spring 1672. Because the controversies over his theory of color so disturbed Newton, he decided against submitting the "Observations" to avoid further controversy. This work contained new observations and measurements, and also eliminated some of the erroneous results of "Coloured Circles." It also abandoned the hypothetico-deductive format of "Coloured Circles" for an inductive one in which a series of observations were presented with conclusions deduced from them. When Newton had regained his equanimity in 1675, he sent a revision of the "Observations" from 1672 to the Royal Society, where it was read and registered, though he would not allow its publication. The "Observations" was the definitive statement of Newton's experimental investigation of the colors of thin films and served as the basis, nearly verbatim, for the corresponding section in the *Opticks*, Book II, Part I through Part III, Proposition 8.[33]

If we turn to the key observation that establishes the periodicity of the colors of thin films we are in for a surprise:

> Obs: 5. To determin the intervall of the glasses or thicknesse of the interjacent Air by wch each colour was produced, I measured the diameters of ye first six rings at the most lucid part of their Orbits, & squaring them I found their squares to be in Arithmeticall progression of ye odd numbers 1, 3, 5, 7, 9, 11. And since one of the glasses was plain and the other sphericall, their intervalls at those rings must be in the same progression. I measured also the diameters of the dark or faint rings between the more lucid colours & found their squares to be in Arithmeticall progression of the eaven numbers, 2, 4, 6, 8, 10, 12. *And it being very nice & difficult to take these measures exactly I repeated them divers times at divers parts of the glasses, that by their agreement I might be confirmed in them.* And the same Method I used in determining some others of the following Observations.[34]

Over the years I had read and studied this passage many times, but until a few years ago I had not realized that Newton does not present a single measurement to support one of the more significant results of 17th-century optics, the experimental demonstration of the periodicity of light. And the passage appears virtually unchanged in the *Opticks*. Where are all the measurements that he had made? The sentence in italics represents all the experimental evidence, and it accords with our knowledge of Newton's travails in "Coloured Circles." There are, I suggest, a number of reasons why he suppressed his data. First, he was using average values, and because there was no acceptable method yet to justify such a procedure, they were scarcely used publicly. I know of no cases up to the time of Newton where averages were used to establish the truth of a law. Hooke and Boyle published tables of data to justify Boyle's Law, but the measurements at different degrees of compression and rarefaction of air represented a single experiment. They were, of course, aware that repeated measurements were necessary to establish the truth of the law, and they made it known that other measurements had been made by themselves and others.[35] Even had he wanted to, Newton could not publish a single set of measurements, as was the style among English natural philosophers, because he knew any one set of measurements was not sufficiently precise to establish the law of arithmetic progression or periodicity.[36] It may also have seemed to him that presenting all his data with an average value would only serve to call attention to the variation of the measurements and not their stability.

Newton's attitude to data in fact was more akin to Galileo's, in the mathematical tradition, than to contemporary English natural philosophers. In his *Discourses and Demonstrations Concerning Two New Sciences* (1638) Galileo had presented a mathematical derivation for the law of free fall (the distance fallen is proportional to the square of the time), which was one of the most important contributions to the mathematical sciences in the seventeenth century. When one of the interlocutors in the dialogue asked Salviati, who represents Galileo, if he had experimental proof for this law, he replied by carefully describing the experimental arrangement, the ball, the inclined plane that the ball rolled down, and the water clock used to time it. He recorded the time for the ball to run the whole length of the incline "repeating the same process many times, in order to be quite sure as to the amount of time, in which we never found a difference of even the tenth part of a pulse-beat." When he rolled the ball one quarter of the length "this was found always to be precisely one half" that required for the full run. When it was tried for other lengths, "by experiments repeated a full hundred times, the spaces were always found to be to one another as the squares of the times."[37] No data were reported, only the assertion that they confirmed the law perfectly, exactly the same approach that Newton took in his "Observations." Like Newton, Galileo made experimental tests of his law, as we now know, and was adept at analyzing data.[38] Since mathematical results aspired to certainty, it was difficult for Galileo to invoke imperfect experimental results to confirm their truth.[39] This expectation naturally brings me to the second reason why I believe that Newton suppressed his data. If the first reason suggested why he could not publish his data, the second one suggest why he would not. Newton too subscribed to the belief that mathematics provides greater certainty than natural philosophy, and he had already developed his ideal of near perfect agreement between theory and mathematical law. Since the data was not neat, he simply suppressed it. It is not that the data did not support this law. It did to a per cent or two, and Newton, as we saw, had worked very hard to attain this agreement, but it was not perfect and had its problems.

The final two reasons can be quickly dispatched. Third, even had Newton chosen to use average values, he had no data for a lens with a 51 foot radius. The average values had been corrected for the shape of the lens and, strictly considered, he had measurements only for a flawed, asymmetric lens of 50 feet.[40] Finally, Newton had used the effect he was investigating to determine the shape of the lens, and this might have seemed like circular reasoning, especially when the effect had not yet been established.

In the next observation Newton derived the value for the vibration length or the thickness of the film required for the reappearance of a given color, the basic parameter for his theory. He chose one measurement – not in "Coloured Circles" – for the diameter of the sixth bright ring at normal incidence and calculated that the vibration length for yellow light was 1/80,047 or "to use a round number"

1/80,000 inch.[41] Although Newton did not directly calculate the vibration length in "Coloured Circles," he always worked with average values. I suspect that he chose a measurement from the many that he had that yielded a value close to the average, in the manner of astronomers who always had a surfeit of observations and chose the "best" of them.[42] Newton's determination of the vibration length, as best as it can be compared with modern values, is about 10% too large. Nonetheless, in the 17th century to be able to measure reliably such a minute distance was as remarkable an achievement as the determination of astronomical distances. Newton had no independent means to determine the reasonableness of his result, at least not until twenty years later. While composing the *Opticks* in the early 1690s, he discovered and successfully explained a new phenomenon, the colors of thick plates.

When sunlight shined on a mirror from a reflecting telescope through a small hole in a screen at the center of the circle from which the mirror was formed, Newton noticed four or five colored circles surrounding the hole that appeared like those in Newton's rings except larger and fainter. The thickness of the mirror, which was silvered on the back, was 1/4 inch, yet Newton had found that the colors of thin films vanished when the thickness of the air film was about 1/10,000 inch. He was able to give a precise mathematical account of the colored rings seen in the telescope mirror using the properties of the vibrations that he had earlier developed for thin films, in particular, the same vibration lengths *I*. This investigation forms Book II, Part IV of the *Opticks*.[43]

When Newton first compared his calculated values for the diameters of the rings with his measurements, he found that they differed by about 5 1/2 %. As should be evident from my account of Newton's measurement of the rings in thin films, this was at the upper limit of the error that Newton would tolerate, and we would not expect him to be content with a difference of this magnitude. He set about seeking the source of the error. After checking the obvious sources he was compelled to reopen his determination of the vibration length, which he thought he had settled twenty years earlier and had already included in the part of the *Opticks* that had been completed. He repeated the experiments with two new telescope lenses that had a much smaller radius of curvature, 7 feet 7 inches and 7 feet 8 inches. While the diameters of these rings were less than half the size of his original measurement, the focal length of about 7 1/2 feet could be much more accurately determined than the original one of about 50 feet, which I suspect is the principal source of the error in his earlier determination of the vibration length. He now found, "to use the nearest round Number" 1/89,000 for the vibration length instead of the old 1/80,000.[44] This was a significant change of 11% that brought the value virtually to its modern one. When he used it to recalculate the diameters of the rings in thick plates, it essentially resolved the discrepancy between measured and calculated values. This episode explains why this is one of the few passages from the "Observations" of 1675 that were significantly revised in the *Opticks*.

Returning to the "Observations," we see that Newton again suppressed the data in his determination of the variation of the rings' diameter with index of refraction:

> Obs: 10. Wetting the Object glasses a little at their edges, the water crept in slowly between them, & the circles thereby became lesse & the colours more faint .... By measuring them I found the proportion of their diameters to the diameters of the like circles made by Air to be about seven to eight, & consequently the intervalls of the glasses at like circles caused by those two Mediums water & air are as about three to four. Perhaps it may be a generall Rule that if any other Medium more or lesse dense y^{n} water be compressed between the glasses, their intervall at the Rings caused thereby will be to their intervall caused by interjacent Air, as the sines are wch measure the refraction made out of that Medium into Air.[45]

Newton found for the ratio of the diameters of the rings in water and air,

$$\frac{D_w}{D_a} = \frac{7}{8}.$$

By equation (2) we have,

$$\frac{d_w}{d_a} \propto \frac{D_w^2}{D_a^2} = \frac{n_a}{n_w} = \frac{7^2}{8^2} = \frac{49}{64} = \frac{3\frac{1}{16}}{4} \approx \frac{3}{4}.$$

Again he avoided average values and chose one value – not even a measurement but a ratio representing measured values – and presented a simplified account of his experimental procedure.

Newton was probably so ready to accept this as a universal rule valid for any medium, because he was able to deduce it from his physical model of light particles and aethereal vibrations. In "Coloured Circles" he had stated this law without derivation in Prop. 4 in terms both of the index of refraction (the "subtilty" of the medium) and "ye motions of ye rays in that medium," but it is readily motivated. If the particles move faster in water in proportion to the increase of index of refraction (as the emission theory required), then they would more quickly reach the lower surface of the film and encounter the first aethereal condensation. The vibration length would then be shorter in the inverse proportion.

Finally, we can look at Newton's determination of the ratio of the pulse lengths or intervals for red and violet rays:

> Obs: 13. Appointing an Assistant to move the Prism to & fro about its Axis, that all its colours might successively fall on the same place of the paper & be reflected from the circles to my eye whilst I held it immoveable; I found the circles wch the Red light made to be manifestly biger then those wch were made by the Blew & Violet. ... The intervall of ye Glasses at any of the rings when they were made by the utmost red light was to their intervall at ye same Ring

> when made by the utmost Violet, greater then 3 to 2 & lesse then 13 to 8. By the most of my observations it was as 14 to 9.[46]

In what is now a familiar story Newton not only gave no measurements, but he did not even explain how he determined this essential ratio. Of course, it involved his hypothetical pulses, but he made no attempt to formulate it in observable or phenomenological terms. In the *Opticks* Newton replaced the hypothetical vibrations of the aether with his fits of easy reflection and transmission, which he intended to be free of hypotheses. However, the fits were added in Book II, Part III well after this observation in Part I, and Newton made no attempt to explain how the measurement was made.

Let me briefly summarize the various factors that I believe led Newton to publish summary rather than real experimental results. His decision to use average values, which we may now recognize as a sound, innovative procedure, was not yet either mathematically or philosophically justified or an accepted scientific practice. Newton thus chose to avoid them. The asymmetric lens that he used in his experiments was the cause of many of his problems and compelled him to use average values. The measurements were subsequently corrected for the lens's asymmetry and no longer had a simple and direct relation to what was initially observed and tested. Since his data did not present immediate and compelling evidence for his laws but rather were messy, he simply chose to suppress them. Finally, his refusal to mix hypothetical entities such as vibrations and light corpuscles with conclusively demonstrated results lead him to suppress the methods by which he established some of his results, such as the ratio of vibration lengths for the red and violet.

Newton's imaginative and fruitful use of physical models and his zealous pursuit of precise measurement and confirmation are largely hidden from the reader of the *Opticks* because he chose to present only the highly-polished fruits of his investigations. He had carried out a quantitative and mathematical experimental research program at the very highest level, something that was rare in the early modern era. Readers of the *Opticks*, however, were not able to learn the nature of such a research program from his text, and there were few exemplars of such a program available for about a century.

CONCLUSION: DIFFRACTION

By briefly comparing Newton's investigation of the colors of thin films with some of his other experimental investigations, especially diffraction, I can indicate which characteristics are common to his experimental investigations and which are specific to the colors of thin films. Newton undertook his investigation of diffraction at the same time as that on thick plates, namely, in the early 1690s when he was completing the *Opticks*. After he initially completed the *Opticks* in 1691, he became dissatisfied with the part on diffraction and removed it from the manuscript. No doubt because he intended to resume his experiments on

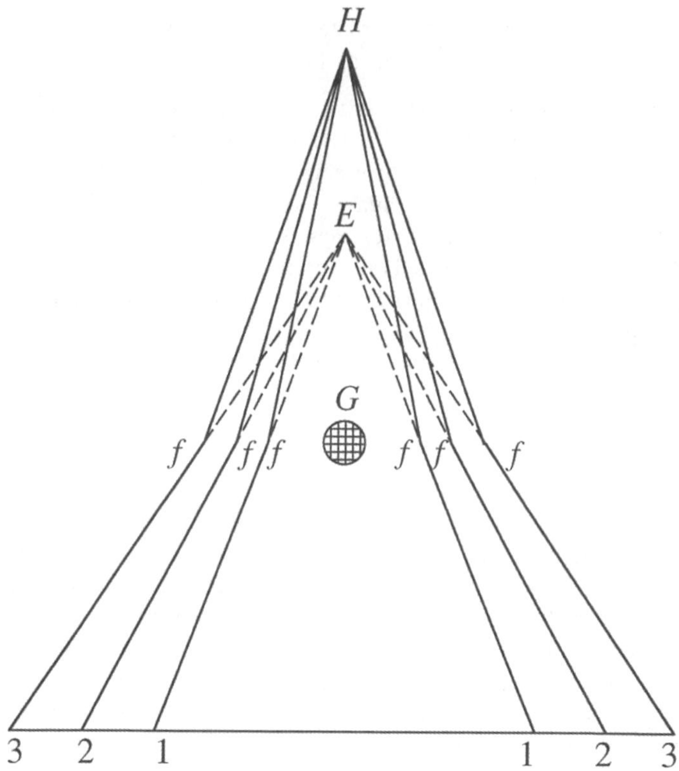

Figure 4. The three pairs of fringes 1, 2, 3 formed by diffraction from a hair *G*, based on Newton's assumption that the fringes propagate from *f*, the point of diffraction, in the same straight lines as the rays.

diffraction at a later date, we have more notebook material for this area of optical research than for any other. This includes a plan of experiments to be undertaken, a sheet of raw data, drafts for the *Opticks*, and pages of calculations and notes on the experiments.[47]

An examination of his notebooks reveals that just as with his experiments on Newton's rings, he used an underlying physical model to guide and interpret his experiments and to deduce laws governing the phenomena. In Figure 4, *H* represents a narrow hole through which light rays (consisting of light corpuscles) proceed to the vicinity of a hair *G* where they are deflected by a short-range force at *f* and, when stopped by a screen, depict three pairs of colored fringes 1, 2, 3. He assumed that the paths of the fringes were identical to, or coincided with, the rectilinear paths of the rays that produced them; that is, that the rays proceeding from *f* to 1, 2, 3 depict the fringes at whatever distance the screen is placed from the hair.

Again, just as in his published account of Newton's rings, Newton did not reveal his model to his readers. I was able to uncover it from his notebooks and then reproduce all of his calculations exactly, so that there can be no reasonable doubt that he actually used this model. Without his notebooks and the keys to Newton's model his investigatory path would scarcely be intelligible.

In the manuscripts of the drafts for the part of the *Opticks* on diffraction we can follow him deducing various laws describing the phenomena and then replacing them with refined laws and new data. His model also apparently guided him in making his measurements, just as it did with Newton's rings. This procedure can present serious problems, such as leading one to ignore other physical effects, as I have already indicated. It can also lead one to tinker with or "massage" the data to agree better with theory (perhaps unconsciously), as seems to have been the case with Newton's diffraction measurements. The problem here is that unlike his measurements of Newton's rings, his linear propagation model was untenable. The reason why Newton had removed the part on diffraction from the first completed state of the *Opticks* was that he had afterwards carried out an experiment (Book III, Observation IX) that refuted his model. He conclusively showed that the fringes and rays did not propagate along identical rectilinear paths and that the fringes followed a curved one. Newton never resumed his experiments on diffraction and when he decided to publish the *Opticks* he simply revised the part he had already written. This required him to go through the earlier version and remove all laws based on the linear propagation model. Some of the data, however, still reflects the good agreement with the rejected model.[48]

Since Newton's use of average values in his research on the colors of thin films was totally hidden from the public and is such a striking innovation, it is natural to ask whether he also suppressed the use of averages in his research on diffraction. Insofar as I can determine, no where else in his optical investigations did he use averages.[49] In his work on diffraction he reported the measured values for his experiments. This difference, I believe, reflects the different nature of the phenomena and measurements, for by refining individual diffraction measurements he had no need for averaging, which was unavoidable with Newton's rings. He was able to reduce the differences between measurements and between measurements and laws to a few per cent, though, as we have seen, the agreement with the laws was spurious.

More than with most scientists, Newton's notebooks are essential to uncovering his investigatory path. Because of his methodology, which required hypotheses and demonstrated principles to be kept separate, he tried to efface all traces of physical models from his published reports of his investigations. And because of his quest to establish a more certain science, he strove to presente his experimental observations as unambiguously agreeing with experimental laws.[50]

NOTES

[1] Unbound notes are in fact typical of 17th century natural philosophy and were used by, among others, Galileo (as described by Jürgen Renn and Peter Damerow in this volume) and Christiaan Huygens.

[2] Cambridge University Library (CUL), MS Add. 3970, f. 373r. Some raw data for Newton's measurements of indexes of refraction survives (Johannes A. Lohne, "Newton's table of refractive powers: Origins, accuracy, and influence," *Sudhoffs Archiv für Geschichte der Medizin und Naturwissenschaften* 61 (1977): 229–47), but that most likely occurred because the measurements are recorded on the versos of sheets, a draft for Book II, Part IV, that Newton otherwise wanted to preserve.

[3] Frederic L. Holmes in "Scientific writing and scientific discovery," *Isis* 78 (1987): 220–235, has perceptively argued that writing up a laboratory investigation for publication is an essential part of the process of discovery and cannot be viewed as a separate process of justification. Holmes was concerned with early biology and chemistry and so did not consider theory, but there can be little doubt that his thesis can be readily extended to experimental and theoretical investigations such as Newton's.

[4] This manuscript, Add. 3970, ff. 350r–353v, consists of two sheets of paper folded in half with one inserted inside the other to form a booklet. Newton wrote with tiny handwriting on both recto and verso, though two pages are blank. The manuscript was published by Richard S. Westfall, "Isaac Newton's coloured circles twixt two contiguous glasses," *Archive for History of Exact Sciences* 2 (1965): 181–196.

[5] Add 3970, f. 352r; Westfall, "Coloured circles," p. 194. The manuscript is badly damaged, and the square brackets here and elsewhere are my restorations or additions unless otherwise indicated. The many alterations that Newton made in his manuscripts will not be indicated, since they do not have significance for this paper. In another example where he is obviously drawing from a larger store of measurements, he states that some measurements "almost ever" fell between particular values, and then buttresses it with "two of them"; *ibid*, in note 11 below I provide evidence that in preparing this notebook Newton was most likely working from raw data and calculations.

[6] I have already treated Newton's investigation of the colors of thin films in *Fits, Passions, and Paroxysms: Physics, Method, and Chemistry and Newton's Theories of Colored Bodies and Fits of Easy Reflection* (Cambridge University Press, 1993), ch. 2, though I was not there particularly concerned with Newton's experimental techniques.

[7] CUL, MS Add. 3975, pp. 6–12, nos. 24–43; in J. E. McGuire and Martin Tamny, eds., *Certain Philosophical Questions: Newton's Trinity Notebook* (Cambridge University Press, 1983), pp. 472–478. Add. 3975 is a bound notebook.

[8] Add 3970, f. 350r; Westfall, "Coloured Circles," p. 192; italics added.

[9] Newton, *The Correspondence of Isaac Newton*, eds. H. W. Turnbull, J. F. Scott, A. Rupert Hall, and Laura Tilling, 7 vols. (Cambridge University Press, 1959–77), vol. 1, p. 100.

[10] Although Newton records that one glass was "convex of a sphaere 100 foot diameter," i.e., with a 50 foot radius, and the other glass was plane (Add 3970, f. 350v; Westfall, "Coloured Circles," p. 191), in the "Observations" he relates that he used the plane surface of a plano-convex lens (Obs. 4, Thomas Birch, ed., *The History of the Royal Society of London, for Improving of Natural Knowledge, from Its First Rise*, 4 vols. [London: 1756–1757; rpt.: Brussels, *Culture et Civilisation*, 1968], vol. 3, p. 274). This has no effect on the observed phenomena. Birch's edition of the "Observations" is reprinted in I. Bernard Cohen, ed., *Isaac Newton's Papers and Letters on Natural Philosophy and Related Documents* (Cambridge, MA: Harvard University Press, 1958).

[11] Add 3970, f. 350v; Westfall, "Coloured Circles," p. 192. There is an error in the second set of measurements in this table that indicates that in composing "Coloured circles" Newton was working from raw data and calculations or perhaps earlier worksheets that are no longer extant. In line 1 the difference of the squares of circles 1 and 2 should be 627 and not 617. Nonetheless, the average value in the next column, 610, is that which results from using the correct 627, which

indicates that 617 was a copying error and not a calculational error. Thus, the data and calculations in "Coloured Circles" are not Newton's original record of his experiments.

12 Westfall omitted this column in his publication, perhaps because the numbers are carried over from f. 350v to the opposite page on the folded sheet, f. 353r, and he was working with a microfilm or photocopy from which it would not be apparent.

13 For some account of averages and, more generally, data analysis in the early modern era see Anders Hald, *A History of Probability and Statistics and Their Applications before 1750* (New York: John Wiley & Sons, 1990), chap. 10; Churchill Eisenhart, "The background and evolution of the method of least squares," Prepared for distribution to participants of the 34th session of the International Statistical Institute, Ottawa, Canada, 21–29 August 1963; R. L. Plackett, "The principle of the arithmetic mean," *Biometrika* 45 (1958): 130–135; Plackett, "Data analysis before 1750," *International Statistical Review* 56 (1988): 181–95; Oscar Sheynin "Mathematical treatment of astronomical observations (a historical essay)," *Archive for History of Exact Sciences* 11 (1973): 97–126. I thank Noel J. Swerdlow for his comments concerning the early history of averages.

14 See Charles Webster, "The discovery of Boyle's law, and the concept of the elasticity of air in the seventeenth century," *Archive for History of Exact Sciences* 2 (1965): 441–502.

15 Westfall, "Newton and the fudge factor," *Science* 179 (1973): 751–758, esp. pp. 751, 753.

16 Add 3970, f. 350v; Westfall, "Coloured Circles," p. 192.

17 I have already treated this episode in *Fits, Passions, and Paroxysms*, pp. 65–68.

18 This sequence of experiments can be found in Add 3970, ff. 350v, 351r,v; Westfall, "Coloured Circles," pp. 192–193. In some of these proportions Newton based his calculations on values that represent earlier measurements rather than the final version in the table, but the differences are small and of little consequence, so I will not indicate them. This first experiment is recorded on the bottom of f. 350v in Figure 3.

19 Newton, *The Optical Papers of Isaac Newton. Vol. 1: The Optical Lectures, 1670–1672*, ed. Alan E. Shapiro (Cambridge University Press, 1984), p. 187. Newton, in fact, found 4.002 to 3 which, he declared, differed from 4/3 "by so small a difference that the error would be insensible if I assumed it to be 4 to 3."

20 Add 3970, f. 351r; Westfall, "Coloured Circles," p. 193.

21 Add 3970, f. 351v; Westfall, "Coloured Circles," p. 194.

22 Add 3970, f. 352v; Westfall, "Coloured Circles," p. 195; the second set of square brackets are Newton's.

23 *Ibid.* The "45 parts greater" should be "40." In the manuscript Newton originally had "575" before changing it to "about 580" and "645" before "640," but he did not return and change the difference. Newton's "about 610" is the mean of the four measured values.

24 Newton's calculation of the corrected radius of the lens does not survive, but he later used "about one & fifty foot" (Newton's draft of his reply to Hooke's critique of his theory of color, spring 1672, Add. 3970, f. 442v; see also Obs. 6 in the "Observations" of 1675 and in the *Opticks*, Bk. II, Pt. I; and note 40 below). We may find the correction to the radius in the following way. The thickness of the air film d must be the same for each surface of the lens for a ring of the same order. Therefore, if we let D_1, D_2 represent the diameters of the colored rings measured with different sides of the lens, and R_1, R_2 represent the radius of each refracting surface of the lens, then from equation (1) we have for the two surfaces,

$$\frac{D_1^2}{R_1} = \frac{D_2^2}{R_2}, \text{ or}$$

$$\frac{D_1^2}{D_2^2} = \frac{R_1}{R_2} = \frac{R+\Delta}{R-\Delta},$$

where Δ represents the variation of each surface from 50 feet, assuming that the difference is equally divided between the two surfaces. We then have

$$\Delta = \frac{D_2^2 - D_1^2}{D_2^2 + D_1^2} R.$$

If we use the measurement where, $D_2^2 = 620$, $D_1^2 = 580$, and $R = 50$ feet, then $\Delta = 1$ ft., 8 in.; using the other measurement gives virtually the same result. (Readers who use the metric system should recall that there are 12 inches to the foot). Westfall determined a correction of 80 in. for the diameter of the lens or $\Delta = 40$ in. for the radius, which is twice my result; "Coloured Circles," p. 185, n. 10. However, he made a simple slip, which, when corrected, agrees with my result. He performed his calculation by using the median figure 600 and then determined what variation in the lens's diameter would yield a variation of 40 in D^2, whereas, he should have calculated with a variation of 20 to produce the requisite 620 and 580. Newton's value of about 51 feet is either crudely rounded off or he had other reasons for choosing 51 rather than 52 feet.

[25] The first corrects for the measured size of the ring. Since the rings are viewed through a glass lens, the true size of the ring is actually larger than the measurement because of the refraction in the lens. The second corrects for the true thickness of the film. Because the rings are viewed slightly obliquely (about 4° off the perpendicular), the path through the film is slightly longer than its actual thickness. The first effect turns out to be significant while the other can be ignored. See the *Opticks*, Bk. II, Pt. I, Obs. 6.

[26] Add. 3970, f. 538v; Newton, *Correspondence*, vol. 1, p. 364.

[27] Newton to Oldenburg for Hooke, 11 June 1672 in Newton, *Correspondence*, vol. 1, p. 174.

[28] "An Hypothesis"; Add. 3970, f. 539v; Newton, *Correspondence*, vol. 1, p. 370.

[29] Add 3970, f. 353v; Westfall, "Coloured Circles," p. 195.

[30] Add 3970, f. 353v; Westfall, "Coloured Circles," p. 196.

[31] *Ibid.*

[32] Add. 3975, p. 22; McGuire and Tamny, *Certain Philosophical Questions*, p. 488. In note 42 I indicate how Newton may have arrived at this number.

[33] What I call the "Observations" was untitled both in the 1672 and 1675 versions, in *Newton's Papers and Letters* it is called "Newton's second paper on color and light," and in Newton's *Correspondence* "Discourse of Observations." I will cite the 1675 version, since it has been published and scarcely differs from the earlier version in the cited passages.

[34] Add. 3970, f. 502r; Birch, vol. 3, pp. 274–275; italics added.

[35] On the changing nature of experimental evidence in the mathematical sciences in the seventeenth century see Peter Dear, *Discipline and Experience: The Mathematical Way in the Scientific Revolution* (University of Chicago Press, 1995).

[36] Narrowly considered, an average value is not an actual measurement but, as Newton wrote in "Of y[e] Coloured Circles," it is a "rectifyed or meane" value, i.e., one that has been corrected and particularly to an (al)chemist refined or purified; Add 3970, f. 350v; Westfall, "Coloured Circles," p. 192.

[37] Galileo Galilei, *Two New Sciences, Including Centers of Gravity & Force of Percussion*, trans. Stillman Drake (Madison: University of Wisconsin Press, 1974), p. 170.

[38] We have come a long way since Koyré's denial of Galileo's experimentation, and there is now an enormous literature on his experiments. While Galilean scholars differ on their interpretations of particular manuscripts, there is little disagreement that he was carrying out a sophisticated experimental and theoretical research program. See, for example, Stillman Drake, "Galileo's physical measurements," *American Journal of Physics* 54 (1986): 302–306; and the paper by Damerow and Renn in this volume. Lest one be sceptical of Galileo's claim to have measured the time for a ball to roll down an inclined plane to better than the tenth part of a pulse-beat, Thomas B. Settle confirmed this claim by an experimental replication in a landmark paper, "An experiment in the history of science," *Science* 133 (1961): 19–23. On Galileo's statistical analysis of the numerous observations of the new star of 1572 in his *Dialogue Concerning the Two Chief World Systems* (1632), see Hald, *History of Probability*, pp. 149–160.

[39] Galileo also had to confront the problem that in the Aristotelian tradition of scientific

demonstration, to which he in part still subscribed, empirical evidence had to be evident and frequently recurring; see Dear, *Discipline and Experience*. This was not a concern in Newton's thought.

[40] In spring 1672, when Newton had intended to accompany his reply to Hooke's critique of his theory of color with his "Observations," he was planning to cover up the problem with the lens. In the draft of his letter to Hooke he described the biconvex lens as "being *on both sides* a segment of a sphere whose semidiameter was about one & fifty foot as I found by measuring it that I might more precisely determin these phaenomena" (Add. 3970, f. 442v; italics added).

[41] Obs. 6, Add. 3970, f. 502v; Birch, vol. 3, p. 275.

[42] If we use the average that Newton determined when correcting for the radius of the lens in "Of y[e] Coloured Circles" (see at note 23), i.e., $D^2 = 610$ *and* $R = 51$ ft., then by equation (1) $d = I = 1/80{,}264$ for the first dark ring.

[43] Newton's investigation of the colors of thick plates is described in my *Fits, Passions, and Paroxysms*, ch. 4.

[44] Newton, *Opticks: Or, a Treatise of the Reflections, Refractions, Inflexions and Colours of Light. Based on the Fourth Edition London, 1730* (New York: Dover Publications, 1952), Bk. II, Pt. I, Obs. 6, p. 203.

[45] Add. 3970, ff. 503r,v; Birch, vol. 3, pp. 276–277.

[46] Add. 3970, f. 503v; Birch, vol. 3, p. 277.

[47] See, for example, Add. 3970, ff. 357–358, 373, 643, and 645–646. I discuss Newton's experimental investigation of diffraction, his notebooks, his data, and give specific citations to the numerous manuscripts in "Newton's experiments on diffraction and the delayed publication of the *Opticks*," in Jed Z. Buchwald and I. Bernard Cohen, eds., *Isaac Newton's Natural Philosophy* (Cambridge, MA: MIT Press, 2001), pp. 47–76. It should be reiterated that Newton's "notebooks" in this investigation consisted of loose sheets of paper.

[48] See Michael Nauenberg, "Comparison of Newton's diffraction measurements with the theory of Fresnel," in Richard H. Dalitz and Michael Nauenberg, eds., *The Foundations of Newtonian Scholarship* (Singapore: World Scientific, 2000), pp. 56–69. It should be added that Newton's measurements generally agree with modern theory, and that his problems occurred with the most sensitive or difficult measurements.

[49] George Smith has told me in a conversation that he has found concealed use of average values in the *Principia.*

[50] Westfall in "Newton and the fudge factor" showed how in the second edition of the *Principia* Newton achieved a spurious precision by fudging his data so that it would agree with his theoretical calculations. While there is no question that Newton fiddled with or massaged some of his optical observations a bit (see my "Newton's experiments on diffraction"), most of the precision that he achieved was legitimate.

MARCO BRESADOLA*

AT PLAY WITH NATURE: LUIGI GALVANI'S EXPERIMENTAL APPROACH TO MUSCULAR PHYSIOLOGY*

Italian scholars seem to share the growing interest of historians of science in research notebooks. A recent sign of this trend, which in Italy has important precedents, especially in the work of Luigi Belloni, is the publication of Lazzaro Spallanzani's diaries of experiments and observations, and Maria Teresa Monti's study of them.[1] Another Italian man of science, who left a conspicuous laboratory register, is Luigi Galvani, whose work – to use the words of Bernard Cohen – "opened up a new field of research and stimulated able minds to produce a kind of revolution within science." Here Cohen is clearly referring to Galvani's investigation into animal electricity, which "opened up" a new field of animal physiology and "stimulated" the study of electrical phenomena, including the work of Alessandro Volta.[2]

Most of the records of Galvani's experimental investigation into animal electricity were published more than fifty years ago, but have not yet received sufficient attention from historians.[3] In my Ph.D. dissertation I devoted many pages to the reconstruction of Galvani's investigation through his experimental records in the decade which preceded the *De viribus electricitatis in motu musculari*, the famous treaty of 1791 in which he published his theory of animal electricity. Taking Frederic Holmes' work on Bernard, Lavoisier and Krebs as my model, I intended to offer a "fine-structure" study of Galvani's science. This was aimed at the clarification of his experimental practice and at the reconstruction of the route taken by his work on animal electricity, as well as on other subjects such as the "chemistry of life."[4]

My focus here will be more limited. I shall concentrate on Galvani's experimentation on animal electricity during an early stage of his investigation for two reasons: first, to account for the emergence of Galvani's experimental approach to this subject, intended as the interplay between the problems, instruments, experiments and interpretations which characterised his laboratory practice; and second, to explore in detail the pathway which led Galvani to one of the earliest outcomes of this approach, his so-called "first experiment." This experiment, which consisted in the violent contractions of a frog's limbs when its

---

* University of Ferrara-I

Frederic L. Holmes, Jürgen Renn and Hans-Jörg Rheinberger (eds.), Reworking the Bench: Research Notebooks in the History of Science, 67–92

crural nerves were touched with a scalpel and an electric spark was discharged from a distant electrical machine, is considered a "fundamental discovery which was to prove revolutionary" in the history of physiology. It has been defined Galvani's "first experiment" because it opened the *De viribus* and marked the first major step in the process of the "discovery" of animal electricity.[5]

Galvani's "first experiment" is recorded in his laboratory diary and is dated 26 January 1781. It was not, however, the first experiment performed and recorded by Galvani, but was preceded by almost three months of experimentation and record-keeping, beginning 6 November 1780. During these three months Galvani frequented his laboratory every week, sometimes more than once a week. Although he eventually decided not to describe the experiments performed in this period in the *De viribus*, they are very significant for the historian interested in the process of scientific investigation. As I shall argue in this paper, it was in this early stage of his investigation that some of the fundamental features of Galvani's approach to muscular physiology emerged. Moreover, it was in the process of developing this approach and exploring its potential as well as its limitations, that Galvani encountered a new and puzzling phenomenon – i.e. muscular contractions at a distance – which marked a turning point in his investigation. My study, therefore, concerns the "private science" of Galvani – to use Gerald Geison's expression – in a twofold sense: it is centred on laboratory notebooks and other unpublished manuscripts, and it deals with a part of his science which he did not make public. However, it can offer some insights for our understanding of the public form Galvani gave to his investigation into animal electricity.

As the introduction to the present volume makes clear, the historical use of research records depends heavily on the form and content of the records. In order to reconstruct Galvani's or any other scientist's investigative pathway through his/her laboratory notebooks, the records need to be complete and chronologically ordered as much as possible. Indeed, this is fortunately the case here. Although Galvani's experimental records are not collected in notebooks, both their form and content suggest that no extensive set of experiments is lacking, at least for the period under scrutiny. Galvani's record-keeping appears very systematic, a feature that, as we shall see, characterises also his investigative practice. Every session of experimentation is dated and every experiment is numbered. There are no blank spaces between the descriptions of each experiment, while sometimes there are additions in the margins, usually referring to the operative details of an experiment. This indicates that Galvani probably used to fill in the record at the end of the experiment or shortly after, sometimes improving or correcting its description (but never its result) later on. It is thus plausible to infer that the spatial arrangement of the records in the diary reflects the chronological order of the experiments. Typically, Galvani's experimental records consist of short narratives – as is the case for other investigators like Lavoisier, Faraday and Bernard – reporting the experimental arrangement, the operations performed and the outcome, but sometimes there are also conclusions, corollaries and "warnings."

We cannot be sure whether the first surviving experimental record corresponds to the first experiment on animal electricity performed by Galvani. He had been investigating the composition and function of nerves and muscles for many years, using frogs as the experimental animal, but we do not have any direct evidence of his use of electricity. Moreover, among his manuscripts there are several pages in which he drew general conclusions from the experiments performed in the period 1780–1782, indicating precisely the date and number of the experiment to which he was referring. All these experiments are reported in the surviving laboratory notes, as if he had them in front of him while writing his remarks. This suggests that at the end of 1780, whether it was the beginning of his experimentation on animal electricity or not, Galvani undertook a new investigative path and decided to fix it on paper in the form of experimental records.

Galvani's research plan was part of an ongoing debate on the effects and role of electricity in the animal economy. An outline of this debate is needed in order to understand what Galvani drew from it and why he decided to enter this field, and is offered in the first section of this paper. Sections 2–5 form the core of my study and concern Galvani's own approach to muscular motion and his experimental activity until the "first experiment." The final section is devoted to a discussion of the published report of this experiment in the light of the investigative pathway that preceded it, and concludes with some remarks on the historical use of laboratory notebooks.

ENTERING THE FIELD

In an unpublished memoir, dated 25 November 1782, Galvani explained the reasons why he had decided to undertake his electrophysiological investigation:

> A not inconsiderable number of anatomists have thought either that the electric fluid constitutes the undefined, highly subtle fluid that is not unreasonably believed to flow in the nerves, or that it is the nervous fluid itself. I therefore felt it would not be useless to perform various experiments with the electric fluid on those selfsame nerves, in the hope that they would lead to the discovery of the truth or at least shed some light on the darkness still shrouding the phenomena of nerves.[6]

As is well known, by 1780 the idea that electricity was part of the "nervous fluid" – traditionally know also as "animal spirits" – or was identical to it, had a long history and went back at least to Newton's *Opticks* and Stephen Hales' *Haemastaticks*. It was supported by several observations and arguments such as the similarities between the alleged properties of the electric fluid and sensation and muscular motion – phenomena traditionally ascribed to nerves – and the effectiveness of electricity in exciting muscular contractions.[7]

Some among the "not inconsiderable number of anatomists" who had claimed such a neuroelectric conception of sensation and muscular motion worked in Bologna, the city of the Papal State where Galvani had been born in 1737, where he had graduated in medicine and philosophy in 1759, and worked as a university professor of anatomy and as a practising surgeon and physician. The most prominent of them was perhaps Tommaso Laghi, who in the 1750s had advanced the theory that "the electric matter, diffused throughout [the animal body], is directed by the nervous liquid secreted in the cerebral glands to flow through the nerves, so to foster sense and motion."[8] This idea generated a lively debate in the Bologna scientific and medical community gathered around the Academy of Sciences. In fact, some of its members were particularly interested in the study of electricity, its relationships with vital phenomena, and its therapeutic application. Giuseppe Veratti, for example, had published a few years before an important book on medical electricity, a growing field by the middle of the 18th century, devoted to the use of electricity in the treatment of some diseases, which were thought to depend on nervous and muscular disorders. Moreover, there were in Bologna several supporters of the Hallerian theory of "irritability" and "sensibility," including Leopoldo Caldani and Felice Fontana, which were to become two of the leading Hallerians. This theory, published by Albrecht von Haller in 1748, rejected the traditional role ascribed to nerves in muscular motion and considered the property of muscles to contract, i.e. "irritability," as intrinsic to muscles and independent of nerves. Haller and his followers also rejected the identification between the nervous and the electric fluids, advancing some powerful objections which were left substantially unanswered by Laghi and the other supporters of a neuroelectric point of view.[9]

Around 1780 the debate over the cause and mechanism of muscular motion was still lively, at least in Bologna. At the beginning of that year, in a public lecture performed in the anatomy theatre of Bologna, Galvani suggested the idea of "the most noble electric fluid, on which motion, sensation, blood circulation, even life itself, seem to depend."[10] This idea was even more comprehensive than that of Laghi, but it was similarly based on conjectures. Neither Galvani, nor other supporters of a neuroelectric conception, offered new evidence beyond that already rejected by the Hallerians. Nor did they undertake a systematic investigation in order to understand the effects of electricity on muscular motion and the laws regulating them. "Everything yet remains to be done," as Fontana remarked in 1781:

> We must first assure ourselves by certain experiments, whether there is really an electrical principle in the contracting muscles; we must determine the laws that this fluid observes in the human body; and after all it will yet remain to be known what it is that excites this principle, and how it is excited.[11]

This an impressive description of the agenda that Galvani fixed for himself at the end of 1780. Realising like Fontana that the neuroelectric conception rested on clay feet and probably spurred by objections to his ideas on the role of electricity in vital functions, raised at his lectures on anatomy, he decided to undertake a systematic and experimental investigation of the effects of electricity on muscular motion.

Galvani carried out most of his experimental activity in a laboratory built in his house (Fig. 1), following a practice that was quite common in Bologna in the second half of the 18th century. In his laboratory he had the complete equipment of the electrician of the period, as it was described, for example, in Tiberius Cavallo's *Complete treatise of electricity*. He knew quite well how to use these instruments, as well as how to deal with electrical phenomena, having a close familiarity especially with the works of Giambattista Beccaria, one of the main supporters of Franklin's theory of electricity. Like Beccaria, Galvani thought that all known electrical phenomena – attractions and repulsions, sparks and lights, the "electrical wind" and other particular manifestations – were produced by a single and imponderable fluid which tended to be equally diffused everywhere. When something altered this equilibrium between adjacent bodies, the electric fluid flowed from the body with the greater amount of it to one next to it, manifesting itself through its characteristic "signs," i.e. the electrical phenomena. This dynamic notion of electricity was at the core of Galvani's investigative approach and is basic to an understanding of his experiments on the effects of electricity on animal motion.[12]

Galvani's decision to use electricity in the investigation of muscular motion partly depended on the general idea that it had a role in this animal function. But it was also due to the specific observation, noted by Caldani and Fontana, that electricity was the most effective external stimulus in order to obtain muscular contractions, even long after the death of the experimental animal. This was very important for Galvani, who wanted to study the "mechanism" of voluntary motion independently of the action of the will, which was "beyond the limits of human sight and understanding." Thus it was necessary to take animals whose voluntary muscles continued to contract after death, as well as a "mechanical agent" still effective under these particular conditions. The frogs, on one hand, and electricity, on the other, were those who best fitted in this picture, and became the main actors, together with Galvani himself, on the laboratory stage.[13]

REPETITION AND VARIATION

On 6 November 1780 Galvani took some frogs "prepared in the usual manner." We are not sure to what investigation he was referring, although he had been experimenting with frogs for several years. It was probably during his investigation on the heart, the nerves and muscles carried out in the previous decade, that he developed a particular preparation of the frog, which resembled very closely the one adopted by Caldani and Fontana in the repetition of Haller's experiments on irritability. The frog was killed and cut just below the upper limbs, leaving only the lower limbs with their crural nerves and spinal cord attached and uncovered (see Fig. 1). Galvani used this preparation, or some modifications of it, throughout his investigation, making it one of the fundamental elements of his method. Indeed, the way the frog was prepared was not a secondary feature of the experiments, but it was crucial for their result and the possibility of reproducing

Figure 1. Preparatory engraving for plate I of Galvani's *De viribus* (AASB, Fondo Galvani. By courtesy of the Accademia delle Scienze di Bologna).

them successfully. As Galvani remarked in the *De viribus*, "... most of the facts we discovered from these experiments are the result of this technique of preparing and separating the nerves."[14]

In the experiments of 6 November Galvani discharged a Franklin square – an electrical condenser made of a glass pane with two metallic foils attached on each surface (called the coatings of the square) – through the spinal cord, nerves and muscles of a prepared frog, and then stimulated these "animal parts" with electricity. Similar experiments had been performed by Veratti some years earlier, and their results communicated to the Academy of Sciences of Bologna. Veratti had been trying to understand the pernicious effects of electricity on the animal and human body, in particular that of lightning. It had been observed that, when struck by lightning, a body stiffened and its muscles remained contracted. Discharging a Franklin square, as the instrument which could best imitate the power and strength of lightning, through different parts of living animals, including some frogs, Veratti concluded that a large amount of electricity destroyed muscle irritability and prevented the flow of animal spirits in the nerves.[15]

Veratti's experiments were among the very few performed on animals with the use of electricity from which Galvani could take inspiration. However, while Veratti's interest lay mainly in the pernicious effects of electricity, Galvani had a different and more general aim, i.e. the understanding of muscular motion through the use of electricity. In his first experiments, in particular, he wanted to compare the effects of electricity when applied to nerves and muscles. He found that in the former case the electrical discharge did not prevent muscular contractions, while in the latter no contractions occurred. He observed, however, a great variety in the results, which depended "perhaps on the various shocks of different strength, and on the different robustness of the animals" being employed. Galvani immediately realised the difficulty of the research he was starting. The study of the effects of electricity on muscular motion was somewhere in between the study of electrical and physiological phenomena, so both these domains had to be taken into consideration when preparing the experiments and interpreting the phenomena. Thus whether or not contractions occurred in an animal which had been given an electric shock could depend either on the power of the shock – i.e., an electrical parameter – or on the specific animal which was being used – i.e, a physiological parameter.[16]

At the end of his laboratory entry for 6 November Galvani made some notes on how to overcome the difficulties encountered in the experiments. When he returned to the laboratory on 15 November, however, he changed direction, repeating some anatomical observations on the spinal cord, the heart, and the brain of a dead frog, that were common in the study of animal economy. Apparently he was proceeding by exploring known facts and debated questions about muscular motion. He continued on the same path in the following weeks, focusing on three main topics, which can be summarised in the following way:

22 Nov.	Effects of the ligature of nerves on muscular contractions
25 Nov.	Electrical properties of nerves and muscles
29 Nov.	Effects of the ligature and severing of nerves on muscular contractions
2 Dec.	Effects of the severing of nerves on muscle contractions
9 Dec.	Effects of the ligature of nerves on muscular contractions
	Electrical properties of nerves and muscles
16 Dec.	Electrical properties of nerves and muscles
	Quantity of electricity needed to excite muscular contractions
19 Dec.	Electrical properties of nerves and muscles

Two of these topics – which I have summarised as the effects of ligature and severing of nerves, and the electrical properties of nerves and muscles – had very important implications from a physiological point of view. In fact, they referred directly to the fundamental objections formulated by Haller and his supporters against a neuroelectric conception of muscular motion.

In his *Elementa physiologiae*, Haller had reported the observation that while ligating or severing a nerve rendered the muscle connected to that nerve "insensible" and incapable of motion, the same muscle contracted if the nerve was stimulated by electricity. He considered this phenomenon as fundamental evidence against the claim that "animal spirits were of an electrical nature." In fact, if the ligature prevented contractions by arresting the flow of animal spirits in the nerve, but did not prevent the stimulating action of electricity from producing contractions, electric fluid and animal spirits had to be different. Perhaps even more powerful was a second objection, also reported by Haller but developed by Caldani and Fontana in their debate with Laghi during the 1750s. Agreeing with Beccaria's statement that there could not be any electrical imbalance between two conductors in contact and thence any "electrical sign," they argued that no contractions could occur if nerves were conductors like muscles. If, on the other hand, nerves were electrical insulators, they would stop the flow of the electric fluid, thus preventing its action on muscles. Whether nerves were conductors or not, in both cases the nervous fluid could not be of an electrical nature.[17]

Although in the following decades some new developments both in electrical theory and in animal physiology, in particular the study of "electric fish," gave new life to a neuroelectric conception of muscular motion, the arguments based on the ligature and the electric property of nerves continued to be an obstacle. In fact those two arguments were reaffirmed by Auguste Tissot in the late 1770s, and Galvani himself discussed the latter in the *De viribus*.[18] Their importance and urgency lay not only in their theoretical implications, but in their potentiality for the understanding of the effects of electricity on nerves and muscles. To determine whether a body was electrically insulating or conducting was in fact one of the primary tasks when considering that body from the point of view of electrical phenomena. It is thus not surprising that Galvani decided to pursue these questions soon after undertaking his investigation.

While the objections elaborated by the Hallerians were stated clearly, how to examine and question them was quite another matter. Galvani did not have a fixed apparatus or procedure to make use of, nor did he have a clear idea as to the fundamental circumstances he had to take into account. Although on his laboratory bench there were essentially only frogs and some electrical equipment, they could be connected in a variety of ways. A large number of decisions had to be taken when preparing an experiment: what part of the frog to consider and how to prepare it, where to put it, which electrical instrument to use, how to produce the electrical stimulus, on which part of the animal to apply it and in what manner, and where to place all these various items on the bench. Each decision produced a different combination of the items, and thus a different experiment. Moreover, as Galvani had learned in his first experiments, different frogs reacted differently to the same electrical stimulus and to different stimuli. How was he to confront such a complex situation?

Galvani approached this situation in a way similar to that of other scientists when beginning a new and original investigation: he varied systematically the arrangement of the experimental objects and recorded what happened, in his case if contractions occurred or not.[19] This mode of experimentation is quite evident in his examination of the problem of the ligature of nerves, the first question on which he focused on 22 November. In the experiments of that day, and again on 29 November and 2 December, Galvani tied up a nerve of a prepared frog near its insertion into the muscle (Exp. 1, 22 Nov.), then near the spinal cord (Exp. 2, 22 Nov.). He made also a ligature in both nerves, with and without the spinal cord (Exps. 1 and 2, 29 November) and compared the effects of the ligature of nerves with those of severing them (Exp. 3, 29 Nov.; Exps. 2–5, 2 Dec.). As for the use of electricity, he stimulated different parts of the frog – muscles, nerves, and the spinal cord – either "giving a spark of the conductor" of the electrical machine to them or "extracting a spark" from them.[20]

Three outcomes emerged from these trials: first, contractions occurred only if some portion of nerves was involved in the electrical stimulus; second, the strength of muscular contractions depended on the portion of nerve being stimulated with electricity; and third, the electrical stimulus did not act beyond the ligature, contrary to what Haller had claimed. Although Galvani did not draw these conclusions explicitly in his laboratory diary, he was to write them down very soon. More importantly, these experiments showed the potentiality and fruitfulness of his systematic approach and helped him to achieve a greater familiarity with the objects involved in the experimental situation. After this set of experiments on the ligature and severing of nerves, Galvani turned to the question of the electrical properties of nerves and muscles with a more confident attitude and a more complex approach.

THE LAWS OF MUSCULAR CONTRACTION

On 9 December Galvani significantly altered the experimental arrangement. Up to then he had simply kept the whole prepared frog isolated, or not isolated, and had charged or discharged its parts by means of the electrical machine or, in one experiment, the Leyden jar. In these conditions, on 25 November he had compared the conductivity of the spinal cord and nerves with that of muscles, obtaining inconclusive results. He probably realised that in order to overcome that failure he had to design an arrangement which contained all the essential elements of the experimental situation (i.e. nerves and muscles), but separated and isolated one from the other. He did this by resorting to the Franklin square he had used on 6 November, placing the frog's limbs on its upper coating, while the nerves and spinal cord were isolated on a glass pane laid on the square (Fig. 2). After performing another experiment on the severing of nerves with the new arrangement, which confirmed that the action of electrical stimulus on the nerve was "local," Galvani turned to the problem of the electrical properties of nerves and muscles.

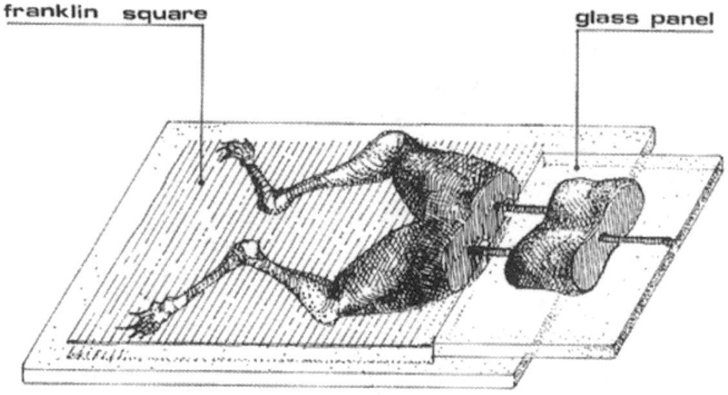

Figure 2. The experimental arrangement of 9 December 1780 (sketch by Ugo Sorrentino).

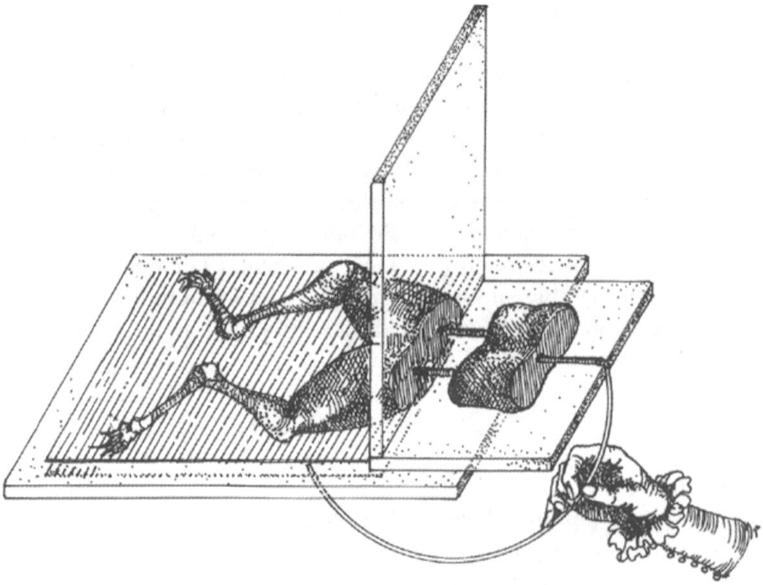

Figure 3. The experimental arrangement of 16 December 1780 (sketch by Ugo Sorrentino).

In the third experiment (the report of the second is incomplete) Galvani charged the Franklin square by connecting the upper coating to the conductor of the electrical machine and giving "60 turns" of the disk. Approaching the spinal cord with a conducting body, he obtained "a small and incomplete discharge," while a greater spark was extracted from the square. Replicating the operation several times he observed that "there was always a small discharge and spark from the spinal cord, a much bigger one from the muscles." In the light of contemporary knowledge of electrical phenomena, as set out for example in Cavallo's textbook, this result was open to various interpretations. In fact, the electric spark, measured by the noise and light produced, "is greater or less, according as the electricity is greater or less; as the parts from which it flies, and on which it strikes, are more blunt or more sharp, and as the conductor is more or less perfect."[21] Assuming that in the experiment the quantity of electricity did not vary (being always the result of 60 turns of the disk of the electrical machine), the smaller spark given off by nerves could depend – Galvani reasoned – either on the fact that "nerves are better conductors than muscles" but "more pointed and of a lesser surface;" or "on the contrary" because nerves were "worse conductors" than muscles.[22]

In order to dissolve this "uncertainty" that, as we have seen, regarded a central problem in the study of animal motion, the following week Galvani modified the experimental arrangement again. On 16 December he fixed a second glass pane perpendicular to the one which supported the nerves and spinal cord at the point where the nerves were inserted in the muscles, but without compressing them (Fig. 3). To fix the second pane, he used insulating materials like wax and sulphur, with which he also entirely covered both nerves. By means of the second pane, and the artificial insulation of the nerves, Galvani aimed at forcing the electric fluid to flow from the upper coating to the spinal cord through the inner substance of nerves, without allowing any dispersion of electricity. It was a sort of "negative control experiment:" if the spark extracted from the spinal cord had been the same as, or greater than, that extracted from the muscles in these particular conditions, it would mean that in normal conditions nerves were not capable of retaining the electric fluid inside themselves. The result, however, was the same as that of 9 December, the spark extracted from the spinal cord being smaller than that extracted from the square and the muscles. In the second experiment Galvani charged the Franklin square again, and then connected the spinal cord and the lower, discharged coating of the square with a "conducting arc." The frog's limbs contracted "five or six or even more times" at every contact, while if he connected the upper coating to the spinal cord, no contractions occurred. This difference was coherent with Beccaria's, and Galvani's, notion of electricity: the spinal cord and the upper coating of the square were equally electrified, so that no flow of electricity and thus no contractions could be expected. More importantly for the problem being investigated, the nerves did not seem to discharge the electricity acquired all at once, as they would have done if they had been good conductors of electricity.[23]

These two experiments marked a decisive step forward in Galvani's investigation. Taken together, they strongly supported one of the two possibilities about the electrical properties of nerves and muscles envisaged on 9 December, i.e. that nerves were worse conductors than muscles. At the end of December he would report this conclusion in his diary as a "corollary," claiming "this shows how difficult the flow [of electricity] through nerves is." In Galvani's eyes, this was a satisfactory answer to one of the Hallerian objections against animal electricity: nerves being bad conductors, electricity could flow through them without being dispersed in the surrounding tissues. And it would eventually become the solution proposed by Galvani in the *De viribus*.[24]

A further step forward was made by Galvani in the last experiments of 16 December. Instead of connecting the upper coating of the Franklin square to the conductor of the electrical machine in order to charge the square, Galvani applied the conductor directly, first, to the spinal cord, and then to the muscles. In the first case (electricity applied to the cord), when the lower coating of the square was touched with a conducting body, there were contractions, while in the second (electricity applied to the muscles) there were not any. This showed that a given quantity of electricity was sufficient to excite contractions if it was given to the nerves, whereas it was not enough if given to the muscles. For Galvani this meant that the action of electricity in stimulating the contraction "depends only on the nerve." In his diary he also remarked that the Franklin square appeared "hardly ever charged," so that the quantity of electricity involved in the phenomenon was very small.[25]

Galvani would very soon be concerned with the problem of the quantity of electricity sufficient to produce muscular contractions. For the moment, however, he was more interested in understanding the flow of electricity between the frog and the Franklin square. On 19 December he resorted to the arrangement used on the 9th (see Fig. 2), and varied the last experiments performed three days before. Having connected the spinal cord to the conductor of the electrical machine, he observed contractions at every turn of the disk of the machine. He obtained contractions also upon touching the upper coating of the square, while no contractions occurred when he touched the lower coating. This result was quite contrary to that of 16 December, and in fact when he returned to the laboratory ten days later, Galvani repeated the experiment, altering again the experimental arrangement.[26]

As Christmas drew near, Galvani interrupted his experimental activity and decided to set down in writing the results he had achieved to date. This decision may have been dictated also by the need to clarify which problems he had solved and separate them from the ones left unsolved, or which had emerged during the first two months of research, such as those of the last experiments. On "Christmas Day, 1780" Galvani wrote down a plan for a treaty on the "neuromuscular force" acting in "cold-blooded animals and in voluntary muscles," under the conditions examined in the November and December experiments. He did not report any experiment; instead, he drew up some "laws" concerning muscular contraction: first, "the contraction of the muscle produced by the irritation of the nerve is proportional to the smallest parts of the nerve set in motion by the stimulus, and to the force by which they are set in motion;" second, "whatever the irritating cause, the irritation of the nerve is basically local, i.e. does not go beyond the point to which the irritating cause has been applied;" and third, "the communication and propagation of that [irritating] action, or of the motion induced, depends exclusively on the nerve."[27]

These laws were generalisations of the experimental outcomes obtained by Galvani in the first two months of experimentation: the first and second derived from the experiments on the ligature of nerves, while the third synthesised the findings of several trials (for example, those of 16 December). More importantly, they were phenomenological laws, in the sense that they reduced the phenomena to general rules without involving a unifying explanation at a more fundamental theoretical level. The terms used by Galvani – "irritation" and "irritant cause" – echoed the Hallerian terminology, but did not imply the concept of "irritability" as distinct from "sensibility" and the nervous force or fluid. These laws could in fact be accepted both within a Hallerian interpretation and within a neuroelectric conception of muscular motion. Their meaning was analogous to that of the laws of electrical phenomena described for example by Cavallo, who in his work had treated separately "such natural laws concerning electricity, as by innumerable experiments, have been found uniformly true, and are independent on any hypothesis."[28] It seems clear, therefore, that although Galvani had suggested a neuroelectric conception some months earlier, at this stage of his investigation he was looking for an understanding of the phenomena of muscular motion, without trying to force the experimental outcomes into that framework.

FROM LAWS TO CAUSES. A FUNDAMENTAL AFTERTHOUGHT

On 29 December Galvani reconsidered the inconsistent results obtained in the last two days of experimentation. He simplified the experimental arrangement, eliminating the Franklin square, placing the whole frog on a glass pane and connecting it to the conductor of the electrical machine. He obtained the contractions at every turn of the disk of the machine and again, after he had stopped turning the disk, whenever he touched the frog's tibias. Upon removing the conductor which connected the frog to the electrical machine, and then re-positioning it, he obtained up to 17 contractions when he touched the animal. He also observed a strange phenomenon he had already noticed on 19 December: when he could not obtain any further contractions, if he "allowed the animal to rest without turning the machine, and then touched it again, there were contractions for two or three times." This phenomenon, which for the moment remained unexplained, was to be interpreted by Galvani in the light of later experiments.[29]

The fact that the charged frog's limbs contracted several times upon being touched suggested that its nerves did not discharge all at once. This confirmed the interpretation of the experiments of 16 December, that nerves were bad conductors of electricity. But it suggested to Galvani also another "corollary," related to the quantity of electricity sufficient to excite muscular contractions. "Very little electric fluid, far from being enough to give the minimum electrical sign – he remarked in his diary – is enough to excite the contractions." In order to test this conclusion, in the second experiment Galvani placed some "light bodies" near the conductor of the machine: while touching the conductor caused the contractions, no motion of the bodies was observed. Electricians considered the attraction of light bodies as one of the fundamental signs through which the electric fluid manifested itself, especially when in very small quantities. This experiment, therefore, showed that a quantity of electricity insufficient to attract light bodies was nevertheless sufficient to produce muscular contractions, i.e. that muscular contraction was the minimum sign of electrical imbalance.[30]

This conclusion was a kind of fourth law regarding the action of the nervous force, following the three that Galvani had formulated some days before on Christmas Day. In the draft of a memoir written a couple of years later he included it among "my discoveries," reporting it in the same terms as in the experimental records.[31] Galvani was right: it was indeed a significant discovery both in the study of muscular motion and in the knowledge of electricity. When Volta read the *De viribus*, one of the first things that attracted him was that the electricity detected by Galvani in the animals was "so weak that we are not able ... to make it sensible with the most delicate electrometer." This fact was particularly important for an investigator like Volta, who had long been concerned with the search for weak electricity, and it may have been one of the insights that pushed him to focus on Galvani's experiments on animal electricity.[32] From a historical point of view, this early outcome of Galvani's investigation strengthens interest in his experimental activity before the "first experiment" and leads to reconsideration of the claim, expressed for example by Marcello Pera, that this activity "did not yield any noteworthy findings."[33]

Although the experiments of 29 December had provided some important evidence on the questions of the electrical properties of nerves and the quantity of electricity in the production of muscular contractions, they did not clear up what happened between the frog, the electrical machine and the Franklin square in terms of the flow of electricity. On 30 December and again on 10 January Galvani focused on that question, resorting to the experimental arrangement with the two glass panes (see Fig. 3). In several experiments he charged the frog through a muscle and touched different parts of the arrangement with the conducting arc, first while turning the disk of the machine, and then after stopping it. From these trials he came up with a "reflection:"

> Therefore it is the frog that retains the electric vapour, which leaves the frog upon touching the conductor [of the electrical machine] and gets to the lower coating [of the Franklin square].[34]

This argument seemed to explain all the phenomena observed in the previous experiments with the same experimental arrangement. Although Galvani felt "sure" of the explanation of 10 January, the following week he re-examined it, according to a habit which was to become an essential part of his experimental practice. One of the guiding principles of his research when he had achieved a result was in fact to repeat the experiment under different conditions, in order to ascertain that the result was independent of those specific conditions.

On 17 January Galvani changed the experimental arrangement again. The elements were the same as in the previous experiments, but this time he placed the whole frog on the glass pane and separated it from the Franklin square. Then he connected both the square and the spinal cord to the conductor of the electrical machine. Galvani sketched this set up on the margins of his record of the experiment, apparently at the end of the trial (Fig. 4). It was the first time that Galvani had represented an experimental arrangement in the process of his investigation, a practice he then continued. Apparently these sketches did not have a mere mnemonic function: they did not represent particularly difficult or elaborate set ups that needed to be fixed on paper in order to replicate them. Rather, they marked important steps in the course of Galvani's investigative pathway, as they related to experimental situations he considered particularly relevant. Their principal function was to isolate the experimental circumstances of a phenomenon from the mess of instruments, animals and apparatus that dominated the laboratory bench. In so doing, they made it easier to identify those very circumstances and their role in a phenomenon, acting as a kind of visual reasoning in the interpretation of the experimental situation.[35] In fact, it was while

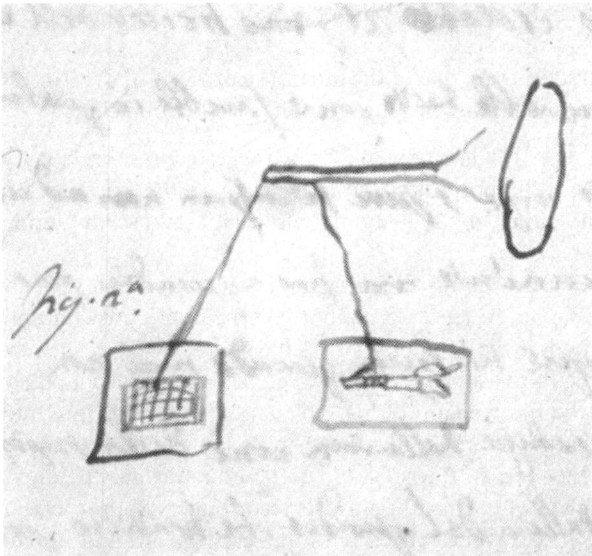

Figure 4. Galvani's sketch of the experimental arrangement of 17 January 1781 (AASB, Fondo Galvani. By courtesy of the Accademia delle Scienze di Bologna).

reflecting on the experimental arrangement set up on 17 January that Galvani realised the possibility of new interpretations of the phenomenon of muscular contractions.

After turning the disk of the machine to charge the square and the frog, Galvani connected the lower coating of the square and the spinal cord with a conducting arc. At each contact, 15 times in a row, there were contractions of the legs. This result seemed to confirm the conclusion of the previous week, and in fact Galvani remarked in his diary:

> This phenomenon seems to depend on a certain amount of electric fluid accumulated in the frog, which flows to the lower coating of the armed [i.e. Franklin] square, considerably discharged by the charge of the upper one and by the touching of conducting bodies.

He then inserted a few more words (here underlined):

> This phenomenon seems to depend either on the extrinsic electric fluid accumulated, held and confined by the frog's nerves and spinal cord, or on that intrinsic to nerves awakened by the extrinsic one, which flows to the lower coating of the armed [i.e. Franklin] square, considerably discharged by the charge of the upper one and by the touching of conducting bodies.[36]

In this last version Galvani claimed two possible explanations for the contractions observed in the experiment:

(1) The contractions are due to the electricity of the machine accumulated in the frog;
(2) The contractions are due to the intrinsic electricity of the frog set in motion by the electrical stimulus produced by the electrical machine.

While explanation (1) had already been suggested the previous week, (2) was totally new. More importantly, the latter claimed the existence of an electricity intrinsic to the animal body – animal electricity, to use the term Galvani later adopted.

It was the first time in his experimental investigation that Galvani realised that he had found some evidence to support the neuroelectric conception of animal motion advanced at the beginning of 1780 – so much so that he recorded it in his diary. It is difficult to ascertain exactly when he realised such a possibility, whether while writing the first version of his explanation or later. From the manuscript it seems likely that he did not make the changes immediately, because he added the new words in the margin. But he did so quite soon, certainly before performing the second experiment of the day, because its record refers both to explanation (1) and (2). Therefore it was while reflecting on the experiment, and in particular on the new arrangement he had adopted, that Galvani added the second explanation. In fact, the problem investigated – i.e. the circulation of electricity between the electrical machine, the Franklin square, and the frog – and the outcome, were the same as 10 January. What was new was the arrangement of the experimental objects, which were still connected but at the same time physically separated from each other. A simplification of previous ones, the new set up allowed Galvani to see more clearly the relationships between the animal and the electrical instruments, and the circulation of electricity within them. In this case, the experiment was not designed as a test of the neuroelectric hypothesis; rather, the hypothesis of an intrinsic electricity (explanation 2) was a sort of afterthought suggested by the experimental arrangement.

The change in the interpretation of the first experiment of 17 January has not been noticed by historians because the published edition of Galvani's laboratory diary reports only the second version.[37] They have rightly emphasised the importance of this experiment, and of its interpretation in terms of an intrinsic electricity, in relation to what followed, in particular Galvani's observation of the contractions at a distance, i.e. his "first experiment" of 26 January. In order to understand Galvani's investigative pathway and the origin of his "first experiment" more thoroughly, however, it is equally important to underline the connections with previous experiments. Although the experiment of 17 or that of 26 January 1781 can arguably be considered decisive turning points, neither was an isolated performance, and even less a flash of insight. Rather, they were the outcomes, expected or not, of an investigative process that consisted in slight shifts, or small variations of the experimental arrangements.

The importance of the experiment of 17 January for Galvani's investigation went beyond the fact, remarkable in itself, of suggesting some evidence to support a neuroelectric conception of muscular motion. It marked also a change in the scope of the investigation. While up to that point Galvani had concentrated on the laws of muscular contractions at a phenomenological level, he now entered the more difficult and hazardous area of theoretical reasoning about the cause of the phenomenon. An indication of such a change of scope is that in the laboratory diary "corollaries" and "warnings," as well as tentative expressions such as "it seems," "one may wonder," and "there is a suspicion," became more frequent. Moreover, although explanations (1) and (2) seemed to Galvani the most plausible, he adopted an open-minded attitude towards the experimental outcomes, taking into account other possible causes of the phenomenon observed:

(3) The contractions are due to a residue of electricity in the lower coating of the Franklin square, which flows to the frog;
(4) The contractions are due to the simple "mechanical" pressure of the conducting arc on the spinal cord;
(5) The contractions are due to the electric fluid accumulated on the glass pane which supports the frog.

As on previous occasions, for example when he had examined the question of the electrical properties of nerves and muscles, Galvani considered the electrical and physiological perspectives as sources from which to draw the interpretation of phenomena, as well as the problems to be investigated. In fact, explanations (3) and (5) derived directly from his knowledge of electricity, while (4) was drawn from his familiarity with the studies on "irritability" and muscular motion.

In the following experiments Galvani put to the test each of the proposed explanations, beginning with (3). He varied the experimental arrangement, connecting only the Franklin square to the electrical machine, while the frog remained uncharged on the glass pane. If contractions were due to a residue of electricity on the lower coating of the square – Galvani may have reasoned – they should occur when coating and frog were connected. But this did not happen; instead, contractions occurred when connecting the frog to the upper coating, positively charged. This experiment, a sort of control experiment of the same type as that of 16 December, seemed thus to exclude explanation (3). On replicating the same experiment with another frog, however, Galvani obtained a different result: at first, when he connected the lower coating of the square to the spinal cord, there were no contractions, but shortly afterwards he noticed some quite evident contractions.[38]

The different result must have puzzled Galvani. In fact, he reconsidered explanation (3) and envisaged the possibility that contractions were not produced by electricity, but were due to the irritation of the cord by means of the arc, i.e. explanation (4). But he also suggested that the delayed contractions could be produced by the electric fluid "naturally existing in the nerves or spinal cord, ... which needed some time to accumulate." This last remark added a new element to the interpretative framework he had established, and may have well been an *ad hoc* hypothesis designed to save the concept of intrinsic electricity. Once it had emerged from the experimental situation, this concept apparently began to influence Galvani's consideration of the phenomena. On the other hand, the remark could have been suggested by the observation, recorded on 19 and 29 December, that contractions could be excited again after they had ceased and a certain span of time had elapsed. Moreover, this did not prevent Galvani from carrying on his examination of the possible causes of muscular contractions.

FACING THE INCONSTANCIES OF NATURE

On 20 January Galvani repeated his most recent experiments, observing no contractions when connecting the frog either to the lower coating of the charged square or to the upper one.[39] The results being so uncertain, four days later he changed direction, and examined whether it was the glass on which the frog was laid that accumulated electricity (explanation 5). In order to exclude such a possibility, he replaced the glass pane with one made of copper, i.e. a conductive material, and charged both the Franklin square and the frog, as in the first experiment of 17 January. Upon contact between the lower coating of the square and the frog, however, no contractions occurred. This was an indication that glass was vital in the experimental situation and that, contrary to what Galvani had believed up to then, it was not the frog which accumulated and held the electric fluid. His puzzlement increased after some experiments performed on the same day, in which he repeated the experiment with the original arrangement of 17 January and obtained different results. At the end of his experimental record of 24 January, Galvani remarked:

> Reflection – In all this year's experiments as well as in many others, a great irregularity, inconstancy and anomaly has been observed, not only when different frogs were used, but also with the same ones.

Compared with similar remarks made in November and December, Galvani now observed such irregularity in the same animal, and not only between different ones. However, instead of being discouraged by a possible source of further confusion, he turned this observation into an argument that by analogy supported the existence of intrinsic electricity. The observed irregularity and inconstancy – Galvani remarked – "show the nervous force to be just as various, inconsistent, and unreliable, as that of the electric fluid; and that in itself would be an argument for believing that the nervous force is the same as the electric."[40]

Analogy plays an important role in Galvani's investigation, as it does for every creative thinker. The most interesting and best-known example is the analogy between muscle and Leyden jar, developed by Galvani at the end of the 1780s and acting both as an investigative tool and as an explanatory model.[41] In the case under examination, however, the analogy was of the same sort as those advanced by other supporters of a neuroelectric conception of muscular motion, who had limited themselves to a similarity between the properties of the electric and nervous fluids. The fact that Galvani adopted a similar argument suggests that at this early stage of his investigation he had not yet completely freed himself from the traditional way of reasoning about the cause of animal motion. On the other hand, such an argument was only one among many developed by Galvani, and did not prevent him from pursuing a systematic experimental investigation of the mechanism of muscular contractions. In fact, two days after this remark he was in his laboratory again, and received an unexpected reward.

On 26 January Galvani focused again on the circulation of electricity between the electrical machine, the Franklin square and the frog, apparently trying to eliminate the "irregularities and inconstancies" observed in the previous experiments. He repeated the second experiment of 17 January, placing the frog on the glass pane and charging only the Franklin square. Having observed no contractions, he connected the lower coating of the square and the frog *while* charging the former. When he touched the frog, or even approached it with a conducting body, the muscles contracted. Evidently, Galvani reflected, "it was the electric vapour, which leaves the lower coating while the upper one is charging, that went to the frog" and produced the contractions. The same was true, however, if he used a "glass cylinder" instead of the conducting body. Galvani had also noticed this last phenomenon two days earlier, calling it "marvellous," because the glass, being an insulating material, should not have discharged the nerves and caused the contractions. Indeed Galvani thought that the glass had become a conductor because of some humidity he observed on it. But after drying the glass and again bringing it close to the nerves, he observed that these were attracted, and "arose" from the pane, as if they were conducting bodies.[42]

In order to test this last suspicion, which contradicted his conclusions of 29 December on the electrical properties of nerves, Galvani resorted to the experimental arrangement adopted then. He eliminated the Franklin square, and placed a frog on the glass pane, connecting the conductor of the electrical machine to the muscles and turning the disk of the machine twelve times. Even though he had removed the connection between frog and machine with a conducting body, which should have discharged the animal, "there were still motions, upon touching the nerves, for five, six, eight or more times." This outcome seemed to confirm that nerves were bad conductors of electricity and that they held the electric fluid inside themselves, a conclusion Galvani had reached on 17 January. In the next experiment, Galvani repeated the operation, but introduced "light bodies" as indicators of the presence of electricity in the conductor of the electrical machine, as he had done on 29 December in order to investigate the quantity of electricity involved in the phenomenon of contractions. As in the previous experiment, he obtained the contractions at every touching of the frog, even after he had removed and reconnected the frog and the machine by hand. This time, however, he noted that contractions occurred as long as the light bodies were attracted by the conductor of the machine, and not afterwards. Galvani therefore concluded: "the end of contractions and the end of the attractions of light bodies in the conductor were simultaneous."[43]

In the light of the preceding experiments, the results now obtained by Galvani were highly problematic. First, they contradicted what he had established as a "corollary" at the end of December, that "a quantity of electricity so small that it does not manifest itself through any electric sign, including the attraction of light bodies, is sufficient to excite muscular contractions." Second, they showed that it was the electricity of the electrical machine that passed to the frog every time it was touched, and caused the contractions. This was strong evidence in support of the first explanation of 17 January, and against the existence of an intrinsic electricity. We do not know which of these two implications Galvani had in mind when he performed the following experiment, or if he was thinking of both of them. What we do know is that it marked a real breakthrough in his experimental pathway.

AN UNEXPECTED AND "MARVELLOUS" PHENOMENON

In the sixth experiment of 26 January, the frog was laid on the glass pane at a certain distance from the conductor of the electrical machine. The experimental arrangement was therefore very similar to the preceding one, except for a small but fundamental change: this time there was no connection between the frog and the machine. Whenever "one, my wife or someone else – Galvani recorded in his register – brought a finger close to the conductor and elicited sparks [from it], and at the same time I rubbed the crural nerves or spinal cord with an anatomical knife . . ., there were contractions even though no conductor was applied to the glass where the frog lay." "This phenomenon – Galvani continued – was consistent and is doubtless marvellous."[44]

Where did Galvani's wonder derive from? At least three meanings of "marvellous" are to be underlined here: the phenomenon of contractions at a distance was in fact unexpected, unprecedented, and its cause was unknown. It was unexpected in the light of the facts established just shortly before: the occurrence of contractions when the animal was unconnected and at whatever distance from the electrical machine conflicted with the idea that it was the electricity of the machine which flowed to the frog and produced the contractions, as well as with the idea that the electrical stimulus excited muscular contractions only until light bodies were attracted. But it was also an unprecedented phenomenon, characterised by new experimental conditions. In all previous experiments, the electric stimulus had been applied directly to the frog, either by connecting it to the Franklin square and to the electrical machine, or by a spark applied to, or elicited from, the animal. This time, on the contrary, the frog was physically separated from any electrical source, and the electric stimulus was provided indirectly, too. It must have been really surprising, "marvellous," to see the movements of the frog's limbs simultaneously with the shooting out of a spark from the electrical machine.

Galvani's puzzlement was increased by his inability to give an immediate explanation for the phenomenon. Although he related it to some previous experiments, especially those of January in which "there was no connection between the conductor of the machine and the glass [where the frog lay]," the "cause" of the contractions was "unknown." Not only did the phenomenon not seem to follow the laws he had established at the end of December, but it did not provide clear evidence to support any of the interpretations advanced on 17 January. Rather, it opened up the experimental situation by adding new conditions in which contractions might occur. In fact, in accordance with his experimental approach, on 26 January and again five days later, Galvani systematically varied the experimental arrangement – the distance between the animal and the electrical machine, the instrument which touched the frog and the animal parts being rubbed, the material on which the animal lay, the position and role of the experimenters, and even the preparation of the frog – in order to find out the essential conditions in which the phenomenon took place. Only after these trials did Galvani conclude that the phenomenon of contractions at a distance was due to "a very subtle fluid existing in the nerves" and that this fluid was "the electric fluid."[45]

These conclusions, established as "corollaries" in his laboratory diary, seemed not only to confirm the neuroelectric conception suggested by Galvani one year before, but seriously questioned the Hallerian theory of "irritability." As he put it in another "corollary" of 31 January, the fact that contractions occurred when the "animal electric fluid" left the nerve and not when it got to the muscles made it very difficult to believe that "the muscle contracts due to an irritation produced in the muscle itself." If this was true – Galvani wondered – "What of irritability?" The interrogative form of this sentence, however, suggests that Galvani did not feel totally sure about the interpretation of muscular motion in terms of animal electricity. Moreover, he realised that the phenomenon of contractions at a distance needed further investigation in order to clear up some circumstances, for example why the spinal cord, when separated from the spinal column, "loses its neuro muscular force more quickly." Indeed, next day Galvani was again in the laboratory, repeating and varying the experimental arrangement of 31 January. Quite unexpectedly, this new set of experiments, which kept Galvani busy for the next two months, would eventually lead him to completely subvert the interpretative framework worked out at the end of January. But this is another story, a different, though strictly connected and equally instructive stage of his experimental venture in the field of muscular physiology.[46]

The reconstruction of Galvani's experimental activity at an early stage of his investigation into muscular motion has allowed us to highlight some aspects of his initial approach to this subject. Although at the beginning Galvani took from "outside" – from the research carried out by other investigators and from his own investigation into related fields – the experimental objects, i.e. the frog and the electrical apparatus, as well as the questions to be investigated, it was *within* his laboratory activity that he developed procedures to deal with them. He did not superimpose a well established set of operative concepts or theoretical presuppositions on natural phenomena; rather, he played with nature in a sort of game of combinations, whose underlying rules (and the implications with regard to the laws of nature – for example those established at the end of

December 1780) could only be determined step by step, experiment after experiment. At the core of this game there were the arrangements of the experimental objects, which functioned as responses to questions or interpretations formulated in the course of the investigation. But they were also prompters of new problems, arguments and phenomena or, more simply, tools to move into the unknown.

The central and complex role of experimental arrangements in Galvani's investigation invites the historian to focus on procedures and operations, which form a fundamental, though often neglected, aspect of scientific activity.[47] It also suggests that the relationships between experiment and theory, thought and action, are more complex than historians sometimes suppose and scientists themselves often convey in their published work. The study of research records, especially in the case of creative scientists like Galvani, can help us to highlight these relationships at a more intimate level and, therefore, to achieve a deeper understanding of scientific activity conceived as an essentially human enterprise.[48]

LABORATORY RECORDS AND THE PUBLISHED REPORT

In the *De viribus*, published more than ten years after the experiments discussed in this paper, Galvani began his narrative "account of the discoveries in the same order of circumstance that chance and fortune in part brought to me, and diligence and attentiveness in part revealed" with the description of the experiment of contractions at a distance. This description differed from the one recorded in the laboratory diary on 26 January 1781, though not in the experimental arrangement adopted or in the phenomena observed, as much as in its circumstances. Galvani wrote that he had placed a dissected and prepared frog on a table where there was also an electrical machine "I having in mind other things." When one of his assistants touched the frog's nerves "by chance," another assistant observed the contractions and informed Galvani, who was "completely engrossed and contemplating other things."[49]

It is difficult not to believe such a vivid and realistic description, which projects us immediately and directly into the laboratory, where the action takes place. In fact, most readers of the *De viribus*, as well as historians, have relied on the story as Galvani told it here, and taken his "first experiment" as a typical case of a chance observation which, in the hands of an attentive and skilful experimenter, can lead to major achievements.[50] Paradoxically, the plausibility of Galvani's public recollection may be strengthened by the allegedly chance character of this experiment, so different from the most common view of scientists, who retrospectively tend to see a surprising result as the "inevitable product of a logical enquiry or of a teleology of the experimental process."[51] If Galvani himself claimed that he had made an observation by chance, or even that he actually did not make the observation because he was "contemplating other things," why not believe him? Usually, and this is true also for Galvani, laboratory notebooks serve as repositories of procedures and results, not of emotional reactions or accidental circumstances, such as those reported by Galvani in the *De viribus*.

And yet, the laboratory diary and the reconstruction of Galvani's investigative pathway suggest that he was personally involved in the experiment and that the experimental situation – the arrangements of the items, the problems investigated and the outcome expected – was directly connected to what preceded it. This is apparently confirmed by a preliminary draft of the *De viribus*, in which the description of the "first experiment" is quite different from that of the published version, and more similar to the laboratory record. In the draft, the phenomenon of contractions at a distance occurred while Galvani was rubbing the nerves with a scalpel in order to "experiment on the power of electricity in the nerves," which is a perfect synthesis of what he had been doing since November 1780.[52] We may thus argue that the story of the chance observation was constructed by Galvani for some reason other than his intention to give a historic narrative of his investigation. This may also lead to a historical analysis of the meaning and role of "chance" and "fortune" in Galvani's thinking.

The discrepancies between the published report of Galvani's "first experiment" and what can be inferred from the laboratory records invite further examination on the role of writing in the investigative process. On the other hand, they should not lead us to dead-end questions such as: Was Galvani telling the truth in the *De viribus* or not? was he reporting some circumstances of an experiment which he had not recorded in his laboratory diary but were in his mind, or was he telling a different story from what had really happened? Even though we have experimental records to compare with the published report, the question remains unanswered. But more importantly, in the case of Galvani's "first experiment" it seems to me to be a mistaken question, even admitting that such questions have any relevance. Much more interesting – and this partly explains my decision to study Galvani's laboratory notebooks – are other questions that Galvani's published recollection poses. Why was there an electrical machine on Galvani's table? What was Galvani doing with frogs? Why was he so "extremely enthusiastic and eager to repeat the experiment so as to clarify the obscure phenomenon and make it known"? These are the questions to which this paper attempts to offer an answer through a detailed study of Galvani's laboratory notebooks.

This is, of course, a partial answer, that needs to be complemented with other sources in order to reconstruct Galvani's scientific and medical education, his professional and research interests, and his work in the context of 18th century science.[53] Nevertheless, laboratory records have allowed us to enter into the laboratory of a creative thinker of our past, to participate in what he was doing day by day, to follow the development of his investigation from the beginning to a significant turning point, and to appreciate the difficulties and obstacles, as well as the successes, that paved his way. For any scholar who undertakes such a venture, the satisfaction and reward that can be drawn greatly exceed the hard work it requires.

NOTES

* This essay is based on my History of Science Ph.D. dissertation on Luigi Galvani (University of Florence, 1999). I wish to thank Giuliano Pancaldi, who has constantly encouraged and sustained my efforts in this field, and Frederic Holmes, who has guided my first steps in the study of Galvani's laboratory notebooks. Thanks to Anna Guagnini and Friedrich Steinle for their comments on an early draft of this paper.

1 Lazzaro Spallanzani, *I Giornali delle Sperienze e Osservazioni*, ed. C. Castellani (Firenze: Giunti, 1994); Maria Teresa Monti, "Metodo e narrazione della scoperta nei diari di laboratorio di Spallanzani," in Walter Bernardi and Paola Manzini, eds., *Il cerchio della vita* (Firenze: Olschki, 1999), pp. 117–135.

2 I. B. Cohen, "Introduction" to Luigi Galvani, *Commentary on the effects of electricity on muscular motion* (Norwalk: Burndy Library, 1953), p. 13. On Galvani's life and work, see Bern Dibner, *Luigi Galvani* (Norwalk, Ct.: Burndy Library, 1971); Giuliano Pancaldi, "Luigi Galvani," in Walter Tega, ed., *Storia illustrata di Bologna, 6* (Milano: Nea, 1989), pp. 281–300; Marco Bresadola, "Medicine and science in the life of Luigi Galvani," *Brain Research Bulletin* 46 (1998): 367–380.

3 Luigi Galvani, *Memorie ed esperimenti inediti* (Bologna: Cappelli, 1937), pp. 233–411. It should be noticed, however, that some experiments described in Galvani's laboratory records have already been examined in J.L. Heilbron, "The contributions of Bologna to Galvanism", *Historical Studies in the Physical and Biological Sciences* 22 (1991): 57–85; Marcello Pera, *The Ambiguous frog. The Galvani-Volte Controversy on Animal Electricity* (Princeton: Princeton University Press, 1992); B.I. Williams, *The Matter of Motion and Galvani's Frogs* (Bletchingdon: Rana, 2000).

4 Bresadola, *Medicina e filosofia naturale in Luigi Galvani* (University of Florence Ph.D. diss., 1999). On the historical approach adopted by Holmes see, e.g., Frederic L. Holmes, "The fine structure of scientific creativity," *History of Science*, 19 (1981): 60–70; "Laboratory notebooks: Can the daily record illuminate the broader picture?," *Proc. of the American Phil. Society* 134 (1990): 349–366; *Antoine Lavoisier–The next crucial year* (Princeton: Princeton University Press, 1998), p. 5.

5 Edwin Clarke and L.S. Jacyna, *Nineteenth-century origins of neuroscientific concepts* (Berkeley: University of California Press, 1987), pp. 165–166. See also Hebbel E. Hoff, "Galvani and the pre-galvanian electrophysiologists," *Annals of Science* 1 (1936): 157–158; Marcello Pera, *The ambiguous frog* (ref. 3), p. 70.

6 Galvani, "Saggio sulla forza nervea e sua relazione coll'elettricità" (25 novembre 1782), in *Opere scelte*, ed. G. Barbensi (Torino: Utet, 1967), p. 123. Engl. transl. in Pera, *The ambiguous frog* (ref. 3), p. 64.

7 On the various conceptions of animal motion in the 18th century and their roots, see Roger K. French, "Ether and physiology," in G.N. Cantor and M.J.S. Hodge, eds., *Conceptions of ether. Studies in the history of ether theories, 1740–1900* (Cambridge: Cambridge University Press, 1981), pp. 111–134; Max Neuburger, *The historical development of experimental brain and spinal cord physiology before Flourens* (Baltimore: The Johns Hophins University Press, 1981); Francois Duchesneau, *La physiologie des lumières* (The Hague: M. Nijoff, 1982).

8 Tommaso Laghi, "De sensitivitate, atque irritabilitate halleriana," in G.B. Fabri, ed., *Sulla insensitività ed irritabilità halleriana, 2* (Bologna: Corciolani, 1757), p. 338.

9 On the Academy of Sciences of Bologna, see the three volumes of *Anatomie accademiche*, ed. Walter Tega and Annarita Angelini (Bologna: Il Mulino, 1986, 1987, 1993). On the Bologna debate on Hallerian irritability see Marta Cavazza, "La recezione della teoria halleriana dell'irritabilità nell'Accademia delle Scienze di Bologna," *Nuncius* 12 (1997): 359–377.

10 Galvani, *De ossibus lectiones quattuor* (Bologna: Compositori, 1966), p. 135.

11 Felice Fontana, *Treatise on the Venom of the Viper* (London: John Cathel, 1795), p. 283 (first ed., Florence: s.n., 1781, *2*, p. 245).

12 Giambattista Beccaria, *Elettricismo artificiale* (Torino: Franzini, 1772). See J.L. Heilbron, *Electricity in the 17th and 18th centuries* (Berkeley: University of California Press, 1979), esp. pp. 365–372.

[13] Galvani, Saggio sulla forza nervea (ref. 6), pp. 123–125.
[14] Galvani, *Commentary* (ref. 2), p. 56.
[15] Giuseppe Veratti, "Sopra l'elettricità riguardo agli animali," Accademia delle Scienze di Bologna (AASB), Tit. IV, Sez. I. Veratti read two dissertations on this topic at the Academy of Sciences of Bologna on 20 April, 1769 and on 4 January, 1770. See *De Bononiensi Scientiarum et Artium Instituto atque Academia Commentarii* 7 (1791): 41–44.
[16] Exps. 1–4, 6 November 1780, in Galvani, *Memorie* (ref. 3), pp. 233–234.
[17] Albrecht von Haller, *Elementa physiologiae corporis humani, 4* (Venetiis: apud A. Milocco, 1769), pp. 254–255; see Roderick W. Home, "Electricity and the nervous fluid," *Journal of the history of biology* 3 (1970): 242–247. The argument based on the ligature of nerves persisted for a long time in the history of neurophysiology, well into the 19th century: see Marco Piccolino, "Animal electricity and the birth of electrophysiology: the legacy of Luigi Galvani," *Brain Research Bulletin* 46 (1998), p. 190 ff.
[18] Samuel Auguste André David Tissot, *Trattato de' nervi, e delle loro malattie* (5 vols., Venezia: Occhi, 1781–1784; French ed. 1778–1780), *1*, pp. 217–218; Galvani, *Commentary* (ref. 2), p. 76.
[19] A similar approach characterises Faraday's experimental investigation into electromagnetism, as discussed in Friedrich Steinle, "Looking for a 'simple case': Faraday and electromagnetic rotation," *History of Science* 33 (1995): 179–202; "Work, finish, publish? The formation of the second series of Faraday's experimental researches in electricity," *Physis* 33 (1996): 141–220.
[20] Galvani, *Memorie* (ref. 3), p. 234–238.
[21] Tiberius Cavallo, *A complete treatise of electricity in theory and practice; with original experiments* (London: Edward and Charles Dilly, 1777), p. 35.
[22] Exps. 1–3, 9 December 1780, in Galvani, *Memorie* (ref. 3), pp. 239–240.
[23] Exps. 1–2, 16 December 1780, in ibid., p. 240.
[24] Galvani, *Commentary* (ref. 2), p. 76.
[25] Exps. 5–6, 16 December 1780, in Galvani, *Memorie* (ref. 3), pp. 240–241.
[26] Exps. 1–2, 19 December 1780, in ibid., pp. 241–242.
[27] AASB, Fondo Galvani, cart. III, plico IIAA, fasc. 1, ss. 1, 9–10.
[28] Cavallo, *A complete treatise of electricity* (ref. 21), p. vi.
[29] Exp. 1, 29 December 1780, in Galvani, *Memorie* (ref. 3), pp. 242–243.
[30] Exp. 2, 29 December 1780, in ibid., pp. 243–244.
[31] Galvani, "Saggio sulla forza nervea" (ref. 6), p. 143.
[32] Alessandro Volta, "Memoria prima sull'elettricità animale," in *Opere, 1* (Milano: Hoepli, 1918), p. 25. See Pancaldi, "Electricity and life. Volta's path to the battery," *Historical Studies in the Physical and Biological Sciences* 21 (1990): 123–160.
[33] Pera, *The ambiguous frog* (ref. 3), p. 69.
[34] Galvani, *Memorie* (ref. 3), pp. 333–334 (exps. of 30 December 1780), 245–247 (exps. of 2 and 10 January 1781).
[35] On the role of the depiction of experiments in animal electricity, see Maria Trumpler's essay in Bresadola and Pancaldi, eds., *Luigi Galvani International Workshop* (Bologna: CIS, 1999), pp. 115–145.
[36] Exp. 1, 17 January 1781, in AASB, Fondo Galvani, cart. V. plico 10, fasc. 1, s. 2.
[37] Exp. 1, 17 January 1781, in Galvani, *Memorie* (ref. 3), pp. 247–248.
[38] Exps. 1–2, 17 January 1781, in ibid., pp. 248–249.
[39] Exps. of 20 January 1781, in ibid., pp. 250–251.
[40] Exps. 1–4, 24 January 1781, in ibid., pp. 251–252 (Engl. transl. in Heilbron, "The contributions" (ref. 3), p. 74).
[41] Galvani, *Commentary* (ref. 2), pp. 73–74.
[42] Exps. 1–3, 26 January 1781, in Galvani, *Memorie* (ref. 3), pp. 252–253.
[43] Exps. 4–5, 26 January 1781, in ibid., pp. 253–254.
[44] Exp. 6, 26 January 1781, in ibid., p. 254: "Il fenomeno fu costante ed è certo maraviglioso."
[45] Exps. 1–2, 31 January 1781, in ibid., pp. 255–258. This conceptual passage is explored in more detail in Bresadola, *Medicina e filosofia naturale* (ref. 4), sect. 3.4.

[46] A reconstruction of this stage of Galvani's experimental pathway is offered in Bresadola, *Medicina e filosofia naturale* (ref. 4), sect. 4.1.
[47] A recent and stimulating discussion of this topic is Hans-Jörg Rheinberger, *Toward a history of epistemic things* (Stanford: Stanford University Press, 1997).
[48] Such a "humanistic" view of scientific activity is a fundamental aspect of Frederic Holmes' studies: see, e.g., Holmes, *Antoine Lavoisier* (ref. 4), pp. 145–146.
[49] Galvani, *Commentary* (ref. 2), pp. 45–47.
[50] Giuseppe Antonio Eandi, "Ragguaglio delle sperienze del Sig. Luigi Galvani accademico bolognese estratto da una lettera diretta al Sig. Conte Prospero Balbo," *Giornale fisico-medico* 2 (1792), p. 95; Jean Alibert, *Elogio storico di Lugi Galvani* (Bologna: S. Tommaso d'Aquino, 1802; French ed. an X), p. 35; Cohen, "Introduction" (ref. 2), p. 57; Clarke and Jacyna, *Nineteenth-century origins* (ref. 5), p. 165.
[51] Rheinberger, *Toward a history of epistemic things* (ref. 47), p. 74.
[52] AASB, Fondo Galvani, cart. II, plico IX, fasc. E, s. 2.
[53] For a thorough discussion of these topics, see Bresadola, *Medicina e filosofia naturale in Luigi Galvani* (ref. 4); see also Bresadola and Marco Piccolino, *Rane, torpedini, Scintille. Luigi Galvani, Alessandro Volta e l'elettricità animale* (Torino: Bollati Boringhieri, 2003).

FRIEDRICH STEINLE*

THE PRACTICE OF STUDYING PRACTICE: ANALYZING RESEARCH RECORDS OF AMPÈRE AND FARADAY

RESEARCH RECORDS AND RESEARCH PRACTICE

My interest in laboratory notes results from my interest in research practice. In emphasizing practice I don't have in mind a contrast between practice and theory. Research practice in fields like theoretical physics or mathematics is of great interest, as Jürgen Renn's and Tilman Sauer's work on Einstein's notebook (in this volume) readily illustrates. It is the contrast between practice and the public picture of research which I address here. The geneticist François Jacob, referring to Ludwik Fleck, highlights that contrast in poignant words.

> Day science calls into play arguments that mesh like gears, results that have the force of certainty. . . . By contrast, night science wanders blind. It hesitates, stumbles, recoils, sweats, wakes with a start. Doubting everything, it is forever trying to find itself, question itself, pull itself back together. Night science is a sort of workshop of the possible where what will become the building material of science is worked out. . . . Where phenomena are still no more than solitary events with no link between them.[1]

Whereas the "day science" appears in the published papers, addressed to a scientific audience or even a more general public, the "night science" is conducted in the laboratory or the study and usually remains hidden. But it is here where experimental results are achieved and new concepts are formed, where steps are taken and innovation is produced. It is this "workshop" of scientific research which is in the focus of interest here – a focus common to the contributions to this volume.

Research practice has to be reconstructed. If personal accounts of the actors exist, they may provide a comfortable starting point of the analysis. Those records often differ, however, from what can be historically reconstructed from other sources. The editors of the present volume point to many examples in their introduction. Particularly helpful are sources of more or less private character which were not intended for communication. Examples are laboratory notebooks,[2] letters, accounts of negotiations with instrument makers and suppliers of

* Max Planck Institute for the History of Science, Berlin

Frederic L. Holmes, Jürgen Renn and Hans-Jörg Rheinberger (eds.), Reworking the Bench: Research Notebooks in the History of Science, 93–118 © *2003 Kluwer Academic Publishers. Printed in Great Britain*

resources, private sketches of concepts, ideas and experiments, but also manuscripts for talks or publications. These diverse research records differ widely in what they can tell. Historical reconstructions will always have to rely on a variety of different types of material. In most cases that is not so much a matter of choice, but rather of what is preserved and accessible.

As the editors highlight in their introduction, one has to be well aware of the inherent limits of any reconstruction of this type. Every reconstruction is inevitably done from a certain point of view. It necessarily focuses on certain questions and leaves others out. A *complete* reconstruction of the historical development is neither possible in principle, nor would it be instructive or helpful. In order to learn something, we have to focus our attention. The general point that any historical narrative necessarily has to take a specific perspective comes sharply to the fore here. Those restrictions have to be kept in mind, and it is necessary to be aware of the interests which drive the historian to undertake such meticulous studies. Within such a frame, however, it should be possible to give fair accounts of the actions and considerations of historical actors.

In any particular case, the possibility of a reconstruction is wholly dependent on the availability of appropriate source materials, which is usually a matter of chance. In my paper I shall sketch two examples which illustrate how different those situations may be and how one may deal with such differences in historical research. My focus is on the practices and techniques which I used as an historian. The historical insights which resulted from those studies and which I presented elsewhere are only touched upon. My cases are taken from the history of electromagnetism and deal with two of its most prominent actors: Ampère and Faraday.

AMPÈRE'S FIRST STEPS INTO ELECTROMAGNETISM

In July 1820, when the Danish researcher Hans-Christian Oersted announced his discovery of an interaction between electricity and magnetism, this news came as a surprise to most researchers in Europe. A wave of investigations ensued across the continent. The activities dealt not so much with theoretical explanations of the effect as with grasping the effect itself and with getting an idea of what was involved here. The effect was puzzling. A magnetic needle changed its position when a nearby wire was connected to the poles of a galvanic battery. It took a position somehow "tranverse" to the wire. Even describing that position appropriately turned out to be a problem. Oersted had reported different angles for different experimental arrangements. What was more, the position of the needle changed or even reversed when the needle was placed above instead of below the wire. Such curious behavior resisted any account in terms of attractive and repulsive central forces. This incompatibility was nowhere felt more sharply than in Paris, where physics was dominated by a powerful tradition of mathematization exactly by means of the concept of central forces – the "Laplacian" physics.[3]

One of the earliest and most important Paris actors in the new field was the mathematics professor André-Marie Ampère, who was not committed to the Laplacian school. His initiative came as a surprise, since he had before not displayed much interest in physics, and had not carried out any experimental activity whatsoever. He threw himself into the new field immediately when he heard the news. Within a few months of feverish work, he presented to the academy new concepts, new effects, new instruments, and new experimental techniques. For another six years, Ampère would be occupied with elaborating the results of that first, intense research period. He came to name his whole account "electrodynamics." In its final version, presented in 1826, it would form the foundation of one of the most important physical theories of the nineteenth century.

There are many questions open on how exactly his early activities developed. In particular, there is the striking fact that, after only three weeks of research, he presented something like a research program in which the main lines of his later theory were outlined. From previous historical studies, we have got a clear idea of his research after that date.[4] But we still don't know much about his very entry into the new field – a field about which Ampère knew even less than most other actors. From Ampère's own few notes the impression emerges that he had started from the beginning with ideas of the same type that he actually came to develop. For several reasons, however, that story appeared implausible. The only way to improve on it was to carefully study his research practice.

Archive Material

The main reason for our lack of knowledge about Ampère's entry in the new field was the problematic state of the sources. Much of the previous historical research relied on public statements by the researcher himself. The very few existing letters of that period provided little help.[5] Not that no material had been kept. The main resource is the *Dossier Ampère* in the archive of the Paris *Académie des Sciences*. It consists of 39 boxes which contain some 400 folders. The folders are roughly classified, but there is no detailed catalogue of their contents. 58 folders are devoted to electrodynamics. They contain roughly a thousand individual documents of various types, such as manuscripts for publications and for lectures, notes for private use, reminders for buying this or that, letters, bills and receipts (some of which are even in other boxes). In many cases Ampère had used his paper successively for more than one purpose, such as for notes on electrodynamic experiments and afterwards (with the paper turned upside down) for calculations in preparing his mathematics lectures.

The problem which makes it so difficult to handle that material is, on the one hand, its sheer bulk, with no catalogue or reliable classification at hand. There was previously even no means available to refer to specific individual documents. In historical studies, references were usually given to complete folders which could well contain up to 30 documents. An even more serious problem is that Ampère almost never gave any indication of the date or period in which the document originated. Since his work in electrodynamics lasted over roughly a decade, the task of dating is essential, but at the same time particularly difficult. The most important start is, of course, a detailed analysis of the content. With a broad knowledge of Ampère's published writings and of the historical circumstances, a first ordering could be achieved. A very important criterion was given by shifts of terminology which Ampère undertook several times. But those criteria did not offer much help once it came to reconstructing shorter periods of research. Further criteria came from more technical points: the type of paper, the handwriting, the contours of separation of different sheets, different writing on both sides of paper etc. Individual documents could be temporally located only by their complex relation to many other documents. Working in the academy archive, I sometimes felt like a detective searching for whatever traces I could find to get more information about a scratched piece of paper. Although such a situation is well known to many historians, the degree to which it was necessary for Ampère's archive is probably unusual.

For the early period in which I was interested, successively and slowly a set of documents could be identified. As an important, negative result of systematically checking the boxes, it became clear that there is nothing like a laboratory notebook or other systematic laboratory records. Ampère most probably never kept records of that type. His style of research was spontaneous and he was anything but a systematic, well-organized actor. If a reconstruction were to be at all possible, it could only rely on other types of material. From the various different things I found, I shall just present two important pieces.

Reconstructing a Lecture Manuscript

In the fall of 1820, Ampère presented a long series of nearly weekly lectures to the academy in which he constantly presented new experiments as well as instruments and theoretical considerations. Many of the manuscripts for those lectures are preserved, particularly for the period from October 1820 on, and Christine Blondel has based her excellent and detailed study on that material.[6] For the very first period in September 1820, however, the situation is more difficult. Ampère gave two lectures to the academy, on 18 and 25 September. The manuscripts are not preserved and were considered as lost. In analyzing a manuscript for a later publication, however,[7] I realized that some pages of it must have been part of the lecture manuscripts of 18 and 25 September. Ampère had taken out those pages from the original version and made them part of his publication. But in editing the text, he had essentially reworked the original version. He had crossed out words, sentences, paragraphs and inserted new ones, cut parts of pages out and glued them in, or put other pages in between.

Any attempt to restore the original version meant, of course, to undo all those changes. Many different procedures had to be applied, such as removing overglued pages, identifying overwritten words, and figuring out the original sequence of pages. Some passages could be restored only after considerable training and several attempts, a few others kept resisting. Even when these procedures worked out, there remained the problem that many parts of the original lecture manuscript had not been included in the later publication. If they still existed, they had to be found and identified in the archive. In searching for them, again a broad range of criteria had to be applied. Some pieces could be identified just by the exact fit of the sheets of paper which Ampère had torn up, or by the fact that the words of the first line continued a (canceled) phrase of another page which already had been identified as part of the manuscript.

With a several months effort, I was finally able to restore the original version of the manuscript as a whole. It came out to be one piece: Ampère's second lecture was just a continuation of the first. An additional difficulty arose when I realized that Ampère had inserted changes in the manuscript even during the very process of writing. With specific and detailed arguments I was in most cases able to separate these changes from those introduced much later.[8] Thus there is now a text available which comes very close to Ampère's original lecture manuscript. It is a special type of source, however. Being the product of a reconstruction, it does not physically exist in the archive, but only in my transcription of the restored text. Only by very detailed arguments for any part of the reconstruction do I feel justified in using it as source material.

This reconstructed text provides many new and surprising insights into the experiments and considerations Ampère conducted in those weeks, into the instruments he designed and used and into his interactions with other researchers. As useful and informative as it may be, however, that source is still a lecture manuscript, a text intended for presentation and not only for private use. Nevertheless I take it very seriously. Ampère wrote it within the same short period of three weeks in which he conducted some most intense research. And phrases like "The experiment ... has been carried out yesterday" or "There are several apparatuses under construction right now"[9] show that it was written in close connexion with ongoing research. Thus the considerations Ampère presented here may not be very distant from those which he actually had in mind when working in the laboratory. One thing which cannot be derived from the lecture manuscript is, of course, the temporal sequence of his activities. It is very probable that Ampère rearranged and regrouped his activities and results when he prepared the manuscript – after all, his lectures were to be delivered at the Paris academy. Thus I was happy to find another document in the archive which provided insight even into how his research developed in time.

Ampère's Private Research Agenda

The manuscript AS 205(b) is one folded page, forming four pages. I give a facsimile in Table 1. The manuscript has text, figures and some calculations. The figures are referred to in the text, but the calculations have no connexion with the text. As I mentioned above, Ampère tended to use paper successively for different purposes. There is no date whatsoever given in the manuscript, but from Ampère's

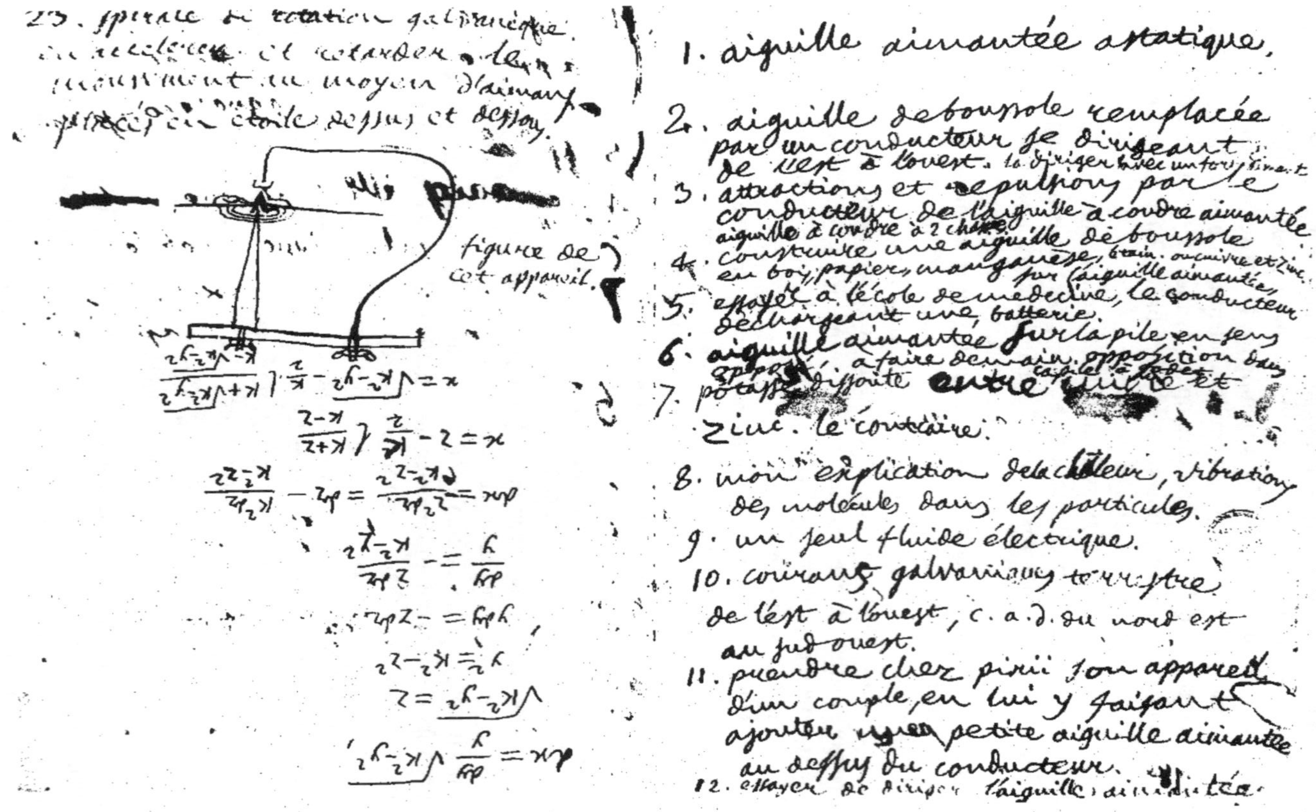
1. aiguille aimantée astatique.
2. aiguille de boussole remplacée par un conducteur se dirigeant de l'est à l'ouest.
3. attractions et répulsions par le conducteur de l'aiguille à coudre aimantée
8. mon explication de la chaleur, vibrations des molécules dans les particules.
9. un seul fluide électrique.
10. courants galvaniques terrestres de l'est à l'ouest, c. a. d. du nord est au sud ouest.
11. prendre chez pixii son appareil d'un couple, en lui y faisant ajouter une petite aiguille aimantée au dessus du conducteur.

Table 1a. Facsimile of Ampère's manuscript AS 205(b). Archives de l'Académie des Sciences, Paris.

Table 1b. Facsimile of Ampère's manuscript AS 205(b). Archives de l'Académie des Sciences, Paris.

terminology it becomes clear that the text was written in the three-weeks period in question. Ampère used the term "courant galvanique" which he did only in that period. Shortly afterwards he would switch to the term "courant électrique."

The text has two lists which I have called A and B. Ampère did not state the purpose of those lists. They differ significantly even in the writing style. List A, which he numbered (1) to (23), is quite clearly written and has not many revisions or changes. The much shorter list B, in contrast (with numbers (1) through (7)), has often been reworked and corrected. Even that first observation made me surmise that the two lists had different purposes.

List A was obviously written in several successive stages at each of which new entries were added. From Ampère's handwriting and the distribution of the text on the paper, I inferred four pieces or blocks of text: (1) – (7), (8) – (12), (13) – (17) and (18) – (23). Ampère used many verbs, mostly in the infinitive mood: "construire une aiguille," "prendre chez Pixii," "essayer de diriger," "mettre des roues," "faire faire un appareil" etc. Obviously that was meant to remind or summon himself to do certain things. Sometimes he wrote down results or thoughts just to record them. In some cases a later point referred to an earlier one. In (12) Ampère came back to experiments proposed in (5), and in (14) to those proposed in (2). In both cases he seems to have carried out the proposals, and on the basis of those results proposed further steps. Hence my view that the list was composed in successive stages gets strong support. The list is a very special kind of research record: not a laboratory notebook, not an ideas notebook, but something like Ampère's private to-do list. It was intended exclusively for private use. From the sequence of entries, we can cautiously infer the sequence of events in the laboratory, at least with regard to the four textblocks.

List B is quite different. Ampère reworked it several times. Again I attempted, with procedures similar to those mentioned above, to reconstruct those various stages. In Table 2, I give a transcription of the present text and the result of the reconstruction of the successive earlier stages. From that reconstruction, I attempted an interpretation of what purpose the list served. It was a list of the experiments and instruments Ampère wanted to discuss in his first academy lecture. In its final version it fits exactly to the reconstructed version of that manuscript. The development of the list shows the various stages Ampère went through in preparing his academy lecture while continuously proceeding in his research.

The two lists A and B were kept parallel and written in the same period as Ampère worked on his lecture manuscript. Even from the distribution of text it becomes clear, for example, that the entries 18 through 23 have been added only after the list B was completed. A close analysis of the specific content reveals many connexions between the two lists. Table 3 gives an overview of the temporal relation between the parts and versions of the two lists.

Table 2. List B of Ampère's Manuscript AS 205(b)

---

1) TRANSCRIPTION

6 appareils

{*séparer les faits.*}
1. aiguille aimantée astatique.
2. appareil pour les attractions et répulsions magnético-galvaniques.

{*réciprocité.*}
~~2.~~ 3. ~~galvanomètre~~ appareil réciproque.

——— ———

4. ~~??conducteur??~~ {aiguille} galvanique. ~~mobile.~~
{5.* galvanoscope.}

{*appareil galvaniq. singeant l'aiguille aimantée.*}
~~5.~~ 6. hélice galvanique mobile.
~~6.~~ 7. hélice galvanique fixe. {moyen le meilleur pour aimanter par le galvanisme.} utilité de ce dernier appareil pour la marine.

{*identification des courans galvaniques.*}
~~7.~~ 5.* galvanoscope

appareils galvaniques pour observer les effets des plans à spirales galvaniques.
appareils pour la direction qu'un de ces plans donne à un aiguille pasacinétique.
appareil pour les attractions et répulsions de 2 de ces plans suivant la direction de leurs courans.
aiguille {aimantée ordinaire} placée dans la direction de la ligne qui joint leurs centres.

2) RESOLVING THE LIST INTO ITS FOUR SUCCESSIVE VERSIONS

Version 1: (up to Ampère's sign in the text)
1. aiguille aimantée astatique.
appareil pour les attractions et répulsions magnético-galvaniques.
2. galvanomètre

Version 2: (new text and a title is added)
6 appareils
1. aiguille aimantée astatique.
2. appareil pour les attractions et répulsions magnético-galvaniques.
3. appareil réciproque.
4. aiguille galvanique (first version presumably: conducteur galvanique mobile.)
5. hélice galvanique mobile.
6. hélice galvanique fixe. utilité de ce dernier appareil pour la marine.
7. galvanoscope

Version 3: (further text in different ink is added; topics 5–7 are renumbered; the text is divided under four headings)
6 appareils
séparer les faits.
1. aiguille aimantée astatique.

2. appareil pour les attractions et répulsions magnético-galvaniques.
réciprocité
3. appareil réciproque.
4. aiguille galvanique.
identification des courans galvaniques.
5. galvanoscope.
appareil galvanique singeant l'aiguille aimantée.
6. hélice galvanique mobile.
7. hélice galvanique fixe. moyen le meilleur pour aimanter par le galvanisme. utilité de ce dernier appareil pour la marine.

Version 4: (point 1–7 as above: new text is added)
appareils galvaniques pour observer les effets des plans à spirales galvaniques.
appareils pour la direction qu'un de ces plans donne à un aiguille pasacinétique.
appareil pour les attractions et répulsions de 2 de ces plans suivant la direction de leurs courans.
aiguille aimantée ordinaire placée dans la direction de la ligne qui joint leurs centres.

Table 3. Correlation Between the Two Lists

List A	*List B*	*Dates of writing*
(1)–(7)		
(8)–(12)	version 1 (two instruments)	
(13)–(17)	version 2 (additions: reciprocal apparatus and galvanic helices)	shortly before 18 September 1820
	version 3 (rearrangement)	around 18 September: Ampère prepared the final lecture manuscript
(18)–(23)	version 4 (additions: instruments with spirals)	after 25 September

Ampère's private research agenda as given in those lists combines very nicely with the reconstructed text of his lecture manuscript. Whereas in the lists he used an extremely condensed language (as is not unusual for such private notes), he explained in the lecture manuscript his instruments, experiments and considerations in much detail. For the first time there is considerable material available which was written in the very period under discussion and not in hindsight. It forms the basis for a new historical analysis.

How Ampère Entered Electromagnetism

The new sources enable the reconstruction of Ampère's activities in the first three weeks of his research. Just to give an idea of the power and limits of that reconstruction, I add a survey of his activities (Table 4). There are, in all cases, specific arguments for ordering the events in the way they are presented here. Only very few of the events can really be dated, however. The survey may not be complete: There may well be other activities which Ampère did not mention in the surviving documents. In some cases, moreover, we do not know whether Ampère actually carried out his proposals. Despite these limitations, the reconstruction provides a lot of previously unknown material. Most of the activities given in the list we did not know before. I will just give a sketch of some of the new things we learn from the reconstruction.[10]

Table 4. Ampère's Research Activities, 3–25 September 1820

	GF	RE	GA	TM	EG	GM	GC
4 September: NEWS OF OERSTED'S RESULT ARRIVE IN PARIS							
Experiments on "separating the actions": "astatic needle."	•						
First version of the regularity of the "directive action": concepts "right-/left hand side" of a current, 'bonhomme.'	•						
Experiments on the reciprocal "directive action": no result with terrestrial magnetism; proposal of using a strong magnet.		•					
Considerations and experiments on the "attractive/ repulsive action."	•						
Formulating a regularity of the "attractive/ repulsive action."	•						
Experiments on the "directive action": varying the material of the needle: wood, paper, manganese, brass, copper, zinc.	•						
Experiments on a possible magnetic effect of ordinary electricity at the École de Medecine: no result.	•						
Experiments on the magnetic effect of the (closed) pile.	•						
Second, more general version of the regularity of the "directive action": concept of a closed galvanic (not yet *electromotive*!) current "circuit."	•						
Chemical determination of the current direction within the pile.	•						
Hypothesis of terrestrial currents.				•			
Considerations on a microscopic theory of electric fluids.					•		
Considering possible *causes* (chemical agency) and further *effects* (heat and light) of terrestrial currents.				•			
Further considerations and proposals of experiments on the magnetic effects of the pile. Proposal of a *galvanoscope* for quantifying the pile's activity.			•				
Considerations on improving the experiments with ordinary electricity.	•						
Designing the *galvanometer* as a measuring instrument for currents in general: introducing a scale beneath the needle.			•				
Considerations on improving the experiments on the reciprocal "directive action" ("aiguille galvanique rectiligne").		•					

Table 4 *continued*

	GF	RE	GA	TM	EG	GM	GC
General hypothesis of circular currents: proposal of "imitating" magnets by helices of wire.						•	
Designing experiments with a moveable galvanic helix ("hélice galvanique mobile"), acted on by magnets and terrestrial magnetism.						•	
Designing experiments for magnetizing iron bars by a fixed galvanic helix ("hélice galvanique fixe").						•	
Successful experiments on the reciprocal "attractive/ repulsive action."		•					
17 September: Repetition of the experiment of the "attractive/ repulsive action" with Arago at the *Obsérvatoire*.	•						
17 September: Improved and successful experiment on the reciprocal "directive action" with Arago at the *Obsérvatoire*.		•					
First successful experiments with moveable galvanic helices						•	
Considerations, in a conversation with Arago, on a possible "reduction" of *all* electromagnetic results to the two general facts.	•						
18 September, afternoon: FIRST LECTURE AT THE ACADEMY.							
19 September: Unsuccessful experiments for directing a straight conductor by terrestrial magnetism.		•					
Elaborate considerations on possible causes of terrestrial currents.				•			
20 September: General considerations on galvanic effects and their relation to the effects of the open pile. Elaboration of the concept of a current circuit: "electromotive circuit."					•		
Considerations on further effects of terrestrial currents: magnetic variation, heat, light.				•			
Further experiments (together with Fresnel and Despretz) on the imitation of magnets by helices.						•	
Unsuccessful experiments (together with Fresnel) on the supposed interaction of two spirals. Ampère surmises insufficient power of the batteries.						•	
22 September: Presentation of the experiments with helices and spirals at a dinner party with Poisson.						•	
24 September: Ampère asks for permission to buy the large battery which Dumotier had built for Dulong.						•	
25 September, morning: Ampère buys the battery and carries out the experiment with two spirals in Dumotier's workshop. For the first time, the experiment works.						•	
Ampère adds a report on the experiment to his manuscript for the afternoon lecture and proposes to do the experiment with straight wires.							•

Table 4 *continued*

	GF	RE	GA	TM	EG	GM	GC
General considerations: The interaction of two spirals is completely equivalent to the interaction of two magnets. Ampère writes a first version of the "Conclusions."						•	
25 September, afternoon: SECOND LECTURE AT THE ACADEMY, ending with a presentation of the "Conclusions." After the session, Ampère presents his apparatus with two spirals. He makes appointments with Alexander v. Humboldt (26 September) and with Laplace (27 September) to present his new theory.							
25 September, evening: Ampère writes enthusiastically to his son: "On ne m'a plus fait d'objection et voilà une nouvelle théorie de l'aimant."						•	

TOPICS

GF reduction to **G**eneral **F**acts
RE **RE**ciprocal effects
GA **GA**lvanoscope/ galvanometer

TM theory of **T**errestrial **M**agnetism
EG theory of **E**lectricity and **G**alvanism (*one* electromotive force)
GM theory of **G**eneral **M**agnetism
GC theory of the interaction of **G**alvanic **C**urrents

The pathway of his research was drastically different from what has been sometimes suggested. Just to give an example, Ampère's discovery of the interaction between straight currents (which would later form the central effect of his whole system) did not happen by chance, and provide the base for his invention of circular arrangements (as has been suggested by Williams and adopted by Hofmann),[11] but the story was exactly the opposite. The background of his research appears in a new light under this perspective. From the reconstruction we learn about whole areas of research which Ampère had conducted, but never mentioned in his own public accounts. He made strong efforts – experiments, considerations, instruments and money – to establish some "reciprocal" effects to those of Oersted: If a fixed galvanic wire was able to move a moveable magnetic needle, a fixed magnet should likewise be able to set in motion a moveable galvanic wire. Although Ampère's efforts to obtain those effects were only partially successful, he learned a lot of new experimental techniques which would become central for his later experiments on electrodynamics.

As the survey shows, Ampère pursued a whole number of different topics in parallel. In the columns which I introduced on the right hand side of the list, I roughly attributed his individual actions to the several topics he himself quite clearly identified and separated. The shorthand notations are explained in the box at the end of the list. It becomes clear, moreover, that Ampère did not start out his research with the ideas in mind that afterward came to dominate his research. Rather those ideas were formed in a quite tortuous and often unsystematic process during the time that he dealt with many different questions at once.

In the first and second weeks of Ampère's work a particular type of research occupied most of his effort. He aimed at establishing empirical regularities and an initial order within the new field. As a result, he succeeded in identifying two "general facts" to which, as he claimed, all other electromagnetic effects could be "reduced." Any thought of his later theory was far away, and the experimental activity he practiced here was of a quite specific, "exploratory" type.[12] Rather than providing anything like a test of theories, such experimentation plays an essential role in the process of forming basic concepts which enable a stable manipulation of the effects. And it was indeed in the context of that activity that Ampère developed and stabilized some of his most fundamental concepts, such as the notion of a current circuit which included both the battery and its connecting wire. By understanding that exploratory activity, we can obtain an appropriate answer to the vexing question of how Ampère, a complete newcomer in electricity and magnetism, could so quickly establish a new line of research. The concepts which were fundamental had not been in his mind before, but were developed in an intense period of exploratory experimentation.

The first three weeks of Ampère's research form a specific research period with its beginning and end, and with a specific type of research being predominant. The mere existence of such a period was hitherto not known, nor could we see its fundamental importance and far-reaching consequences. Ampère's second academy lecture of 25 September 1820 was not only, as has been realized in previous studies, the starting point of a long and well-directed research program. It was likewise the culmination and end of a (previously unknown) foregoing research period whose results provided the framework for Ampère's later research. Those points may illustrate my claim that the reconstruction of Ampère's practice presents a new picture of his research and of the roots of his later development.

FARADAY'S RESEARCH ON ELECTROMAGNETIC INDUCTION

My second case deals with a part of the copious experimental research of Michael Faraday. He had entered the field of electromagnetism for the first time in 1821, when he was "chemical assistant" to Humphry Davy at the Royal Institution of London. At that time, he had obtained a spectacular success by discovering electromagnetic rotations. Due to other obligations, he disappeared again from the scene for ten years. Only in 1831, when Davy had died and Faraday was director of the laboratory, was he in a position to choose his own research agenda. He moved quickly to electromagnetism. In some of his first researches he obtained the new and exciting effect of electromagnetic induction. He worked intensely for several months, finally presenting the effect by means of the concept of magnetic curves. Such an approach differed fundamentally from Ampère's concepts, which had, at that time, been widely accepted. Only some decades later, however, did Faraday's approach (by then called field theory) become widely recognized as a possible alternative to the usual action-at-a-distance-theory. In my research, I aimed at a better understanding of the formative period of 1831/32, when Faraday developed his dissenting approach for the first time.

The setting was quite particular. Although Faraday had deliberately sought for an induction effect (as many others had done,[13]) the effect he actually had obtained looked very different from what he had expected. The induced currents were only transient, and there occurred an induced current in the opposite direction when the cause of induction was taken away. Faraday was puzzled and had no idea how to account for those properties. In that respect, the situation was analogous to the situation in which most researchers had been thrown eleven years earlier by Oersted's report of an electromagnetic interaction. Faraday was well aware, moreover, that a report of the effect would be as spectacular as Oersted's had been.

In other respects, however, the situations could not differ more sharply. Faraday knew himself to be alone with the mysterious and fascinating effect. He could be nearly sure that nobody else had a similar result. Doing research in such a private and quiet atmosphere was very different from knowing that actors all over Europe were starting investigations at the same time. Moreover, Faraday was in a most favorable position. He had one of the best-equipped laboratories of Europe at his disposal. Faraday was eager to preserve that situation for a long period. Although he immediately started research into the new effect, he did not communicate his activities to others. He had obtained the new effect on 29 August 1831, but his first notice to anyone else did not come until three months later, on 21 November, when he handed over an elaborate account to the Royal Society. In a printed form, his first papers would appear only in April 1832. During that whole period, he continued research, and developed and changed his views, sometimes even drastically. At the time of publication, his papers looked significantly different from their initial form.

Archive Material

The sources for a reconstruction of Faraday's research are totally different from what I found with Ampère. There is, first, a very detailed laboratory diary. Faraday kept it carefully over several decades in the form of successive bound volumes. For each experiment he noted the date, the instruments and experimental arrangement (in considerable detail) and the outcome. Often he added sketches of the experimental arrangements. In some rare cases, Faraday wrote down considerations on what a certain experimental result implied, or ideas where to go further. In even fewer cases are there signs of excitement, such as underlining or exclamation marks.

The diary is completely preserved and has been edited and printed in seven volumes.[14] The transcribed text is completed by a facsimile of the figures. While the edition is mostly reliable, it does not give any corrections to the original text (of which there are only few). Likewise it does not reflect Faraday's custom of marking with vertical lines those paragraphs that he had included in a published text. For those points, which may well be important for reconstructing his research practice, recourse to the manuscript version is necessary. For most of the reconstruction work, however, the printed version serves perfectly well.

Faraday's handwriting in the diary is very clear and does not have a provisional character. There are only few corrections, crossing-outs or cancellations. Moreover, there are no physical traces of laboratory events such as stains by chemicals, bleaching or effects of heat. Thus it seems probable that the diary was not written in the lab, but at a separate place. Some days' entries start with a general remark about the whole day, such as "To-day went still more generall to work ... and obtained I think very satisfactory and reconciling results."[15] Those remarks suggest that the diary entries were written in the evenings on the base of provisional notes that Faraday had taken during the day's research (and which are not preserved). Although the diary thus is a well edited text, the order of events does not appear to have been much rearranged. There are sudden and unsystematic switches of topics, as are typical for everyday laboratory work. Faraday also reported failed attempts, bad measurements and so on. Thus the order of the entries within one day may well reflect the temporal sequence in which the experiments were conducted. Sometimes Faraday emphasizes that point explicitly.[16]

The second main source for reconstructing Faraday's research practice has only recently come to scholarly attention. For some of his published papers there are manuscript versions preserved, in particular for his two papers on electromagnetic induction.[17] Those manuscripts are in a specific state. They have many corrections, erasures and rearrangements. In some parts there are three series of different paginations. As a result of a closer analysis which also included other sources, it became clear that the manuscripts went back and forth several times between Faraday and the editor of the *Philosophical Transactions*. Each time, Faraday made some alterations. This was his way of keeping the text up to date with his research. During the process of publication, which went on for several months after submitting the paper, he continued his laboratory work and continuously produced new experimental results and insights. As long as the text of his manuscript was not yet typeset, he called it back repeatedly from the editor to insert these developments.

What we find in the archive is the final version. Since all those changes were done with one and the same physical document, that document bears traces of most of them. In order to find out the sequence of Faraday's research, I reconstructed the successive states of the texts. This work resembled what I described above with respect to Ampère's lecture manuscript. Again, a broad variety of procedures and criteria had to be applied, including technical points (paper, ruptures, handwriting, ink/pencil), considerations of the topics discussed by Faraday, comparison with the diary (dates, formulations), and analysis of the numerous figures (which had been renumbered, rearranged and sometimes redrawn by Faraday). As a result of some weeks' detective work, a reliable reconstruction of the successive versions became possible. By comparing them with the diary and letters it was possible, moreover, to establish a detailed chronology of the versions and their journeys between Faraday, the editor of the *Philosophical Transactions* and two referees of the text.[18]

The two main sources complement each other perfectly. The diary tells what Faraday did, not so much what he thought. Taken alone, it leaves essential questions open. His successive manuscript drafts of his publications, in contrast, give much of his thoughts, but no full record of his research, and in particular not the temporal sequence. Since both texts were produced at the same time, the mutual references can be identified in considerable detail. Often the intention for certain experiments becomes clear from the manuscript draft, and paragraphs of those drafts can be directly related to the results of experiments in the diary. That nearly ideal situation is further improved by an excellent edition of Faraday's letters which is complete, reliable and well commented.[19] With those sources, the fine structure of research can be reconstructed in considerable detail.

A Glance of the Results

From that wealth of material, a survey of Faraday's activities can be composed resembling that given above for Ampère. It is much longer, both due to the detailed character of the sources and the duration of the period which can be regarded as a more or less self-contained research period. As a consequence, a narrative of the whole story, as I have given elsewhere, is quite long.[20] It throws new light on Faraday's achievements and revises certain received views on that development. Even Faraday's general background appears in a new light.

I mention just two points. We are now in a position to study in detail the various roles of experimentation in Faraday's research. It becomes clear that the most important and lasting innovations were introduced by Faraday exactly in the context of exploratory work. This holds in particular for his new use of "magnetic curves." That term was not new, but widely used as a name for the patterns formed by iron filings on a paper above a magnet. Faraday, in contrast, turned it into an essential tool for formulating a regularity, a law of electromagnetic induction. The main problem was to find a conceptual means for expressing such a regularity. He started out with classical concepts, such as the (imagined) Ampèrian currents within magnets. But that attempt ran into problems. The complexity of effects, and in particular the essential role of motion of which he progressively became aware, could not be accounted for in that conceptual framework. Only when Faraday definitely realized that failure, did he switch to other concepts, trying to see whether magnetic curves would serve better. And indeed he could formulate, by means of those curves, the law for many experimental arrangements.

In order to embrace more and more phenomena, he had to widen the concept of magnetic curves, e.g. to subsume the direction of the earth's magnetic dip under that concept. He began to speak of magnetic curves in a circular shape even round a current-carrying wire. In a decisive step, he finally imagined those curves as moveable: expanding from a wire in the moment when the current was switched on, and contracting towards the wire when the current was switched off. By means of such a broad concept of magnetic curves, he was finally able to formulate *all* induction effects he had obtained at that point – and he had a large number and variety of those – in one single law. It is significant and characteristic that the whole development took place in a framework of intense exploratory experimentation.[21]

The reconstruction may well serve, and that is my second point, to study how Faraday's laboratory activities were connected to his communication and to the reaction he received from others. As a brief illustration, a condensed list of Faraday's activities within the two-month period of February and March 1832 (Table 5) may be instructive.

Table 5. Chronology of Faraday's Research Activities in February and March 1832

Dates indicated by * are only indirectly inferred from the sources. When periods of time are given that means "on some date within this period."

9 February 1832	Christie and Bostock give the report on Faraday's first paper. The paper is ordered to be printed. Faraday's second paper (V_1) is handed over to Christie and Bostock as referees.
9–27*:	Conversation with Christie on the difficulties implied by Faraday's previous result concerning the dependency of the inductive effect on the material of the wire. Faraday gets back the manuscript of his second paper (V_1) from Christie.
10	Faraday has got back the manuscript of the first paper. He adds a footnote to (79), concerning his priority with respect to induction effects.
17	Faraday gives a Friday Evening Discourse: "An account of the first two parts of his recent researches in electricity; namely, volta-electric induction and magneto-electric induction."
22	Experiments {342–3}: explanation of the Arago-effect.
23	Experiments {344–54}: Arago-effect obtained by an electromagnet. **Puzzling result!**
23.2.–12.3.	Faraday adds several passages on the results {344–8} to the first paper (111, 117, 119).
27	Experiments {355–67}: Arago-effect; dependency of the inductive effect on the material
29	Experiments {368–73}: dependency of the inductive effect on the material. Faraday adds a passage on this point to the first paper (139).
1 March	Experiments {374–82}: Arago-effect; dependency of the inductive effect on the material.
2	Friday Evening Discourse: "Explanation of Arago's Phaenomena."
5	Experiments {383–5}: dependency of the inductive effect on the material.

5	Faraday receives an offprint of the paper of Nobili and Antinori on inductive effects.
5–8*:	Faraday finishes the manuscript version V_2 of his second paper, hands it over to Christie.
8	Experiments and considerations {386–97}: inductive effects in general. Christie and Bostock give the report on Faraday's second paper. The paper is ordered to be printed.
8–12*:	Faraday adds a passage to the manuscript of the first paper (56), concerning experiments {386–92}, and hands it over to the editor of the *Phil.Trans.*
10	Christie returns the first part of the manuscript V_2 of the second paper to Faraday.
11–12*:	Christie returns the remaining part of the manuscript V_2 of the second paper to Faraday.
12	Faraday deposits a sealed note at the Royal Society.
12*	Faraday withdraws the report and the discussion of the puzzling result {347–8} from the manuscript version V_2 and hands the manuscript over to the editor of the *Phil.Trans.*
13–26*:	Faraday recalls the manuscript of the second paper from the editor. He adds paragraphs (231–42: moving magnetic curves) and gives the altered version V_3 back to the editor. He adds small alterations and a footnote to the proofs of the first paper withdrawing the notion of an 'electrotonic state'.
24	Report in the Literary Gazette, claiming the priority of Nobili and Antinori with respect to electromagnetic induction.
26	Experiments and considerations {398–407}: "mutual relation of electricity, magnetism and motion," Harris' results.
27	Faraday protests to the editor of the Literary Gazette.
30	Friday Evening Discourse, reported by the title "On the Evolution of Electricity, naturally and artificially, by the inductive action of the Earth's magnetism."
31	Report in the Literary Gazette, giving the correct order of events with respect to electromagnetic induction.

In that period Faraday did a lot of laboratory research and, at the same time, was busy with preparing final versions of his papers for print. As a consequence of his first communication in November, his results had become widely known and occasioned much interest. In Paris, the question of priority was raised immediately, since Ampère had also obtained an induction effect in 1822, but had never published it nor pursued the research.[22] Thus Faraday, being very sensitive in matters of priority and plagiarism, was much interested in printing his results as soon as possible, thereby showing his independence of others. Both of the two papers he had delivered to the Royal Society (in November and January) had first to be refereed. In February, the first paper was already in the hand of the referees, Christie and Bostock, whereas the second paper lay at the Royal Society and awaited review.

But it was just in that period that he stumbled upon a highly problematic experimental result (23 February). The result made it quite probable that some basic classifications of his account of the induction effect were weak or even mistaken. Thus he found himself forced to decide between quickly publishing a text which he had just realized contained serious problems, or to stop publication in order to gain time to analyze those problems. The latter option would have led to a further delay in clarifying the priority claims. That situation deeply affected the course of his research. Although he continued to work intensely, he could not solve the problem, and finally decided to give the manuscripts to the printer (12 March). Before he did so, however, he excised the report on the puzzling experiments which he already had formulated and inserted in the manuscript. It was only by happy chance that he obtained, shortly afterwards, a solution to those problems and was, in a last minute action, able to call back the manuscript of his second paper a third time. He introduced extended passages in which he not only reinserted the problematic experiments but presented, at the same time, his new solution. In those newly added passages, Faraday presented a fundamental revision of his classification of induction processes. It was only here that he introduced moving magnetic curves. In the light of his new insights, moreover, his previously introduced theoretical concept of an "electrotonic state" lost all of its functions and thus its justification. At that time, Faraday already had the page proofs of his first paper in hand, a paper in which he had devoted a whole chapter to this concept. The only thing he could now do was to add a footnote to that chapter, announcing that its very subject – the "electrotonic state" – had lost any significance! For his readers, his account of induction looked fundamentally different in the light of those corrections than it would have without them. In the printed version of his papers, there are only few traces of those turbulent developments. It was only by means of a close analysis of his notebook and his successive manuscripts versions that the whole complexity and fragility of the research process could be brought to light.

DIFFERENT RESEARCH RECORDS

The cases of Ampère and Faraday illustrate the variety of research records addressed by the editors in the introduction of this volume. The two cases are extreme. In Ampère the historian has a hard time finding anything whatsoever, whereas in Faraday she finds a nearly ideal situation. Those differences depend on a variety of circumstances. Local and personal styles are an essential aspect, of course. Faraday was trained as a bookbinder and businessman and got his research training as an assistant in Davy's chemistry laboratory. By far the most of his own early research was in the field of quantitative chemical analysis, often for commercial purposes or even as expert at court. In those activities, detailed record keeping had been essential. Ampère, in contrast, had never done experimental work before. He closely cooperated with the Paris instrumentmaker Pixii. Many details of the apparatus and instruments came from Pixii, and Ampère could trust that they were well documented in Pixii's workshop.

Ampère and Faraday had very different ideas of what purpose their laboratory records were to serve. Ampère envisaged just a short-term use until his next talk was given or paper written. He felt no need to keep detailed laboratory records. In the texts intended for the public, he would lay out the details of those experiments and considerations which he regarded as important, and he would do so within a period of several days, when he could still remember them. The idea that he might come back much later to experiments which he had initially regarded as not too important, and that he therefore should leave a detailed record of his activity as a whole, seems to have been absent from his mind. The contrast to Faraday could not be starker. From early on, Faraday intended his diary to be a repository of experiments and instruments. For the sake of easy reference, he gave each entry of his diary a number. He described carefully every arrangement and instrument and gave a detailed description of the outcome so that he could, at a later time, come back to the experiment and replicate it, if necessary. In his research on induction he would, even at the end of the eight months period, often come back to experiments which he had done in the very beginning. When, after some years of research, his diary became too voluminous for him to keep its structure in mind, Faraday started to compose directories, indices and superindices.[23] He regarded that careful way of keeping notes as appropriate to all research and suggested it to students.[24] Faraday put much effort into formulating his notebook entries as clearly and unequivocally as possible. Thus he could use his diary even as a direct textual source for his published papers. Quite often, particular formulations of the diary appeared unchanged in the printed text.

Those differences point, among others, to fundamentally different attitudes towards research in general. For Faraday, research was always provisional. He was very much prepared to take a step back and to revise even earlier decisions, interpretations and concepts.[25] He actually did so very often, and could do so only because he had kept his diary as he had. Ampère, in contrast, was prepared to commit himself quickly to insights he had developed at a given point of time. In his electromagnetic research, there are several episodes of that sort. Some of them were much to his advantage, as was his early and premature commitment to his electrodynamic theory – a step which he later regarded as guided by some sort of divination. In other cases he was less lucky, as with his above-mentioned discovery of the induction effect which he had "swept under the rug" just because it might possibly force him to revise certain theoretical views to which he had already committed himself in public.[26] Those different attitudes have to do with personal characteristics and biographical situations. The extraordinary degree of Faraday's resistance against commitment had to do, among other things, with his religious background, which enforced modesty in the face of nature.[27] But, as I suppose, local cultures which were so different in Paris and London come essentially into play here. It would be most instructive to study under that perspective the way in which other researchers in those environments kept laboratory records. Those studies may considerably enrich and sharpen the local characteristics of research.

SOME PERSPECTIVES

Along with other papers of this volume – one may think, e.g., of Hans-Jörg Rheinberger's study of Carl Correns – my examples illustrate how drastically studies of research practice often change our picture of specific historic cases. Those studies may, moreover, go far beyond micro-history and lead to new views on larger historical developments and on more general aspects of the development of science. In order to illustrate that point, and at the same time to suggest those aspects in which I was particularly interested in my studies, I shall mention three aspects.

An important problem in both the history and philosophy of science is the question of the various roles of experimentation within the process of scientific research. It is now widely agreed that the standard view according to which the prime role of experimentation is the testing of theories or hypotheses, is inappropriate or at least extremely restricted.[28] Experimentation has more and not less important roles. But to spell out those roles has proved to be difficult and is still a task to be done.[29] It is not by chance that the new discussion was triggered by a closer look at what scientists actually did or do in the laboratory. And a better idea of the role(s) of experimentation in the process of forming and stabilizing knowledge cannot be obtained by just studying researchers' published records. It is necessary to focus on research practice. Exploratory experimentation, for example, comes only into view when studying practice. Only here we can find material which enables us to analyze the characteristics and the fundamental importance of that specific experimental activity. What is more, studies of practice can even indicate how and why exploratory experimentation so often disappears from the picture when actors write down their results and rearrange them for public presentation. If it holds true that there is a systematic tendency of certain roles of experimentation to be concealed in public accounts, that would well explain why they escaped attention for so long a time.

That leads me to a second point: the difference between research practice and self-representation. It is obvious that there is a gap – the above quote of Jacob is just a very pointed formulation. But, as the editors of this volume emphasize in their introduction, the very process of researchers writing down their results for publication forms an integral part of research. Sometimes it is in the course of that process that new insights or shifts of emphasis come about. Any publication is, moreover, necessarily written with a certain audience in mind. The expectations of that audience inevitably play a role in writing the paper (be it in order to meet, or, on occasion, not to meet them), and so do general philosophical opinions and particular historical, sociological and biographical settings such as career interests. No wonder that in most cases the published record of research looks different from the original path. The point in focusing on practice is not to tell the 'truer story,' but just to add a part of the story which often has been overlooked. The public presentation is more easily accessible and has for a long time been the main source. By concentrating on practice, an essential part of research comes into view which otherwise remains hidden. The full picture may well provide new historical insights. In studying Ampère, for example, we are now in a position to see in detail which parts of his research he presented in his public accounts and which he left out.[30] Tracing in detail how and to what degree he successively shaped the public image of his research to fit the standard expectations of a Paris physicist is highly instructive, and tells us as much about Ampère as about the Paris scientific culture of the 1820s.

My third point is that insights into research practice do impressively, and in great detail, reveal the inexorable complexity and fragility of research processes. Those who engage in a systematic search for knowledge – in science – are human actors, after all. The interests, motivations and restrictions which they have as humans in culture, history and society cannot be separated from their scientific activities. How those interests matter and sometimes interfere with standards and norms of the search for knowledge, can be brought out particularly sharply by accounts of research practice. The above example of Faraday's introduction of moving magnetic curves is an illustrative example. It shows, moreover, the eminent role mere chance may play in those developments – as it does in everyday life. Accounts of research practice make particularly clear that neither the concentration on epistemic aspects and their dynamics, nor on social or historical aspects can be sufficient for an appropriate understanding of scientific development.

NOTES

1 (Jacob 1998), 126.

2 As was discussed at the workshop, there are cases in which laboratory notebooks do indeed serve for communication within a research group, but not beyond its borders.

3 For an overview, see the still classical study of (Fox 1974).

4 In particular (Blondel 1978) and (Blondel 1982), see also (Williams 1983), (Hofmann 1987a) and (Hofmann 1987b). For Ampère in general, see (Hofmann 1995).

5 Collected and edited (Launay 1936–43).

6 (Blondel 1982).

7 The publication is (Ampère 1820), the manuscript is AS 208bis(f). In order to enable an unequivocal reference to individual documents, I introduced an additional letter behind the number of the folder.

8 In (Steinle 2000a), Anhang A and B, I present all the material and give a detailed discussion of the reconstruction. At the same place, I give the full text of the reconstructed version.

9 Paragraphs r25 and r33 in my numeration of the reconstructed version of the lecture manuscript.

10 In more detail, I discuss the results in (Steinle 2000b) and (Steinle 2000a), chs. 3 and 4.

11 (Williams 1983), (Hofmann 1995).

12 (Steinle 1997), (Steinle 1998).

13 (Ross 1965).

14 (Martin 1932–6).

15 Entry 209, 9 December 1831, (Martin 1932–6) I, 394.

16 "... and the following experiments made ... in the following order": Entry 36, 1 October 1831, (Martin 1932–6) I, 372.

[17] The first to use material of that type were Romo and Doncel: (Romo and Doncel 1994). The manuscripts are located in the archive of the Royal Society of London, RS MS PT 20.4 and 20.5.
[18] I discuss the procedure and criteria for the reconstruction in Appendix C of (Steinle 1996).
[19] (James 1991), (James 1993).
[20] (Steinle 1996).
[21] For a more detailed account, see (Romo and Doncel 1994) and particularly (Steinle 1996). In (Steinle 1994), I analyze the variety of experimental work in that period.
[22] (Williams 1986).
[23] (Tweney 1991).
[24] As emphasized by (Tweney 1991), 304 with reference to (Faraday 1827), 546.
[25] Faraday's hesitation of too rash committment is excellently described by (Crawford 1985).
[26] (Williams 1986), quotation on p. 310.
[27] This becomes particularly obvious in Faraday's public talk "On mental education": (Faraday 1854). That impact of Faraday's religion to his experimental work provides an additional and essential aspect to those worked out by (Cantor 1985), (Cantor 1991) or (Gooding 1991).
[28] The starting point of the recent discussion, and still an excellent introduction to the general problem is (Hacking 1983).
[29] There have been attempts of a more general taxonomy: (Hacking 1988), (Galison 1988).
[30] I give a detailed discussion in ch. 4 of (Steinle 2000a).

REFERENCES

Ampère, André-Marie (1820), "Suite du Mémoire sur l'Action mutuelle entre deux courans électriques, entre un courant électrique et un aimant ou le globe terrestre, et entre deux aimants," *Annales de Chimie et de Physique* 15 (octobre): 170–218.

Blondel, Christine (1978), "Sur les premières recherches de formule électrodynamique par Ampère (octobre 1820)," *Revue d'Histoire des Sciences* 31: 53–65.

Blondel, Christine (1982), *A.-M. Ampère et la création de l'électrodynamique (1820–1827)* (Paris: Bibliothèque Nationale).

Cantor, Geoffrey (1985), "Reading the Book of Nature: The Relation between Faraday's Religion and his Science," in D. C. Gooding and F. A. J. L. James, eds., *Faraday Rediscovered: Essays on the life and work of Michael Faraday, 1791–1867* (Basingstoke: Macmillan), 69–81.

Cantor, Geoffrey N. (1991), *Michael Faraday: Sandemanian and Scientist. A study of Science and religion in the nineteenth century* (Basingstoke: Macmillan).

Crawford, Elspeth (1985), "Learning from Experience," in D. C. Gooding and F. A. J. L. James, eds., *Faraday Rediscovered: Essays on the life and work of Michael Faraday, 1791–1867* (Basingstoke: Macmillan), 211–227.

Faraday, Michael (1827), *Chemical manipulation; being instructions to students in chemistry, on the method of performing experiments of demonstration and research with accuracy and success* (London: Phillips).

Faraday, Michael (1854), "Observations on Mental Education," *Lectures on Education Delivered at the Royal Institution of Great Britain* (London), 39–88.

Fox, Robert (1974), "The Rise and Fall of Laplacian Physics," *Historical Studies in the Physical Sciences* 4: 89–136.

Galison, Peter (1988), "Philosophy in the laboratory," *The Journal of Philosophy* 85: 525–527.

Gooding, David C. (1991), "Michael Faraday's apprenticeship: science as a spiritual path," in R. Ravindra, ed., *Science and Spirit* (New York: Paragon House), 400.

Hacking, Ian (1983), *Representing and Intervening: Introductory topics in the philosophy of natural science* (Cambridge: Cambridge University Press).

Hacking, Ian (1988), "On the stability of the laboratory sciences," *The Journal of Philosophy* 85: 507–514.

Hofmann, James Robert (1987a), "Ampère's Invention of Equilibrium Apparatus: A Response to Experimental Anomaly," *British Journal for the History of Science* 20: 309–341.

Hofmann, James Robert (1987b), "Ampère, Electrodynamics and Experimental Evidence," *Osiris (2nd series)* 3: 45–76.

Hofmann, James Robert (1995), *André-Marie Ampère: Enlightenment and Electrodynamics*. (Oxford: Blackwell).

Jacob, François (1998), *Of files, mice and men* (Cambridge: Harvard University Press).

James, Frank A.J.L., ed. (1991), *The Correspondence of Michael Faraday, Volume 1, 1811 - December 1831, Letters 1 - 524* (London: Institution of Electrical Engineers).

James, Frank A.J.L., ed. (1993), *The Correspondence of Michael Faraday, Volume 2, 1832 - December 1840, Letters 525 - 1333* (London: Institution of Electrical Engineers).

Launay, Louis de, ed. (1936–43), *Correspondance du Grand Ampère* (Paris: Gauthier-Villars).

Martin, Thomas, ed. (1932–6), *Faraday's Diary. Being the various philosophical notes of experimental investigation made by Michael Faraday, DCL, FRS, during the years 1820–1862* ... (London: G. Bell & Sons).

Romo, J. and Doncel, Manuel. G. (1994), "Faraday's Initial Mistake Concerning the Direction of Induced Currents, and the Manuscript of Series I of his Researches," *Archive for History of Exact Sciences* 47: 291–385.

Ross, Sydney (1965), "The Search for Electromagnetic Induction 1820–1831," *Notes and Records of the Royal Society of London* 20: 184–219.

Steinle, Friedrich (1994), "Experiment, Speculation and Law: Faraday's analysis of Arago's wheel," in D. Hull, M. Forbes and R. M. Burian, eds., *PSA 1994. Proceedings of the 1994 Biennial Meeting of the Philosophy of Science Association* (East Lansing: Philosophy of Science Association), vol. 1, 293–303.

Steinle, Friedrich (1996), "Work, Finish, Publish? The formation of the second series of Faraday's Experimental Researches in Electricity," *Physis* 33: 141–220.

Steinle, Friedrich (1997), "Entering New Fields: Exploratory Uses of Experimentation," *Philosophy of Science* 64 (Supplement): S65-S74.

Steinle, Friedrich (1998), "Exploratives vs. theoriebestimmtes Experimentieren: Ampères erste Arbeiten zum Elektromagnetismus," in M. Heidelberger and F. Steinle, eds., *Experimental Essays - Versuche zum Experiment* (Baden-Baden: Nomos-Verlag), 272–297.

Steinle, Friedrich (2000a), *Der Einstieg in ein neues Feld: Forschungspraxis im frühen Elektromagnetismus bei Ampère und Faraday*, Habilitationsschrift, Fachbereich Kommunikationswissenschaften (Berlin: Technische Universität).

Steinle, Friedrich (2000b), "'... et voilà une nouvelle théorie d'aimant': Ampères Weg zur Elektrodynamik," in R. Thiele, ed., *Mathesis: Festschrift zum siebzigsten Geburtstag von Matthias Schramm* (Berlin: GNT-Verlag), 250–281.

Tweney, Ryan (1991), "Faraday's notebooks: the active organization of creative science," *Physics Education* 26: 301–306.

Williams, Leslie Pearce (1983), "What were Ampère's earliest discoveries in electrodynamics?," *Isis* 74: 492–508.

Williams, Leslie Pearce (1986), "Why Ampère did not discover electromagnetic induction," *American Journal of Physics* 54: 306–311.

OHAD PARNES*

FROM AGENTS TO CELLS: THEODOR SCHWANN'S RESEARCH NOTES OF THE YEARS 1835–1838

INTRODUCTION

Within three years, between 1835–1838, the young Theodor Schwann revolutionised physiology. Assisting Johannes Müller in Berlin, and working through a series of seemingly unrelated topics like muscle contraction, stomach digestion and the microbiology of putrefaction, Schwann in 1838 identified the elementary unit of life. Both animals and plants, he declared, are made solely from cells and of cells. Moreover, Schwann considered cells not only as structural units, but also as physiological causal factors. At a stroke, Schwann got rid of the centuries-old life-force, and made physiology a science of specific agents and their corresponding specific effects.

Historians have hitherto hardly accounted for this revolutionary move. Although Schwann has been the subject of two full-length biographies and numerous discussions in the historiography of the cell theory, no attempt has been made to understand the specific conceptual presuppositions and experimental practices which enabled Schwann's recognition of cells.[1] Similarly, very little attention has been given to the deeper epistemological consequences of a theory of cells, and to its far-reaching impact on the further development of an experimental science of life.[2] In the following, I will analyse Schwann's discovery of cells in relation to his other physiological investigations, conducted between 1835 and 1838. For such a historiographical endeavour, Schwann's research notes of this period are of special interest.[3]

SCHWANN'S SCIENTIFIC DIARY

Theodor Schwann (1810–1882) studied at the University of Bonn, where he became acquainted with Johannes Müller. In 1833 Schwann went to Berlin in order to complete his medical degree. Johannes Müller had meanwhile been appointed professor of anatomy and physiology at the University of Berlin, and in 1834 Schwann submitted to him his doctoral dissertation, dealing with the necessity of atmospheric air for embryonic development. Müller persuaded

* Max Planck Institute for the History of Science, Berlin

Frederic L. Holmes, Jürgen Renn and Hans-Jörg Rheinberger (eds.), Reworking the Bench: Research Notebooks in the History of Science, 119–140

Schwann to devote himself to research and offered him a position as his assistant at the "anatomical museum" of the university. It was there, in the very short period of 1835–1838, that Schwann arrived at most of his scientific discoveries. Immediately after publishing his work on cells Schwann moved to Belgium, where he was offered a position at the Catholic University of Louvain. This was, in fact, also the end of his active scientific career. In 1849 Schwann was appointed professor of anatomy at the University of Liège, where he stayed until his retirement in 1879.

Fortunately, several volumes of Schwann's unpublished notes from his Berlin period have survived.[4] Schwann entitled them "diaries of scientific and medical observations and experiments" ("Tagebuch über naturwissenschaftliche und medizinische Beobachtungen und Versuche"), but essentially they served as a general writing pad, on which Schwann registered occasional notes and thoughts, as well as some experimental procedures and the results obtained. I retrieved four volumes of notes: the first (henceforward TB 1835) covers the year 1835. The second volume (TB 1836) begins in June 1836 and runs to the end of that year. The third volume (TB 1837) begins January 1837 and ends October of that year. Then there is a fourth, less systematic volume (TB 1838) with sporadic notes and drafts from the period January–July 1838.[5]

On first reading, and from the point of view of the standard historiography of the cell theory, the notebooks seem totally unrevealing. At least, they offer no straightforward insight into the context of Schwann's discovery of cells. The most interesting period, as far as the discovery of cells is concerned, does not seem to be documented in the available notebooks: the third volume (TB 1837) corresponds to the period ending October 1837, that is, shortly before Schwann began working on his first paper on the cell theory (published January 1838),[6] while the next volume, beginning January 1838, already contains drafts for one of Schwann's preliminary articles on the topic. One looks in vain for earlier programmatic declarations of a wish to search for the elementary building block of life. Nor do these notes include reports of systematic microscopical investigations of the structure of tissues. Cells, or any comparable elementary microscopic units, do not appear at all in the first three volumes of notes (1835–37).

The same notebooks acquire a completely different historiographical coherence, however, when they are considered not in terms of a 'history of the cell theory,' but as a documentation of a persistent and consistent attempt to make sense of life in terms of causal agencies.

AGENTS

Agents play the central role in the modern biomedical sciences; "agents" being material objects, exerting specific forces, and bringing about specific physiological processes. Typical such agents are cells, specific enzymes, as well as viruses and genes. This is one of the most powerful and constitutive conceptions of the life sciences, and as such is commonly considered as obvious, almost trivial.

Moreover, it is usually assumed that this modern, etiologically-oriented biomedical thought has its origins in the "bacteriological revolution" which took place in the last two decades of the nineteenth century and is associated with the work of Louis Pasteur and Robert Koch.[7] However, it is commonly overlooked that this so-called revolution, in which the idea of the etiologic agent was, of course, central, was itself already based on some deeper, preceding conceptual changes that took place several decades earlier, between approximately 1840 and 1870. Changes in conceptions of causality and new observations of microscopic structures were distinct but intertwining aspects of one and the same scientific development. The modern biological experiment, where a strict causal relation is established, or rather constituted, between the experimentally defined "cause" and its (again, experimentally defined) "effect" appeared first in German physiology, properly in the period after 1840. Similarly, the conception of micro-biological entities as agents of life processes was intimately connected with the establishment of the idea of the animal cell and its reception after Schwann's publication of 1839.

FROM "PRINCIPLES" TO CELLS: SCHWANN'S SEARCH FOR THE AGENTS OF LIFE

In the following, I will tell the story of Schwann's physiological investigations between 1835 and 1838 as composing a coherent and systematic research programme. I will contend that practically from the launch of his experimental work, Schwann aimed to introduce new experimental methods into physiology, and to arrive at explanations of life processes which would not require recourse to vital forces. At the time practically everybody still considered physiological and pathological processes to be manifestations of an underlying life-force. Life was typically conceived of as a contingent phenomenon, dependent on a very delicate equilibrium of the life-force of the individual and its environment. Normal, healthy life was a result of the complicated interaction of the vital capabilities of the organism with the variety of impulses and stimuli it absorbed from its surroundings. Schwann, in contrast, believed that if one could construct the right experimental setting, it should be possible to discern specific agents for specific physiological phenomena, and to show strict causal relations between the agent and its effect.

Where did the motivation for the undertaking of such an ambitious project come from? There is little in Schwann's published or unpublished work which would suggest a plausible context for this radical break. Still, there is ample evidence, especially in the unpublished notes, of Schwann's preoccupation with the foundations of physiology, and of his wish to establish a new physiological epistemology. In March 1835, shortly after he had begun to conduct a scientific diary, he enclosed in it excerpts from a letter he wrote to his brother.[8] Here he insisted on the need to make a distinction between mind and brain, "thus banishing the question of the seat of the mind from physiology." Having adopted this view, he wrote, it becomes conceivable to claim "that each individual is endowed with his own, peculiar and new *psychic* principle" and still avoid the need to assume a similar individual *life-principle* (a view he ascribed to Johannes Müller).[9] The latter, he adds, could then be conceived of as dependent only upon organisation. Physiology without the mind could thus become a rigorously lawful science, based on strict causal dependencies. "I believe," Schwann writes in February 1835, "we should direct our effort towards the introduction of calculation to physiology." A plausible beginning for such an endeavour, he adds, may be the study of muscle contraction.

The investigation of muscle contractility, made between February and October 1835, was never fully published, and can be retrieved only from the unpublished notes.[10] Essentially it consisted of the measurement of the extent to which a (frog's) muscle contracted under a given (galvanic) stimulus. Historians have considered this experiment one of the earliest examples of the successful quantification of muscle activity. And although Schwann's work was duly credited for its important contribution to muscle physiology, it was never considered relevant to Schwann's later microscopical work.[11] But an essentially different interpretation of this experiment emerges when it is read not simply as a very successful endeavour of measurement, but as an attempt to discern the physiological agent of muscle contraction. Indeed, in his notebook Schwann explicitly stated that the mere quantitative aspect of the muscle experiment was trivial and predictable. His actual aim was more ambitious: the calculation ("Rechnung") was to serve as a tool, helping to discover regularities and hence *laws* of physiological processes. The experiment was conceived as follows (April 16, 1835[12]):

> In order to study the contractility of muscles in a scientific way, we should proceed in the following manner: let us imagine one primitive muscle fibre, located vertically with its lower end connected to a fixed point, its upper end connected to one of the arms of a simple balance.
> At the other side of the balance a scale hangs. When everything is in equilibrium, the muscle fibre is not pulled apart, but is simply compensated for its gravity. Then one puts a weight *G* on the scale and stimulates the fibre to contract. The fibre contracts to the length *L*. The force which the muscle performs under the length *L* may thus be expressed by the weight *G*.
> We could term this force the Magnitude of Contraction, analogous to the magnitude of motion which is employed in physics, and which is commonly expressed by *C*.

Schwann's imagery, even at this early stage, is remarkably different from the prevailing physiological imagery of the time, and also has direct consequences for his experimental practice. Instead of the "contractility" of the whole muscle, Schwann imagines here a "*force*" of an elementary physiological agent – in this case the muscle fibre. This he wishes to represent ("ausdrücken") by activating the muscle against a controlled resistance, namely metal weights, thus defining the experimental entity "magnitude of contraction" ("Kontraktionsgröße"). This magnitude, in turn, depends on the "peculiar force of contraction" of the muscle ("eigentümliche Kontraktionskraft"). The latter cannot, of course, be measured directly but, Schwann notes:

> . . . [the specific contraction force of the muscle] can be calculated from a series of measurements of the magnitude of contraction which the muscle expresses under a constant stimulus but a changing degree of the life-force.

Schwann contends here that by conducting a series of muscle experiments under various conditions, and by assuming that one can control and measure all the relevant magnitudes *apart* from the life-force of the muscle, one can then conclude that the remaining variation of the extent to which the muscle contracts must be ascribed to the variation of its life-force. Consequently, this (presumably regular) rate of change could indeed be considered as the *numerical representation of the life-force*. It is important to note the very peculiar manner, for the time, in which the term "life-force" is used here, namely as a concrete trait of an elementary microscopic unit, in this case the muscle-fibre. Indeed, Schwann was even contemplating the underlying mechanism which could explain this causal relation (April 20, 1835):

> Maybe one can explain the contraction of muscles if one assumes that the light-coloured parts of the primitive fibres act merely by their elasticity, whereas the dark points possess the ability to exert a force of attraction upon each other under a stimulus. If one assumes this force of attraction to be inversely proportional to the square of the distance, then the resistance of the light-coloured parts increases the closer the dark points are to each other. That is why the magnitude of contraction decreases relative to the shortening of the muscle and finally arrives at a point where it is zero, i.e. the force of attraction of the dark points and the resistance of the light-coloured parts are in equilibrium.

Furthermore:

> The given explanation fits the phenomenon of linear contraction of muscles, but not quite their rippling. I don't see any way of explaining both kinds of contraction out of one single principle, as long as microscopical anatomy does not supply the appropriate data for such an explanation.

Note how the concept of a "Prinzip," or 'principle,' is employed here in a microscopical–anatomical context. Schwann is using (and conceiving) it in a unique, ambiguous sense: both as an explicatory "principle" of a physiological phenomenon, and as an anatomical-microscopical (material) entity. Four days later, on April 24, Schwann notes:

> An independent field of study which could be pursued would be "comparative general anatomy." In the same way as the special comparative anatomy is helping physiology to understand the essential form of the organs, this new science would aim to understand the essential structure.

It would be another two and a half years before Schwann observed cells but, in a sense, it is here that Schwann's work on the cell theory begins – that is, in the endeavour to account for physiological processes in terms of the lawful effects of force-exerting microscopic units.

The particular attempt to make sense of the life-force of a single muscle fibre eventually failed.[13] The only conclusion Schwann was able to derive from the experiment, which was conducted in October 1835, was that there exists a lawful relation between the initial length of a muscle and its corresponding force of contraction.[14] Still, the peculiar use of "Prinzip" will become the cornerstone of Schwann's physiological project, the understanding of physiological processes through microscopic organisation.

The period of December 1835–June 1836 was predominantly devoted to the investigation of stomach digestion, culminating in the identification of the active digestive substance in the stomach, which Schwann termed "pepsin." This work is considered a milestone in the history of physiology and biochemistry; not only was it the first digestive process to be fully explained (and artificially simulated) but it was also the first identification of a human enzyme. This investigation, in which no use was made of a microscope, attains direct relevance to the cell theory when considered in terms of physiological agencies.

As we learn from the notebooks, the problem of digestion, which intrigued Johannes Müller for several years, was handed over to Schwann in October 1835 after another pupil of Müller's, Jacob Gerson, had failed to solve it.[15] Gerson was only able to repeat, with partial success, experiments published two years earlier by Johann Eberle on the possibility of exciting an artificial digestion using a mucus extract of an animal stomach ("Magenschleimhaut").[16] In November 1836 Müller and Schwann conducted a similar series of experiments on artificial digestion with *Magenschleimhaut*, this time fully corroborating Eberle's work but failing to identify the mechanism underlying the activity of the mucous extract.[17] Müller considered the investigation completed at this stage but Schwann had not yet conceded. His aim, as he wrote, was to understand the "essence" ("Wesen") of the digestive process.[18]

At the time, digestion was typically considered in terms of a chemical reaction, taking place within a vital environment.[19] Schwann, in contrast, attempted right from the start to explain this process not as part of a complex 'vital' activity, but as if it were brought about by one single, specific material agent. This was far from straightforward. There was no hint of the existence of such a material agency: it could neither be plausibly attributed to any of the known anatomical components of the stomach, nor could it be observed microscopically. Where should one look for such an agent? On November 30 Schwann noted in his diary:

> In my experiments on digestion I have discovered that if one prepares, in the usual way, artificial stomach-liquid, and then dilutes it with water and filters it through linen and then again through paper, one gets a clear, yellowish liquid which still possesses the ability to bring about a chemical process.

Within this "yellowish liquid" Schwann conjectured a principle of digestion. But how can one determine the material nature of such a "digestive principle"? "The path of the common course of analysis," says Schwann in his published report, "is in this case blocked." The only real proof of the existence of this substance, he argues, is by its digestive power, which, in turn, can be demonstrated only in association with an acid. Heat and alcohol, on the other hand, destroy the digestive capability of the substance and thus cannot be used as means of extracting or distillation. Therefore, Schwann tells us, he decided to go another way:

> The idea was to establish the characteristic reaction of the principle of digestion, in the digestive solution, against a series of reagents, without actually isolating this substance. I.e., simply by observing whether the digestive force is lost or not when a specific reaction takes place. A full analytical process and an isolated demonstration of the digestive principle may be possible at a later date.[20]

The epistemological step conveyed in this sentence can hardly be overestimated. It is, actually, one of the most basic presuppositions of modern experimental life sciences. What Schwann is proposing is to employ a set of experimental procedures to characterise a *physiological* entity which cannot be discerned in any other way (e.g. directly observed, or precipitated, or defined chemically). The demonstration of the existence of such an entity would then be substantially contingent upon the experimental context, i.e. upon its ability to bring about a specific effect under specific conditions. Schwann, in other words, is claiming full epistemological sovereignty for the physiological experiment: chemistry and physics are employed in the experimental procedure, but the nature of the agent discerned does not have to be either chemical or physical; it can be determined merely by its peculiar experimental profile. Accordingly, Schwann's investigation consists mainly of a series of reactions of the solution, suspected to contain the *digestive principle*, with a long list of standard chemical and physical agents. He summarises his results as follows:

> It is soluble in water and in diluted chloric acid and in acetic acid; it is decomposed by alcohol but I cannot tell whether it is soluble in it or not. It is changed by a strong heat, but it is not clear whether it is also precipitated by it. Lead makes it precipitate both from an acidic and from a neutralised solution.[21]

The exact details of each result in this long series of reactions are of little importance. The choice of reagents was anyway largely coincidental. Even so:

> Through these reactions the digestive principle can be characterised as a *peculiar, specific substance*.[22]

Thus a "digestive principle" (named *pepsin*)[23] is defined which is not merely one of many conditions which have to be fulfilled for digestion to take place, but is considered as a factor with a *preferred* epistemological status. This was a completely new way of conceiving of physiological processes: instead of the endless chain of material transformations, dependent upon the organism's vitality and a complex web of interrelations with its environment, the physiological process was reduced to a specific material agency. Schwann immediately realised how useful this stance could be for a general theory of life. It may be imaginable, he writes in his notebook on June 28, 1836, that many other physiological processes follow an analogous course:[24]

> Similarly, one could imagine the process of nutrition: if we consider, for example, a muscle, then one could understand its process of nutrition in similar terms to that of the principle of digestion. The role which the acids and the albumen play in digestion are taken by the arterial blood in the nutrition of a muscle. Thus the muscle-matter has a similar effect on the blood which is supplied to it to the effect the principle of digestion has on albumen, with one difference, namely that the substances which are generated from the blood through the decomposing force of the muscle-matter are themselves muscle-matter. If we assume such a process, then we could explain the constant renewal of the muscle-matter, analogous to the way the digestive principle loses its digestive force through the process of digestion.

Schwann is applying the new concept of a *digestive principle* to construct a general model of nutrition: the tissues possess a multitude of "nutrition principles," able to generate new tissue (out of substances supplied by the blood, for example) which, in turn, possesses more of these principles. If one further assumes that one of the by-products of this process is a substance which blocks growth, Schwann notes, then one can even explain the cease of generation in older age (still June 28, 1836):

> The newly generated substance is able to bring about exactly the same process of decomposition of the blood, whereas the 'earlier' substance which brought about the generation of the substance which is now active, is changed and dissolved. If we now add a further hypothesis, that this process of dissolution is not complete, but that each time the active substance of the muscle-matter is dissolved one of its constituents remains active and that this constituent is able to suppress the process of nutrition (in the same manner as sulphuric acid carbonate suppresses the process of digestion), then one could explain the whole developmental process of life in all organised bodies.

Thus Schwann is conceiving an all-encompassing physiological model: generation, embryonic development and nutrition must be explicable in a unified way, in terms of these nutritive principles, i.e. active material agents within the tissues. Schwann even envisaged several experiments which could corroborate his nutrition theory. He hypothesised that the mammary gland, and possibly the liver, contain such nutritional agents, and noted that a possible experimental test of this hypothesis would be to introduce pieces of these tissues to the blood stream of a living animal and to monitor the changes they exerted on the blood (e.g. on the level of bile).

This imagery is quite reminiscent of what Schwann would later term the "metabolic" activities of the cell.

At this point of his work, Schwann was already thinking in terms of active physiological agents, and the digestion investigation provided him with a first successful employment of this imagery. The next step would be the location of such agencies within observable microscopic structures.

After completing his work on digestion, Schwann returned to a subject which preoccupied him from the start of his scientific career: the problem of putrefaction. As the notebooks reveal, this investigation was originally motivated by an attempt to discern another conjectured physiological agent, that of *respiration*. Indeed, Schwann was convinced for several years that respiration was a process brought about by a peculiar microscopic entity, floating in the air, and triggering the respiratory process in the lungs. The respiration hypothesis is actually the earliest evidence of Schwann's search for physiological agents. It does not appear in any of Schwann's publications, but is extensively discussed in the unpublished notes. Only years later, after Johannes Müller's death in 1858, did Schwann mention publicly his respiration theory, claiming he had been thinking about it as early as 1831:

> As a student in Bonn, in 1831, while having a walk with Müller, I had told him my idea that respiration may involve some kind of organic substance contained in the air in the form of vapour. This substance would not be discerned in a standard analysis of the air (as it was conducted at the time), because of its vapour nature, and maybe it would not react in any typical way with other substances.[25]

One of the first entries in Schwann's diary, on February 5, 1835, discusses ways of testing this hypothesis. Like most of his contemporaries, Schwann considered motile microscopic organisms to be little animals ("Infusoria"). Thus if he could show that these microorganisms cannot live when supplied with air devoid of the respiration principle, he might be able to prove his hypothesis. This was his initial motivation to construct an experiment in which air was pre-heated, as heating was assumed to destroy the organised principle of respiration. In order, Schwann writes in his diary, to decide whether an organic substance plays a crucial role in respiration, one should provide a vessel containing putrefying organic substances and infusoria with air which has been heated to boiling point before its introduction to the vessel. The vessel itself would also be heated, in order to destroy all the existing respiration principle in it. After a while, with a constant supply of preheated air, one would check whether the infusoria in the vessel were still alive. Assuming one would not find any living infusoria in this solution, then:

> – If a later analysis of the gas shows that the air was not changed through the heating, then a two-fold explanation would be required for the infusoria, which were either destroyed by the heating, or really required the substance in

question, in the atmospheric air, for their respiration. Assuming the first case, then the claim that the infusoria are generated by spontaneous generation would be out of the question.

In January 1837 Schwann finally succeeded in constructing an appropriate experimental apparatus. He could demonstrate that an organic substance (a meat infusion) would not putrefy when put in a vessel which was boiled and then exposed only to air streaming through a glowing metal tube. Needless to say, the experiments did not prove the existence of a respiration principle. But while conducting them, Schwann gradually realised that they were very suitable for the study of spontaneous generation. The experiment proved that it was not the mere exposure to air which enabled the decomposition of organic matter, but rather the living organisms floating in the air that caused this decomposition. Thus putrefaction was not, as traditionally believed, a "spontaneous decomposition" of organic matter when deprived of a life-force, but merely a specific process brought about by "infusoria," i.e. specific, living microscopic organisms.[26]

This investigation had a further important influence on Schwann's physiological imagery. A series of similar experiments, conducted in February 1837 and using currant juice instead of a meat infusion, led Schwann to identify *yeast* as the agent of alcoholic fermentation.[27] Moreover, Schwann recognised yeast to be a microscopic plant, namely a fungus, and located the agency of alcoholic fermentation within its 'globules' or cells.[28] That many plants appear to be composed of 'cells' was quite a common view at the time; Schwann's investigation, however, was the first to locate a physiological activity (the transformation of sugar into alcohol and carbonic acid) within such a morphological unit.

NOTE ON THE RELATION BETWEEN SCHWANN'S PUBLISHED AND UNPUBLISHED WORK

The attentive reader has probably noticed that my analysis has indiscriminately incorporated quotations from both Schwann's unpublished diaries and his publications. This may be the place to elaborate briefly the role Schwann's notebooks may have played in his research activities, and on the relation between the published and the unpublished documents in his work.

One of the most striking things about Schwann's diaries is the inconsistency of the entries. While some investigations are recorded in much detail, others are only skimpily described and discussed. Moreover, some of the most interesting and successful experiments, described in detail in the published accounts, are not documented at all in his scientific diary. The work on digestion, for example, is only very partially discussed in the notes, with much less experimental detail than in the published reports; and in the notes on spontaneous generation some of the most crucial experiments are missing. Especially puzzling is the fact that the investigation on muscle contraction, conducted in 1835, which Schwann finally *did not* publish, is covered in great detail in the notebook – including general theoretical considerations as well as the full registration of the actual experiment. Indeed, this is the only investigation appearing in the notes where Schwann did not simply write down the experimental results, but actually structured them in a communicable, in fact publishable form. Why would Schwann compose, in his diary, a publishable account of an experiment which was never to be published? And why would he avoid recording successful experiments that were obviously worth publishing? One plausible explanation would be that Schwann occasionally used his notebooks for making drafts of publications. He may even have handed over those drafts directly out of the notebooks in case he did decide to publish them, which would explain why some published investigations are missing from the diary. In that case, the reason the muscle experiment appears in such detail in the notebook would be that it was also meant for publication, but for one reason or another Schwann decided to leave it unpublished.[29] At the same time, this would suggest that Schwann hardly edited and reorganised his accounts before submitting them for publication. This, I believe, worked both ways: while the entries in his scientific diary occasionally served as drafts for publications, the published reports were quite authentic accounts of his experimental trail, and can be considered as almost chronological reports of his actual experimental work.[30]

This assumption proves especially compelling when applied to the 1839 treatise on cells, the *Microscopical Investigations into the Accordance in the Structure and Growth of Animals and Plants*.[31] There is additional historical evidence to support such a hypothesis. We know that Schwann was under considerable time pressure to finish this last work in Berlin, as he was quite nervous about his job situation. Finishing the book was a main prerequisite for a "Berufung" for a university professorship. Moreover, we know that the work was written in the same sequence as its three published parts, each completed and submitted separately: the first part of *MU* (an investigation of cartilage) was sent to the French Academy of Sciences in August 1838, the second part (the generalisation of the scheme) in December 1838, while the third and concluding part (and the whole treatise) was completed in February-March 1839.

Thus I would make the very unconventional suggestion that we consider this treatise as a chronological documentation of Schwann's investigative trail. I would contend that the published report represents, with minor changes and minimal editing, Schwann's actual account of his research towards a cell theory. Admittedly, this historiographical assumption is a very useful one – as noted, there is scarce unpublished documentation of Schwann's work on the cell theory. Still, I believe it is also a very compelling and plausible assumption, enabling a coherent understanding of Schwann's discovery of cells.

CELLS: THE *MICROSCOPICAL INVESTIGATIONS* AS A LABORATORY NOTEBOOK

Part I: October 1837–February 1838

Schwann repeatedly claimed that his initial impulse to study cells had been an informal report by the botanist Matthias Schleiden on his investigations into the development of plant cells. Years later Schwann wrote:

> One day, when I was dining with M. Schleiden, this illustrious botanist pointed out to me the role that the nucleus plays in the development of plant cells. I at once recalled having seen a similar organ in the cells of the notochord, and in the same instant I grasped the extreme importance that my discovery would have if I succeeded in showing that this nucleus plays the same role in the cells of the notochord as does the nucleus of plants in the development of plant cells.[32]

This report, as noted, was written with forty years' of hindsight, and must be considered accordingly. As I would argue, Schwann's conception of cells, as well as their consequent entrenchment in the modern biomedical scheme, were not originally the result of a purely anatomical–microscopical endeavour. The cell theory was but a plausible expansion of the conception of specific agency. When in October 1837, over lunch, Schwann was informed by Schleiden about his latest investigations into the development of plant cells, he merely realised the usefulness of these ideas for his own programme.[33]

Let us recall my historiographical presupposition, that the *Microscopical Investigations (MU)* is a more or less chronological account of Schwann's actual experimental work. In the first part, completed in the first half of 1838, Schwann merely attempted to demonstrate that cellular-generative processes similar to those Matthias Schleiden had demonstrated for plants existed in *some* animal tissues too. To this end he studied in detail the generative characteristics of one kind of tissue, namely cartilage. Indeed, in 1839 Schwann claimed that Schleiden's ideas had merely helped him understand "the previously enigmatical contents of the cells in the branchial cartilages of the frog's larvae," as he was now able to "recognise in them young cells, provided with a nucleus."[34] But as he writes in his notebook, this was not enough for a demonstration of the cellular nature of all animal tissues. In January 1838 Schwann was obviously still not quite sure how sweepingly he should formulate his generalisation. Drafting his first paper on cells he writes:

> Considering the life-cycle ("Entwicklungsgeschichte") of other tissues it is very probable that they too ~~are generated originally out of cells~~, in part and maybe all are generated out of cells, and that they show some analogies with plant-cells. If a rigorous proof could be delivered, that in the animal body there exist corpuscles analogous to the plant cell, then we could parallel these

> two and then we could consider the generation of animal organisation as nothing other than a modification of plant organisation.[35]

Schwann realised that the criteria used by Matthias Schleiden in his work on the plant cell did not suffice for a general cell theory; i.e. that the mere cellular appearance of animal tissues was not a sufficient criterion for such a generalisation. Still, in January 1838 he wrote in his notebook:

> Not every cell is a corpuscle analogous to the plant cell and even the polyhedral form, which is a necessary characteristic of cells which are densely pressed together, is not a really useful characteristic, as it tells us nothing more about a possible similarity than that these cells are indeed densely pressed together.

Instead:

> If one wants to parallel an animal structure with the plant cell then one must prove not only that this structure is a cell, but also that similar forces to those of the plant cell are active in it or, as this is impossible to do directly, that the phenomena through which the activity of these ~~cells~~ forces manifests itself, namely nutrition and growth, take place in a similar manner to those in the cells of plants.[36]

Note the crossing-out of "cells," which Schwann seems to conceive of as almost interchangeable with "forces." Microscopical structure is considered as an expression of a physiological process; a process brought about by specific agencies and manifested by specific (generative) effects. Thus, while completing the first part of *MU* Schwann realised that the proof for the cellular nature of all kinds of specific tissues could not consist merely of 'pure' observations. Instead, he had to argue that all these tissues could be considered as the physiological effects of the activity of the same material principles. Accordingly, Schwann set out to show that the generation of (animal) tissues is brought about by a specific agent within the tissues. He thus considered cells to be generated out of an amorphous generative substance, which he termed "Cytoblastem." Within this substance the nuclei were assumed to form, and around the nuclei the cells, in turn, were supposed to build up.[37] For each tissue Schwann had to corroborate his hypothesis anew; each observation, therefore, would actually constitute an independent, isolated experiment. This he attempted in the second part of his book, written in autumn and winter 1838.

Part II. ca. March–August 1838

The historical sequence becomes manifest when we compare the way Schwann discussed the cartilage in the first part of the book (i.e. probably towards the end of 1837), with the way it is discussed in the second part (i.e. almost one year later). When investigating the cartilage in 1837, Schwann's aim had been to demonstrate an analogy between the structure of the tissue and that of plants, based on Schleiden's theory of (plant) cell formation. Now his project had substantially changed; it was no longer the demonstration of an accordance with the plant model which he was pursuing, but the demonstration of an accordance with his theory of cells. At the first stage of his observation, in winter 1837/38, he was still relying on the traditional conception of "cell" as a morphological attribute:

> The simplest form of cartilage is exhibited in the cartilages of the branchial rays of fishes. ... The structure of this cartilage is very simple ... it perfectly resembles, in its whole appearance, the parenchymatous cellular tissue of plants. ... little polyhedral cell-cavities with rounded corners are seen lying closely together.[38]

The adaptation of Schleiden's watch-glass theory for the explanation of the generation of this tissue proved to be fairly strained. The argumentation was mainly morphological; the "appearance" was one of the indications Schwann lists here for the cellular nature of the tissue. The emphasis placed on the presence of the nucleus ("Kern") in each cell is also a relic of Schleiden's scheme, according to which new cells are generated within existing ones by the inflation of the membrane out of this "Kern" (which Schleiden called "Cytoblast"):

> The process of formation of this cartilage is as follows. It consists originally of cells, which lie in very close contact, but every one of which has its special, very thin cell-membrane. This follows, firstly, from the complete accordance in appearance, of cartilage in its earliest stage, with vegetable cellular tissue; secondly, from the presence of the nucleus in the young cells of cartilage, a structure which, as will subsequently be seen, occurs in almost all the cells proved to exist in other tissues; thirdly, from the fact, that a separation of the cell-walls is often distinctly perceptible in instances where they are thickened.[39]

Several months later, in the second part of the book, Schwann returned again to the cartilage. This time he argued for its cellular nature as follows:

> When we examine the stripes at the parenchymatous membrane ("Schleimhaut") of the cartilage, we can see that the inter-cellular substance between the cartilage cells protrudes over the outermost cells and, moreover, that it constitutes a kind of thin coating over these outer cells so that the external border of the cartilage is not made directly of cartilage cells. Therefore the cartilage cells lie entirely embedded in this intercellular substance, which is their *cytoblastem*. And it is in this cytoblastem, *not* in already existing cells, that new cells are generated.
>
> ...
>
> One can observe, namely, a series of structures, of which some are naked cell nuclei, smaller than the nuclei of the mature cells, others are nuclei which are

> densely covered with a cell; in short all the transitional stages beginning with merely naked cell nuclei, passing through nuclei covered with a small cell, and ending with the mature cells. Thus we can conclude that development is taking place here in a similar manner to that in most cells.[40]

Instead of describing a structure, Schwann was here describing a process. The effect is the one hypothesised – the postulated generative process, beginning with the generative substance and ending with the full-fledged tissue. The emphasis this time is mainly on the "intercellular substance," which is identified as the *cytoblastem*, i.e. the amorphous substance within which the actual agents of cell formation are embedded.

The cartilage was just one tissue which Schwann examined in the second part of *MU*. His aim was to demonstrate that "all remaining tissues" could be similarly considered as derived from cells, and this he did by systematically surveying tissues according to the scheme described, repeating the same pattern of analysis employed for the cartilage. Thus the aim of demonstrating the cellular nature of all animal tissues had actually already been achieved in the second part of *MU*.

Part III. ca. September 1838–January 1839

The third part of the book was written at the end of 1838 and in the first month of 1839. Schwann entitled this part "Theory of Cells" (in contrast to the "cell-theory" put forward in the first two parts of *MU*). It is usually considered by historians as an unfortunate addendum to Schwann's actual work, a theoretical-speculative appendix to his cell theory.[41] But, when considered under the historiographical perspective I have applied in this paper, it becomes strikingly integral to the whole work. Indeed, read from that perspective, this last piece composed in Berlin becomes a recapitulation of the whole physiological programme Schwann had employed for the previous five years. From the muscle experiment through stomach digestion and fermentation, it was the heuristic agent which enabled him to demonstrate the "principles" of physiological processes without recourse to a vital force, and it was the same heuristics that had finally led him to the cell theory. Now, a kind of a reverse logic was taking place. Having demonstrated and envisioned cells, Schwann arrived at what he believed to be the ultimate agent of life, and he could re-consider his previous observations and experimentation in this light. Cells, thus, became the realisation of his programme:

> We have seen that all organised bodies are composed of essentially similar parts, namely cells; that these cells are formed and grow in accordance with essentially similar laws; and, therefore, that these processes must, in every instance, be brought about by the same powers.[42]

The idea of the physiological force-exerting agent, already employed for his muscle experiment, has only to be slightly modified in order to fit the cellular theory:

> Now, if we find that some of these elementary parts, not differing from the

> others, are capable of separating themselves from the organism, and pursuing an independent growth, we may thence conclude that each of the other elementary parts, each cell, is already possessed of power to take up fresh molecules and grow; and that, therefore, every elementary part possesses a power of its own, an independent life, by means of which it would be enabled to develop itself independently.[43]

His considerations of the principles of nutrition, for example, elaborated in mid 1836, are similarly incorporated into the cellular language:

> We must ascribe to all cells an independent vitality. ... The cause of nutrition and growth resides not in the organism as a whole, but in the separate elementary parts – the cells.[44]

Cells are endowed with the ability to differentiate and specify; they are able to make the specific out of the non-specific. And this is what Schwann defines as the "metabolic":

> The cytoblastem, in which the cells are formed, contains the elements of the materials of which the cells is composed, but in other combinations; it is not a mere solution of cell-material, but it contains only certain organic substances in solution. The cells, therefore, not only attract materials out of the cytoblastem, but they must have the faculty of producing chemical changes in its constituent particles. Besides which, all the parts of the cell itself may be chemically altered during the process of its vegetation. The unknown cause of all these phenomena, which we comprise under term metabolic phenomena of the cells, we will denominate the *metabolic power*.[45]

It is a radical change in the way life processes are conceived of, and it is difficult to overestimate the extent to which Schwann changes the prevailing imagery of physiology in these lines. Cells are defined, in the most explicit way, as the fundamental, force-exerting principles of life. In other words, life is conceived of as being composed of a series of specific material processes, and the agents of these processes are the cells.

CONCLUSION: AUTHENTIC NOTES OR ORGANISED ACCOUNTS?

My argument for the close relation between Schwann's published and unpublished notes can be turned around. Perhaps, it could be argued, Schwann's research notes were not notes at all, at least not in the strictest sense of the word? Perhaps these notebooks were in practice the place where Schwann drafted his publications, and thus represent a very high level of narrative organisation, a filtered selection of those results which he considered worthy of publication, partially already rewritten in a rhetorical, public-oriented way? The alternative, it seems, would be to assume that Schwann was such an extraordinary investigator that he never needed to revise or change anything in his drafts before publishing them. But then:

did he make no mistakes? Did he never embark an investigative avenue which ended up as a dead end? Could it really be that no observation, no experimental result which he had arrived at was ignored, considered problematic, evaded because it did not fit his expectations? Did Schwann produce camera-ready texts even at the stage of conducting his experiments?

Of course not. The nature of the notebooks leaves no doubt about their personal role. Schwann also incorporated into his diary thoughts and experimental results which were not to become publications, as the example of the "respiration principle" clearly shows. The diary is full of wild speculations, fragmentary investigations and reports of unsuccessful experiments. The nature of the diary does seem to indicate, however, that it was not an immediate "logbook" of his work, and that Schwann often selected what should be registered in it and what not. Indeed, the title "Diary" fits the role of these notebooks very well; it seems Schwann occasionally set out to recapitulate his scientific activities of the previous day, sometimes of the previous several days or even weeks. In this sense the notebooks do represent a first stage of reorganisation and selection. At the same time, I believe one could consider these notes as very 'close' to his actual intellectual and experimental activity. In a sense, the partially organised nature of the diary may even be viewed as a further corroboration of the programmatic stance of Schwann's work in those years.

The story of Schwann's scientific diary, like practically all the case studies presented in this volume, is a peculiar and idiosyncratic one. There is, however, one historiographical lesson to be drawn from this case: that we should not let our over-enthusiasm for retrieving hitherto unknown research notes lead us astray. At least in some cases we should be allowed to believe the author, and follow the principle of charity. Read in the appropriate way, published accounts may be no less useful for the reconstruction of an implicit investigative pathway than the unpublished research notes.

NOTES

1 Recent accounts of the history of the cell theory are Cremer (1985), Duchesneau (1987). Two biographies of Schwann exist, one in French (Florkin 1960), one in German (Watermann 1960).

2 The numerous recent works dedicated to the 'experimentalisation' of life in the 19th century pay little attention to the relation between the new experimental stance and the cell theory. E.g. Rheinberger & Hagner (1993), Coleman & Holmes (1988).

3 This paper derives from my extensive study of Schwann's work and the establishment of the conception of the physiological agent in the modern life sciences (Parnes 2000). See also Parnes (2000a).

4 The original notebooks were kept by the Schwann family in Düsseldorf, and were duplicated and transcribed by Schwann's biographer Marcel Florkin in the 1960s. I retrieved them from Florkin's Nachlass, kept at the *Centre d'Information et de Conservation* of the library of the University of Liège, Belgium. The notebooks may be incomplete.

5 To the best of my knowledge, only very few historians have seriously consulted the notebooks,

and nobody has hitherto studied them systematically. Marcel Florkin, Schwann's biographer, discusses a few passages of the notebooks in his book (Florkin 1960), as does Schwann's second biographer, Rembert Watermann (Watermann 1960). In addition to these two, Larry Holmes made some use of Schwann's notes concerning stomach digestion for his book on Claude Bernard (Holmes 1986). Richard Kremer analysed two pages of Schwann's notes, reprinted in Florkin (1960), in which Schwann registered some of his experimental results concerning muscle contraction (Kremer 1990).

[6] Schwann (1838a). This was the first in a series of three short papers (followed by 1838b in February, and 1838c in April) which preceded the comprehensive treatise (Schwann 1839).

[7] See, for example, Evans (1993) for such an account.

[8] The same letter also appeared in Florkin (1960a).

[9] Italic added.

[10] The results of this investigation were reported very briefly in Johannes Müller's *Handbuch der Physiologie* (Müller 1834–40), pp. 59–62, and at the "Versammlung deutscher Naturforscher und Ärtze" in Jena in November 1836 (Schwann 1837c).

[11] E.g. Kremer (1990).

[12] Unless otherwise stated (see Note 38), all translations from Schwann's writings are my own.

[13] The attempt failed, as Schwann himself acknowledged, mainly for two reasons: 1. the inability to isolate and experiment with single muscle fibres and 2. the problematic role of the nerves in the process, i.e. the dependency of the 'magnitude of contraction' upon the extent to which the nerves conduct the galvanic stimulus. It was probably this failure to achieve his more ambitious aims which eventually dissuaded Schwann from publishing this investigation.

[14] See footnote 11.

[15] Gerson (1835).

[16] Eberle (1834).

[17] The results were published in a joint paper (Müller & Schwann 1836), in fact written solely by Schwann.

[18] Schwann 1836, p. 90.

[19] The nature of the acid present in the stomach had been known at least since the work of William Prout (Prout 1824) and the very extensive investigation by Friedrich Tiedemann and Leopold Gmelin (Tiedemann & Gmelin 1826–27), but at the same time it was recognised that acid alone does not suffice for a full digestive process, and that additional, presumably vital factors take part in the process. A rather detailed historical account of digestion research in the first half of the nineteenth century, including Schwann's work, can be found in Holmes (1974).

[20] Schwann 1836, pp. 113–114.

[21] Schwann 1836, pp. 122–123.

[22] Schwann 1836, p. 123. Emphasis added.

[23] It was only in a somewhat later publication (Schwann 1836a), that Schwann re-named the "Verdauungsprinzip" "Pepsin."

[24] Schwann's reasoning also involved the employment of a "effect by contact" model, which he borrowed from Eilhard Mitscherlich (e.g. Mitscherlich 1834). This model was later transformed by Berzelius into "catalysis" (e.g. Berzelius 1836). I will not enter into that aspect of the story here.

[25] Watermann 1960, pp. 173–174.

[26] Schwann (1837). "Infusoria" was used at the time for all animal-like microscopic organisms, in contrast to microscopic plants like fungi or algae. The distinction between animal-like 'protozoa' and plant-like 'bacteria' was established gradually between 1850 and 1870.

[27] The results of this investigation were read by Johannes Müller at the "Gesellschaft naturforschender Freunde zu Berlin" in February 1837 (Schwann 1837a), and later published in the *Annalen der Physik* (Schwann 1837b).

[28] At the time, yeast was considered to be a peculiar chemical substance, characterised by its fermenting qualities.

[29] Another logical possibility would be to assume that almost all the experiments Schwann reported

in his publications had actually never been (successfully) conducted. This is highly improbable. We have ample evidence of Schwann conducting some his experiments in public (at the "Versammlung deutscher Naturforscher" in Jena in 1836), as well as demonstrating them to his colleagues (e.g. Johannes Müller, Eilhard Mitscherlich).

[30] Of course, this does not mean that Schwann published everything he wrote in the diary, nor do I suggest that Schwann never changed and edited his papers. I would, however, contend that large parts of his publications are almost unedited excerpts from his research notes. Indeed, the style of his major publication supports this assumption, e.g., his reporting of experimental dead ends and failures.

[31] *Mikroskopische Untersuchungen über die Übereinstimmung in der Struktur und dem Wachstume der Tiere und Pflanzen* (Schwann 1839). Henceforward cited as "MU."

[32] Schwann 1879, p. 51. (Translation from Hughes (1959)) p. 38.

[33] Schleiden's work was later published in Schleiden (1838).

[34] Schwann 1847, p. 7.

[35] The crossing out is in the original manuscript (TB 1838, in January).

[36] The crossing out of "cells" is in the original. This passage, somewhat modified, was later incorporated into Schwann's *Microscopical Investigations* (Schwann 1847, p. 6).

[37] Thus, in Schwann's theory, cell formation is a process akin to crystallisation; the recognition of division as the sole way of cell proliferation had to await the work of Rudolf Virchow and Robert Remak, some two decades later.

[38] Unless otherwise stated, the following English translations are from the original English translation by Henry Smith (Schwann 1847, here p. 15). The original reference will be added in square brackets [MU, here p. 17].

[39] Schwann 1847, p. 18 [MU, p. 19].

[40] MU, pp. 112–113. Interesting enough, the translator Henry Smith found this second set of observations on the cartilage superfluous, as "all these tissues have been already treated of" (Schwann 1847, p. 97). The reader was referred to the observations in the first part of the book. Accordingly, this translation is mine.

[41] E.g. in Hughes (1959)

[42] Schwann 1847, pp. 191–192 [MU, p. 227].

[43] Schwann 1847, p. 192 [MU, p. 228].

[44] Schwann 1847, p. 192 [MU, pp. 228–229].

[45] Schwann 1847, p. 197 [MU, p. 234]. Emphasis added. Note how similar this description is to Schwann's speculation about a 'principle of nutrition' following his work on digestion.

REFERENCES

Berzelius, Jacob (1836), "Einige Ideen über eine bei der Bildung organischer Verbindungen in der lebenden Natur wirksame, aber bisher nicht bemerkte Kraft," *Jahres-Bericht über die Fortschritte der physischen Wissenschaften* 15: 237–245.

Coleman, William, & Frederic L. Holmes, eds. (1988), *The Investigative Enterprise: Experimental Physiology in Nineteenth-Century Medicine* (Berkeley: University of California Press).

Cremer, Thomas (1985), *Von der Zellenlehre zur Chromosomentheorie* (Berlin, Heidelberg, New York, Tokyo: Springer Verlag).

Duchesneau, François (1987), *Genèse De La Théorie Cellulaire*. Collection Analytiques, vol. 1 (Montreal: Bellarmin).

Eberle, J. N (1834), *Physiologie der Verdauung, nach Versuchen auf natürlichem und künstlichem Wege* (Würzburg: C.C. Etlinger'schen Verlagsbuchhandlung,).

Evans, Alfred S.(1993), *Causation and Disease. A Chronological Journey* (New York and London: Plenum Medical Book Company).

Florkin, Marcel M. (1952), "Le journal des expériences de Théodore Schwann sur la génération spontanée, la fermentation alcoolique et la putréfaction," *Bulletin de l'Académie Royale de Médecine de Belgique* (6. Ser.) 17: 164–167.

Florkin, Marcel (1960) *Naissance et déviation de la Théorie Cellulaire dans l'oeuvre de Théodore Schwann* (Paris: Hermann, 1960).

Florkin, Marcel, ed. (1960a), *Lettres De Théodore Schwann (1810–1882)*. Mémoires de la société royale des sciences de Liège vol. 2 (5. Ser.) Nr. 3 (Liège).

Gerson, Jacob (1835), *Experimenta de chymificatione artificiosa* (Berlin: Nietack).

Holmes, Frederic Lawrence (1974), *Claude Bernard and Animal Chemistry. The Emergence of a Scientist* (Cambridge, MA: Harvard University Press).

Holmes, Frederic Lawrence (1986), "Claude Bernard, The Milieu Intérieur, and Regulatory Physiology," *History and Philosophy of the Life Sciences* 8: 3–25.

Hughes, Arthur (1959), *A History of Cytology* (London and New York: Abelard-Schuman).

Kremer, Richard L. (1990), *The Thermodynamics of Life and Experimental Physiology, 1770–1880* (Harvard dissertations in the history of science), (New York and London: Garland Publishing Inc.).

Mitscherlich, Eilhard (1834), "Ueber die Aetherbildung," *Annalen der Physik und Chemie* 31, no. 18: 273–282.

Müller, Johannes and Theodor Schwann (1836), "Versuche über die künstliche Verdauung," *Archiv für Anatomie, Physiologie und wissenschaftliche Medicin*: 66–89.

Müller, Johannes. *Handbuch der Physiologie des Menschen für Vorlesungen*, vol. 2 (Coblenz: J. Hölscher) 1834–40.

Parnes, Ohad (February 2000), *Agents of Life and Disease* (Ph.D. Dissertation, submitted to Tel-Aviv University).

Parnes, Ohad (2000a), "The Envisioning of Cells," *Science in Context* 13, no. 1: 71–92.

Prout, William, (1824), "On the nature of the acid salien matters usually existing in the stomaches of animals," *Philosophical Transactions of the Royal Society of London* 114: 45–49.

Rheinberger, Hans-Jörg, and Michael Hagner, eds. (1993), *Die Experimentalisierung des Lebens. Experimentalsysteme in den biologischen Wissenschaften 1850/1950* (Berlin: Akademie Verlag).

Schleiden, M. J. (1838), "Beiträge zur Phytogenesis," *Archiv für Anatomie, Physiologie und wissenschaftliche Medicin*: 137–176.

Schwann, Theodor (1936), "Über das Wesen des Verdauungsprocesses," *Archiv für Anatomie, Physiologie und wissenschaftliche Medicin*: 90–138.

Schwann, Theodor (1936a), "Ueber das Wesen des Verdauungsprocesses," *Annalen der Physik und Chemie* 38 (2. Ser.): 358–364.

Schwann, Theodor (1837a), "Untersuchungen des Herrn Theodor Schwann über Fäulnis und Weingährung," *Mittheilungen aus den Verhandlungen der Gesellschaft naturforschender Freunde zu Berlin.*

Schwann, Theodor (1837b), "Vorläufige Mittheilung betreffend Versuche über die Weingährung und Fäulnis," *Annalen der Physik und Chemie* 41 (2. Ser.): 184–93.

Schwann, Theodor (1837c), "Auszug seiner Untersuchungen über die Gesetze der Muskelkraft (Versammlung der Naturforscher und Aertze zu Jena am 18. September 1836)," *Isis von Oken* no. 5–7: 523–524.

Schwann, Theodor (1838a), "Ueber die Analogie in der Structur und dem Wachstum der Thiere und Pflanzen," *Neue Notizen aus dem Gebiete der Natur- und Heilkunde* 5: 91 (no. 3 of vol. 5): 34–36.

Schwann, Theodor (1838b), "Fortsetzung der Untersuchungen über die Uebereinstimmung in der Structur der Thiere und Pflanzen" *Neue Notizen aus dem Gebiete der Natur- und Heilkunde* 5: 103 (no. 15 of vol. 5): 225–229.

Schwann, Theodor (1838c), "Nachtrag zu den Untersuchungen über die Uebereinstimmung in der Structur der Thiere und Pflanzen," *Neue Notizen aus dem Gebiete der Natur- und Heilkunde* 6: 112 (no. 2 of vol. 6): 21–23.

Schwann, Theodor (1839), *Mikroskopische Untersuchungen über die Übereinstimmung in der*

Struktur und dem Wachstum der Thiere und Pflanzen, (Berlin: Sander'sche Buchhandlung [G. E. Reimer]).
Schwann, Theodor (1847), *Microscopical Researches into the Accordance in the Structure and Growth of Animals and Plants*, trans. Henry Smith (London: The Sydenham Society).
Schwann, Theodor (1879), *Manifestation En L'honneur De M. Le Professeur Th. Schwann. Liége, 23 Juin 1878. Liber Memorialis Publie Par La Commision Organisatrice* (Düsseldorf: L. Schwann).
Tiedemann, Friedrich and Leopold Gmelin (1826–27), *Die Verdauung nach Versuchen*, (Heidelberg and Leipzig: Karl Groos).
Watermann, Rembert (1960), *Theodor Schwann. Leben und Werk* (Düsseldorf: L. Schwann).

H. OTTO SIBUM*

NARRATING BY NUMBERS: KEEPING AN ACCOUNT OF EARLY 19TH CENTURY LABORATORY EXPERIENCES†

In this paper I will discuss the meaning of laboratory notebooks in science and in history of science by concentrating on an early Victorian "gentlemen specialists" practice of notebook writing, just at a time when a public discourse about how to accurately account for the often private and messy bench work events was initiated. Of particular interest is the numerical technique of keeping a "laboratory book" of James Prescott Joule written in the years between 1843 and 1858.[1] It captures the beginning of an intensive research period in which he contributed most importantly to effect a change in our understanding of the nature of heat and shows a specific mode of keeping an account of laboratory experiences. However, in order to provide a clearer sense for the historical significance of Joule's mode of narrating by numbers we will briefly compare this practice with contemporary works of Charles Babbage and Michael Faraday. Michael Faraday's detailed notebook entries have attracted the attention of historians of science for a long time and their investigations certainly have moulded our image of experimental investigations in this period. For our concerns here two characteristics of Faraday's notebook writing are important to mention, firstly, his strikingly different literary technology of recording laboratory events and, secondly, his attempt to establish notebooks as a research technology in science. In 1827 Faraday had just published his "Chemical Manipulation, being Instructions to Students in Chemistry, on the Methods of Performing Experiments of Demonstration or of Research, with Accuracy and Success" which we might regard as one of the earliest student manuals in science which aims at teaching the art of experiment. There he states:

> The Laboratory notebook, intended to receive the account of the results of experiments, should always be at hand, as should also pen and ink. All the results worthy of record should be entered at the time the experiments are made, whilst the things themselves are under the eye, and can be re-examined if doubt or difficulty arise. The practice of delaying to note until the end of a train of experiments or to the conclusion of the day, is a bad one, as it then becomes difficult accurately to remember that succession of events. There is a

* Max Planck Institute for the History of Science, Berlin

Frederic L. Holmes, Jürgen Renn and Hans-Jörg Rheinberger (eds.), Reworking the Bench: Research Notebooks in the History of Science, 141–158 © *2003 Kluwer Academic Publishers. Printed in Great Britain*

> probability also that some important point which may suggest itself during writing, cannot be ascertained by reference to experiment, because of its occurrence to the mind at too late a period.[2]

Faraday described the notebook as the experimenter's companion of the bench, a tool always at hand to immediately account for important laboratory events. His attempt to provide general rules for how to use this tool properly indicates that he wanted to transform the hetereogenous practices of memorising laboratory experiences into a trustworthy, efficient and communicable research technology. By using the term research technology I partly refer to the term "literary technology" which circumscribes the powerful establishment of writing techniques to communicate experimental practice to readers of early modern natural philosophy. Laboratory notebooks produced at the bench, however, are not used for communication in the first place but to memorise personal or collective experiences. The recording techniques, however, could and did vary according to the kind of research experimenters undertook and the particular forms of life of the experimenters.[3] The distinction made here between technique and technology of tool use is important for our historical understanding of the practice of notebook writing and is strongly informed by the work of the French anthropologist Marcel Mauss. He defined technique as "traditional effective actions" and technology as a term designating a scientific discourse about "techniques."[4] Following this line of thought it is reasonable to regard Faraday's "Chemical Manipulation" amongst other things as intending to spark off a disciplinary discourse about the proper techniques of recording laboratory events and to establish the practice of notebook keeping as a research technology in science.[5] But in practice his book was not widely read nor did the author himself strictly follow the rules provided there.[6] If we look for example at Faraday's own notes we may conclude that many of his pages look like the diary of a Victorian gentleman, written at the conclusion of an exciting day (Figure 1).

The rich narrative about experimental procedures allows the reader to take part in a literary exploration of "new provinces of science." However, the reflexive historian might wonder to what extent the literary coherence – having been constructed only retrospectively – correlates with the working order of the laboratory event.[7] Furthermore instead of assuming notebook writing as a culturally shared and promoted practice, i.e., a research technology of early Victorian science, we are confronted with varying techniques of writing notebooks of which Faraday's mode is only one kind. Joule's numerical technique of keeping accounts of laboratory experiences is another one. It stands in stark contrast to Faraday's practice and hasn't been discussed in the literature as to its historical significance.

In the next section I therefore introduce the reader into the way Joule used his laboratory book by discussing some characteristic pages. The following section discusses approaches to and problems in making sense of Joule's real time recordings of laboratory experiences. I then investigate which "traditional effective

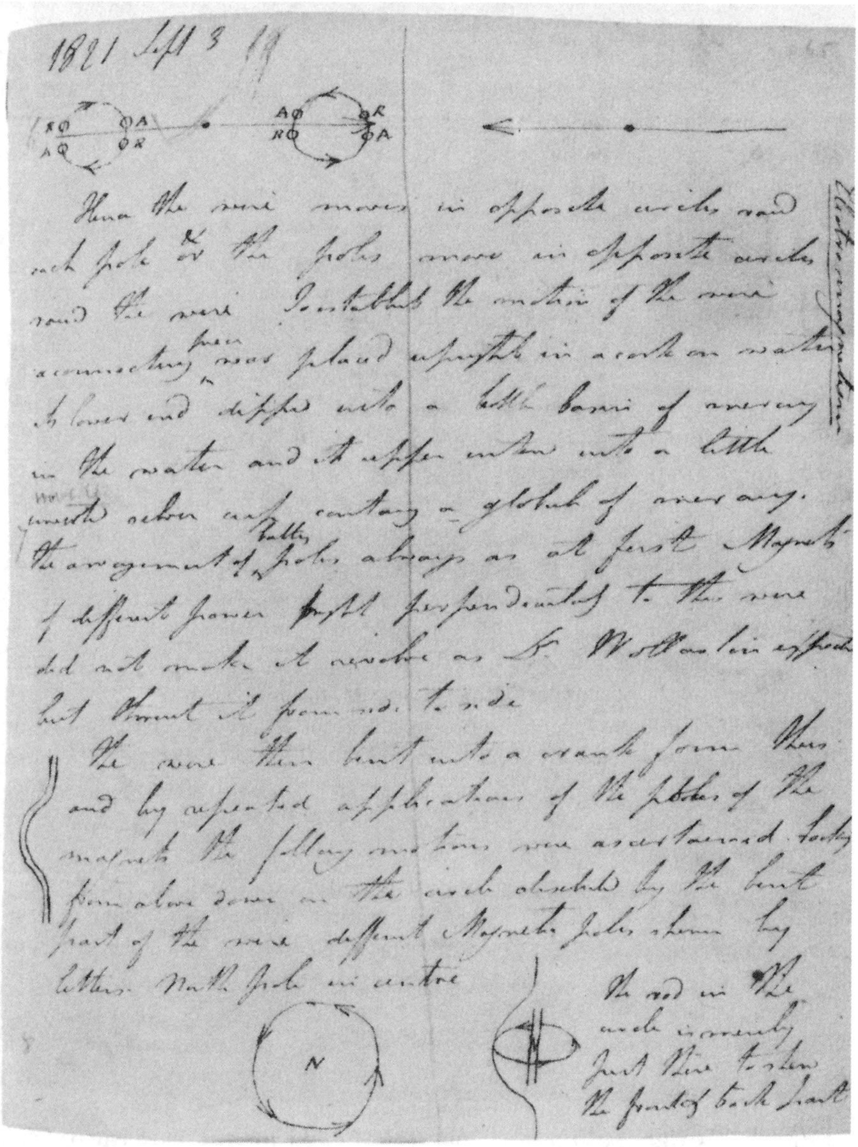

Figure 1. Faraday's notebook page on electromagnetism Sept 3, 1821, in David Gooding, *Experiment and the Making of Meaning* (Dordrecht: Kluwer Academic, 1990).

actions" moulded Joule's practice of memorising laboratory events and even effected the layout of his heat experiments. Finally, I conclude by briefly comparing Joule's mode of keeping an account of laboratory experiences with Charles Babbage's similar practice and suggest that their techniques of "narrating by numbers" have to be seen as part of the new culture of precision broadly taking command in the early 19th century.

JOULE'S LABORATORY BOOK

Joule's laboratory book[8] (Figure 2) consists of over four hundred pages, most of them covered with *real time entries*, jotted down during the experimental procedures.

The page shown here is exemplary: On the top left corner you find the date of experiment "June 16th 1843. Evening," sometimes followed by a statement about the weather conditions "fine & ...." Then the theme of the experiment in question is given: "Experiments to ascertain the heat evolved by the iron alone. 10 cells in series of 5 cell charged." Below this entry one mostly finds real time entries of data accompanied by brief specifications of the experimental procedure "Battery in contact," "contact broken," calculated numbers and comments on results like "gain" or "loss" or "not good." Without here going into details about the specific experiment, let me just go through a few other pages. Page 10 accounts for Joule's thermometric work of comparing thermometer readings (Figure 3).

We see here different layers of writings (some entries are written in light colour, others are in dark pencil marks) which indicate that this page was used at different times during his research. It served as a reference when he checked the accuracy of his temperature readings, as documented on pages 65, 82 and 106 (see entry on this page at the top). Although the book entries follow a calendar chronology over a period of ten years, we find many ruptures, periods of weeks or even months in which no entries were made. Besides many singular entries we also find day by day entries which indicate a series of experimental runs, often performed shortly before Joule presented a paper at a BAAS meeting, or when preparing a publication (Figure 4).

But even those sequences of entries of several experimental runs just provide minimal real time entries with hardly any additional information about how to perform the experiment. Again the first line of the page just gives the date, the type and location of the experiment. "June 1 AM Mechanical Equivalent of heat – Expts in cellar." In the right top corner you see a sketch of the experimental set up used which is detailed in Joule's publication of 1850. According to the publication, the experiment involved a water-filled copper vessel, with a brass paddle wheel, which was set rotating inside the jar by means of falling weights suspended over two wooden pulleys. After the weights fell a total of 1255 inches – achieved through 20 successive trials – the water temperature increased half a degree Fahrenheit by means of mechanical friction. The water and room temperature was taken at the beginning and end of the experiment. Joule called the ratio of the

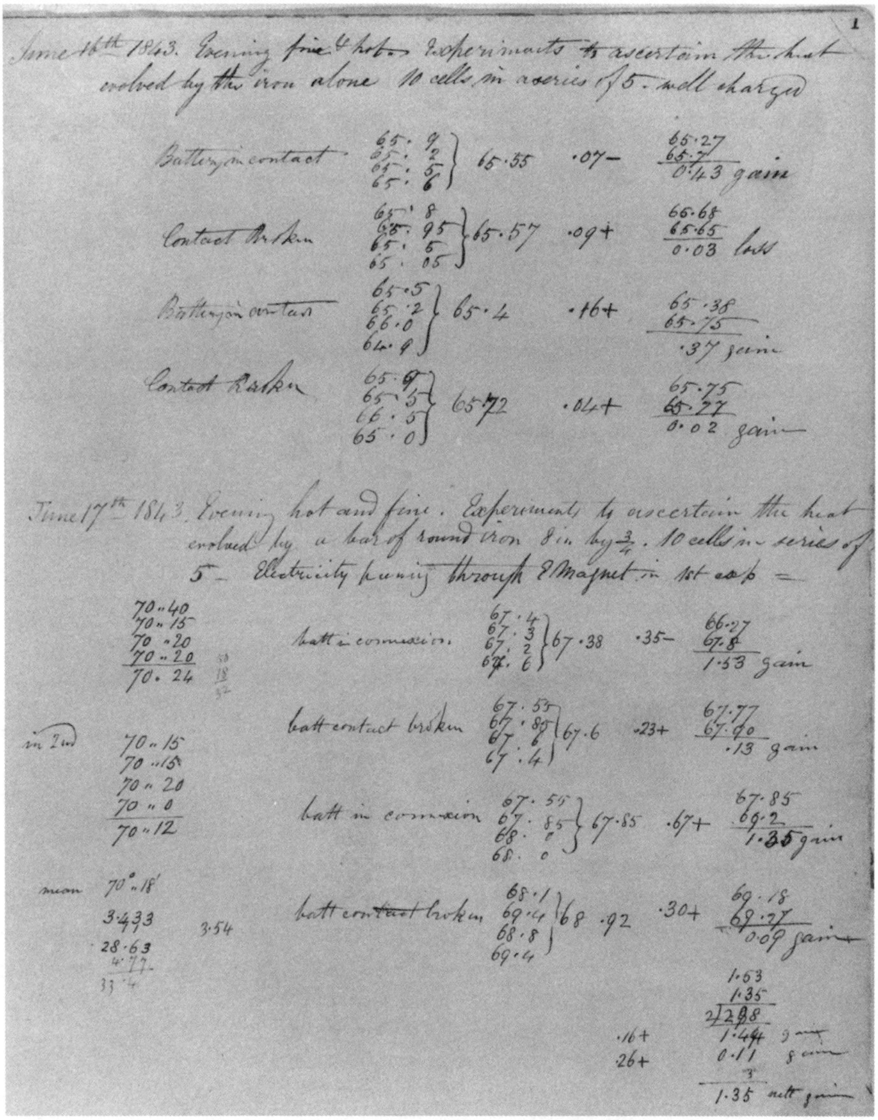

June 16th 1843. Evening fine & hot. Experiments to ascertain the heat evolved by the iron alone 10 cells in a series of 5 – well charged

Battery in contact 65·9, 65·2, 65·5, 65·6 } 65·55 ·07– 65·27 / 65·7 / 0·43 gain

Contact broken 65·8, 65·95, 65·5, 65·05 } 65·57 ·09+ 65·68 / 65·65 / 0·03 loss

Battery in contact 65·5, 65·2, 66·0, 64·9 } 65·4 ·16+ 65·38 / 65·75 / ·37 gain

Contact broken 65·9, 65·5, 66·5, 65·0 } 65·72 ·04+ 65·75 / 65·77 / 0·02 gain

June 17th 1843. Evening hot and fine. Experiments to ascertain the heat evolved by a bar of round iron 8 in. by 3/4. 10 cells in series of 5 – Electricity passing through E Magnet in 1st exp =

70..40, 70..15, 70..20, 70..20 / 70·24

batt in connexion 67·4, 67·3, 67·2, 67·6 } 67·38 ·35– 66·27 / 67·8 / 1·53 gain

batt contact broken 67·55, 67·85, 67·6, 67·4 } 67·6 ·23+ 67·77 / 67·90 / ·13 gain

in 2nd 70..15, 70..15, 70..20, 70..0 / 70..12

batt in connexion 67·55, 67·85, 68·0, 68·0 } 67·85 ·67+ 67·85 / 69·2 / 1·35 gain

mean 70°..18' 3·433 28·63 3·54

batt contact broken 68·1, 69·4, 68·8, 69·4 } 68·92 ·30+ 69·18 / 69·27 / 0·09 gain

1·53 / 1·35 / 2)2·88 / 1·44 gain ·16+ / 0·11 gain ·26+ / 1·35 nett gain

Figure 2. Joule's notebook page 1, June 16th 1843 (James Prescott Joule, *Laboratory Book from 1843–1858*, University of Manchester Institute of Science and Technology [UMIST], Manchester).

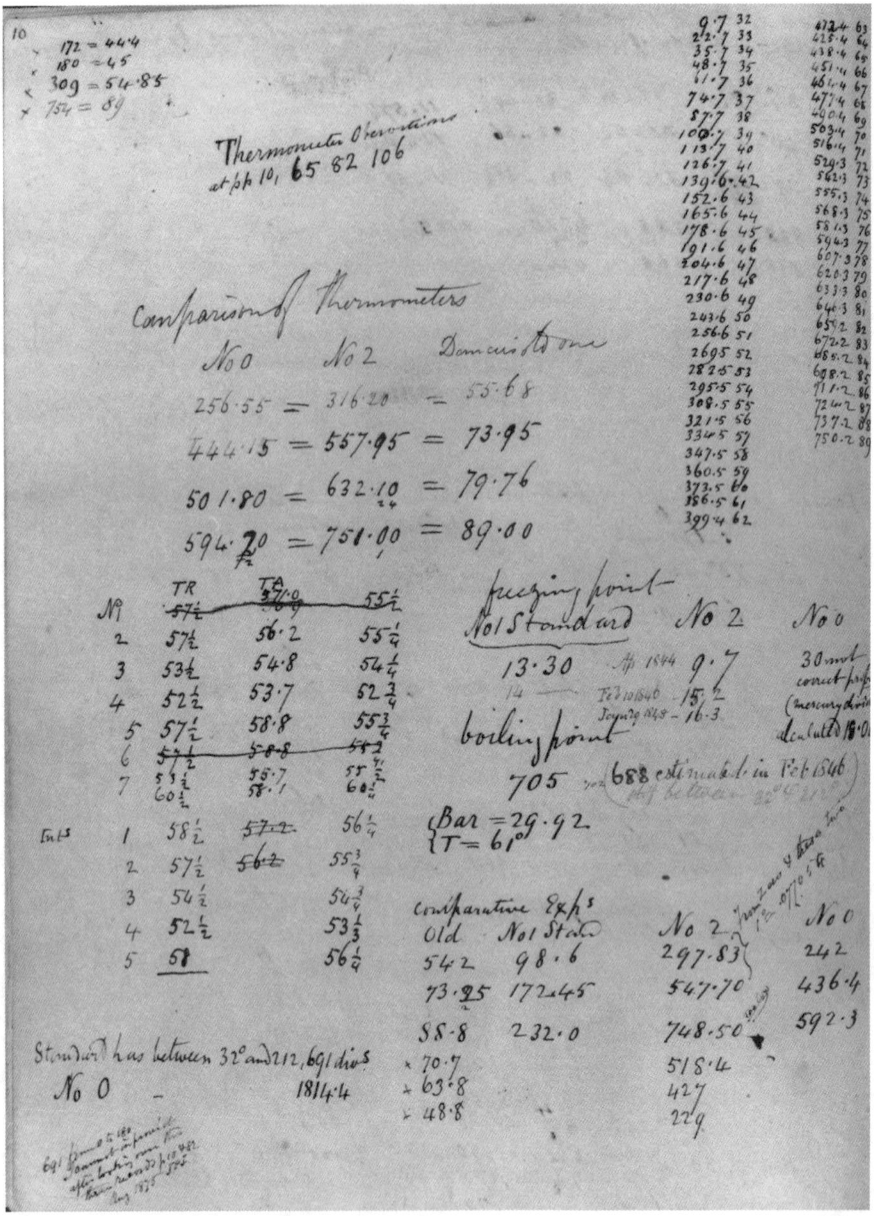
Thermometer Observations at pp 10, 65 82 106

Comparison of Thermometers

No 0	No 2	Dancer's
256·55 =	316·20 =	55·68
444·15 =	557·95 =	73·95
501·80 =	632·10 =	79·76
594·70 =	751·00 =	89·00

freezing point

No 1 Standard	No 2	No 0
13·30	9·7	

boiling point

705

Bar = 29·92
T = 61°

Comparative Exps

Old	No 1 Stand	No 2	No 0
54·2	98·6	297·83	242
73·25	172·45	547·70	436·4
88·8	232·0	748·50	592·3

Standard has between 32° and 212, 691 divs

No 0 - 1814·4

Figure 3. Joule's notebook page 10, *Comparison of Thermometers*.

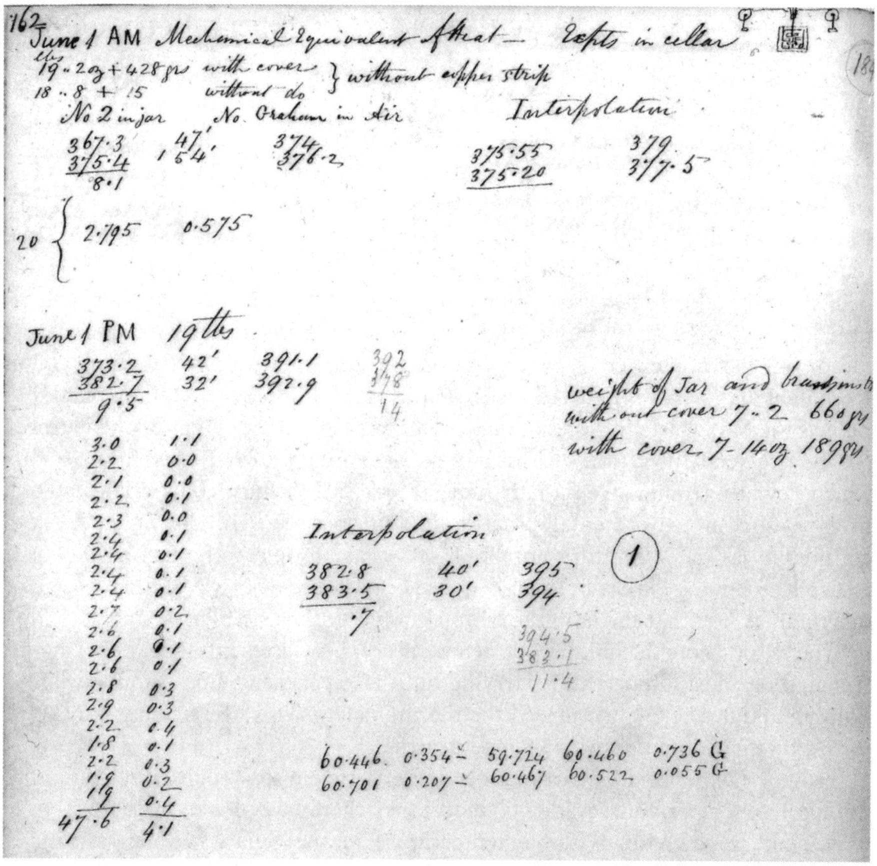

Figure 4. Joule's notebook page 162, June 1, 1847, *Mechanical Equivalent of Heat.*

work done to stir the water and the temperature difference the "mechanical equivalent of heat." The notebook pages offer minimal information and nothing about the machine is retrievable. If we try to classify the kind of experiments documented in this laboratory book we identify heterogeneous themes ranging from the determination of the quantity of heat produced through electrical currents, experiments on the combustion of hay, how to make good beer, as well as experiments on the heating effects of the friction of fluids. Moreover we find entries on experiments to check the accuracy of a balance and to determine the specific heat of flint glass or of copper, brass and water. In contrast to Faraday's mode of keeping a notebook Joule's pages do not show any literary coherence of experimental actions. His technique of keeping an account of the spatio-temporal events was in the first place a means to express everything important *numerically*.

MAKING SENSE OF LABORATORY EXPERIENCES

Historians who have made the effort to investigate Joule's notebooks have not answered the question why Joule chose this way of recording his work. The laboratory book has most often served as a resource book that provides "empirical" evidence for the historians' narratives; but has not yet been treated as a specific research technology with its own history. Joule's biographer Donald S.L. Cardwell initially sought an answer and suggested the possibility "that the requirements of brewing technology and the accountancy needed to run a business helped to mould Joule's scientific attitude."[9] But finally he situates Joule's practical reasoning about inputs and outputs in technical systems like the electro motor in the context of Manchester engineering. Older studies from Leon Rosenfeld and Gordon Jones explicitly tell us that Joule's book-keeping mentality reflects the Manchester business climate.[10] A further and more fine-grained answer can be provided through a historiographical approach of *reworking experiments* which deviates from conventional studies because it includes a material-aesthetic investigation of experimental practice in that period.[11] In the light of these considerations the laboratory notebook serves as an object of study as well as a research technology in order to make sense of past laboratory experiences. One of our findings was that the laboratory notebook even had its blind spots: it yielded no traces of those techniques of temperature measurement that we came to recognise as most important in carrying out the experiment. For example we find nothing about where Joule placed the thermometers. This is extremely important for estimating room and water temperatures.

From the experience of carrying out the experiment in a comparable building, noting the temperature gradients detectable there, we could infer that it is impossible to maintain a stable temperature in the cellar. We can assume a constant temperature fluctuation of about 1 to 1.5 degree Fahrenheit because of the glassless windows which gave open access to the outside temperature conditions. In the water vessel we also identified a temperature gradient which indicated that just after finishing the friction experiment the temperature distributed in the water is not homogenous. Therefore Joule had to know how deep to immerse the thermometer and for how long to wait before he could read off the result. Notebooks most often remain silent about this working knowledge of how to begin and end a proper measurement because the authors were not writing for an audience but for themselves. This circumstance makes the notes often extremely interesting but at the same time cryptic. The detailed reconstitution of the working gestures and course of experimental practices through performing the experiment with a replica, however, made visible experimental techniques and forms of knowing whose origin we could locate in the brewing culture of which Joule was an integral part. Elsewhere I have described this connection in detail.[12]

Through further historical explorations we have meanwhile reconstituted a knowledge space consisting of a network of collaborators whose existence and importance has not yet been recognised by historians.[13] Joule's laboratory book with its aesthetics of precision expresses a scientific attitude emerging in this small community of gentlemen specialists. In the following I will investigate the extent to which the brewing culture with its techniques of mensuration moulded Joule's practice of memorising laboratory events and how this even effected the layout of his heat experiments.

CULTURAL TECHNIQUES OF MENSURATION

On the level of the techniques employed it is reasonable to argue that Joule's mode of keeping a laboratory book parallels very closely the brewers' practice of keeping their excise books.

> Before putting any water upon his malt for brewing, the brewer is to enter in an excise book or paper, the *date of such entry*, the *quantity of malt* intended to be used, and the *date of the brewing*; and he is not to remove any such malt from the mash-tun after all the worts have been drawn off, until the officer shall have gauged the same, ...[14]

This is precisely the format structuring nearly every laboratory book page (see Figure 4): The *date of entry* is given for example as "June 1 AM Mechanical Equivalent of Heat – Expts in cellar" followed by the *quantity of materials intended to be used* here the mass of the copper vessel "19lbs 2 ounces plus 428 grs with cover ..." as well as the real time entries of temperature readings and time recordings. With regard to the increasing demands of precision in beer production and the extreme control of the brewing practices by the Excise officers it is reasonable to say that the brewing premise was one of the most important sites for reaching a trust in numbers. Joule's first laboratory book (which disappeared under obscure circumstances) has been described as "originally a partly used unruled manuscript book in which are written out with great neatness and accuracy the rules of arithmetic (with worked examples), the properties of conic sections, and exercises in book-keeping." The exercises chosen by Joule's tutor clearly addressed important practical problems of mensuration in brewing.[15] In the second laboratory book we see some hybrid forms of the brewer's and natural philosopher's world: the book contains research on brewing as well as his better known research themes. In October 14th, 1846, for example he started a series of brewing experiments of which the pages are the only ones documented in a more literary form, like a cook book recipe but peppered with numerical readings (Figure 5).[16]

Joule recorded all his extensive research experiences in this laboratory book and we have to regard entries on brewing as an integral part of his scientific agenda.[17] But there is more to say. We even can identify a brewer's experimental technique hardly noticed in his laboratory book and in its importance for several investigations on the nature of heat. During many years of work in directing the brewing processes Joule developed a technique in which he "weighted" his heat

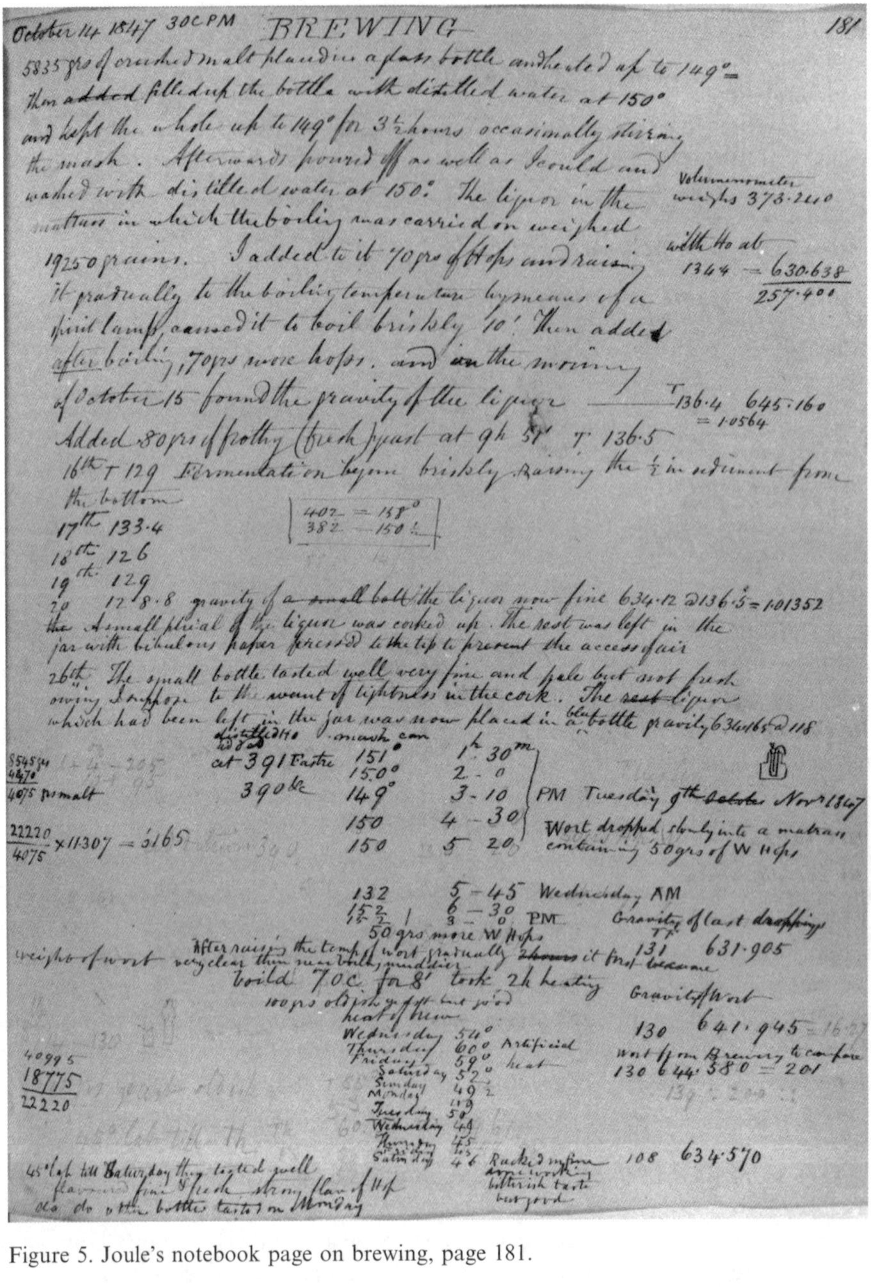

Figure 5. Joule's notebook page on brewing, page 181.

measurements. In order to guarantee stable brewing temperature conditions particularly the gains and losses of heat from the mash tun had to be controlled very carefully and Joule knew a great deal about the effects of the cooling of worts in relation to the air temperature. Irrespective of the controversies scientists had about the nature of heat, in their daily routine brewers had to find practical solutions to control heat exchanges. Attention had to be paid constantly to the heat flows going out or into the mash tun according to the actual weather (air temperature), the material conditions of the tun and the temperature difference between the mash and the air. It therefore became habitual for Joule to observe brewing phenomena in relation to the temperature conditions of the environment. And we can see this even in those notebook entries which have nothing to do with brewing and do not require such information. He often wrote comments like "Evening hot & fine" etc. (see Figure 2).

Heat was nature's main instrument, as Joule said, constantly interchanging in the material world. Thermometers were helpful devices – at least for scientific brewers like Joule – to detect these processes. But how reliably could they indicate these heat exchanges characteristic for the brewing process? During the course of his research he followed different lines of inquiry, like the improvement of thermometers as shown in the notebook page "comparison of thermometers" and which has left only a few traces of an intensive research project conducted together with John Benjamin Dancer.[18]

But in order to trust his numbers Joule often "weighted" his heat measurements by a practice which he indicated in several notebook pages as "Interpolation" (see Figures 4, 6).

This technique had its origins as a practical solution to judge heat conditions in brewing but Joule applied it in many of his research experiments. It was even formative for the layout of his paddle-wheel experiments. The notebook page clearly shows this: The left column "No. 2 in jar" gives the data of the water temperature "No Graham in Air" lists the air temperatures. At the right side you will read "Interpolation" over two columns each representing the water and air temperatures. During the performance of the experiment with a replica we found that after finishing off the friction experiment the water temperature still increased as indicated for example under "June 1 PM" where Joule wrote down the values 373.2 and 382.7, a difference of 9.5 degrees which equal 0.73 Fahrenheit. Under interpolation we read that the water temperature had further increased from 382.8 up to 383.5 which equals 0.7 degrees on Joule's temperature scale. The subsequent change of temperature was also recorded. Without going into further detail here we see that Joule has laid out his experiment in such a way that he firstly measured water and air temperature at the beginning and the end of his friction experiment. In a subsequent session his technique of "Interpolation" accounted for the fact that a temperature measurement taken at a time $\mathbf{t}$ in a water vessel did not represent the actual temperature even at the subsequent time $\mathbf{t_1}$ because the temperature state may have altered already according to the continuous exchange with the environment. In order to be able to judge what amount of heat finally

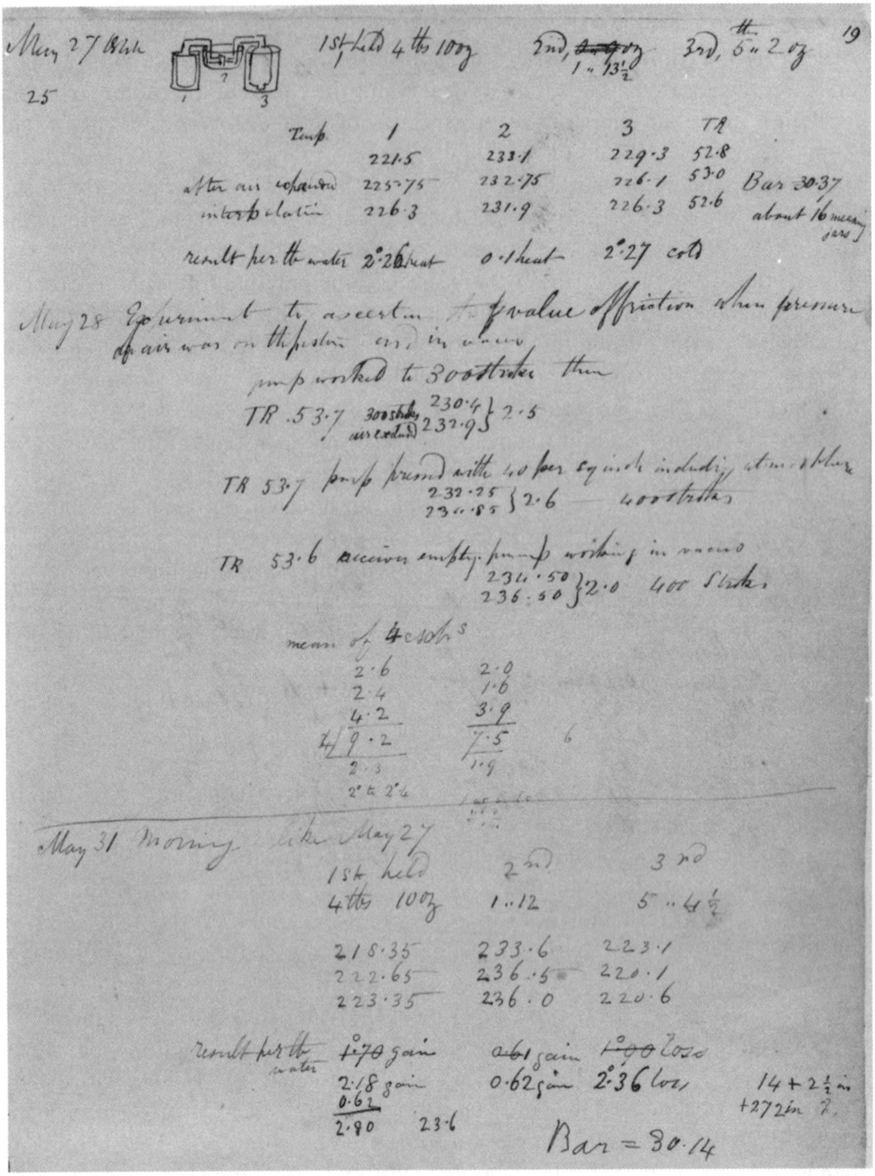

Figure 6. Joule's notebook page 19, May 27 (see Leon Rosenfeld, "Joule's Scientific Outlook," *Bulletin of the British Society for the History of Science* 1 (1952) 7: 169–176).

went into or out of the vessel during the trial a second temperature measurement was executed precisely after the time span the previous friction experiment had lasted. By doing so Joule could account for the gains and losses which might have occurred through radiation and conduction during the friction experiment effected by the temperature difference existing between water and air. These quantities had to be deduced or subtracted from the recorded temperature increase in the water in order to achieve the net result of heat produced through mechanical friction. Hatton and Rosenfeld also identified this technique when studying Joule's notebook pages documenting experiments mentioned in the paper "On the Changes of Temperature by the Rarefraction and Condensation of Air," in which two copper vessels were connected by a pipe and a valve, the whole being completely immersed in water (Figure 6). The one was charged with air at high pressure and the other was evacuated. When the valve was opened air flowed from one vessel to the other until the pressures equalised. Joule argued that "no change of temperature occurs when air is allowed to expand in such a manner as not to develop mechanical power." However, Hatton and Rosenfeld wrote that:

> The main difficulty is in the understanding of the temperature readings. Joule took three temperature values for each test and for each vessel, namely: before expansion, after expansion, and "interpolation." Joule does not explain how he took his final observation, but it is clear that it corresponds to what he describes in his paper as "alternation," "to eliminate the effects of stirring, evaporation, &c." It is assumed that after the air expansion was completed the apparatus was left for a time equal to the time between the first two temperature readings and then the "interpolation" reading was taken.[19]

This technique, or, to be more precise, naming it "Interpolation" was so unconventional that even the referees of his 1850 manuscript "On the Mechanical Equivalent of Heat" suggested that Joule should explain it to the reader by providing further information. Finally Joule wrote:

> Previously to, or immediately after, each of the experiments I made trial of the effect of radiation and conduction of heat to or from the atmosphere in depressing or raising the temperature of the frictional apparatus. In these trials the position of the apparatus, the quantity of water contained by it, the position of the experimenter, in short every thing, with the exception of the apparatus being at rest, was the same as in the experiment in which the effect of friction was observed.[20]

Furthermore for the final publication Joule was also asked to delete the notion "Interpolation" and to set "Radiation" instead.[21] Joule agreed to it – as he did to other important changes – but it partly obfuscated his experimental approach. His whole strategy had been to avoid the discussion of radiation as well as conduction effects by providing an experimental set up and applying a technique of "interpolation" reading which could display the pure effect of converting mechanical power into heat.

NOTEBOOKS AND THE CULTURE OF PRECISION

Joule's numerical mode of keeping an account of his laboratory experiences was not a singular practice. Applying numerical techniques as a means to record laboratory experiences as well as investigating nature was a commonly shared practice amongst British gentlemen specialists. Abstaining from the conventional literary form of narration of experiences as Joule did, in contrast to Faraday's diary writings marks an important step in a broader cultural development of science which we may call the rising culture of precision – an expert culture based on instruments of precision and new gestures of accuracy. Without providing a complete survey of this transformation, the case of Charles Babbage and recent historians' contributions to this topic will suffice to make this clear. In the 1820s Babbage started a series of electrical experimental investigations which he recorded in his notebook – an accountant's ledger book. Figure 7 depicts a characteristic page which shows striking similarities to Joule's mode of accounting.[22]

For a start, it is a real-time account of experimental runs expressed in numbers with minimal written information about the experimental procedure. As in Joule's notebook, we find documented an experimental technique of systematic variation of the experimental conditions, i.e., the successive variation of one parameter, "all the other circumstances being the same." In the case of Joule, scholars describe this mode of working as an integral part of the business of engineering and accounting. Most recent studies on early 19th century British astronomy make a similar claim for Babbage.[23] The works of Babbage, Herschel *et al.* demonstrate that leading astronomers regarded "astronomical book-keeping" as the most successful and perfectly formed system of maintaining records.[24] Babbage's use of a ledger book and accounting techniques at work in his electrical experiments is not astonishing in this context. It is reasonable to argue that the practices of narrating by numbers as exemplified in Joule's and Babbage's notebooks represent core techniques of the rising analytical world.[25]

CONCLUSION

Laboratory notebooks have an ambiguous status in science and in history of science. Scientists use them day by day but the history of this research technology is mostly unwritten. Historians who study them conventionally regard these books as one source revealing individual pathways of important discoveries. But as this paper has shown, irrespective of their richness and private character they have their blind spots. The notebook therefore remains a necessary but insufficient means to study research processes, and the historian's craft is required to make sense of these notes. Reworking past experiments is one way of deciphering and complementing the often cryptic signs left to the historian.[26] Secondly, this paper has shown that what seems to be the private, personal touch of a laboratory book has clear cultural origins. Joule's technique of interpolation originated in the

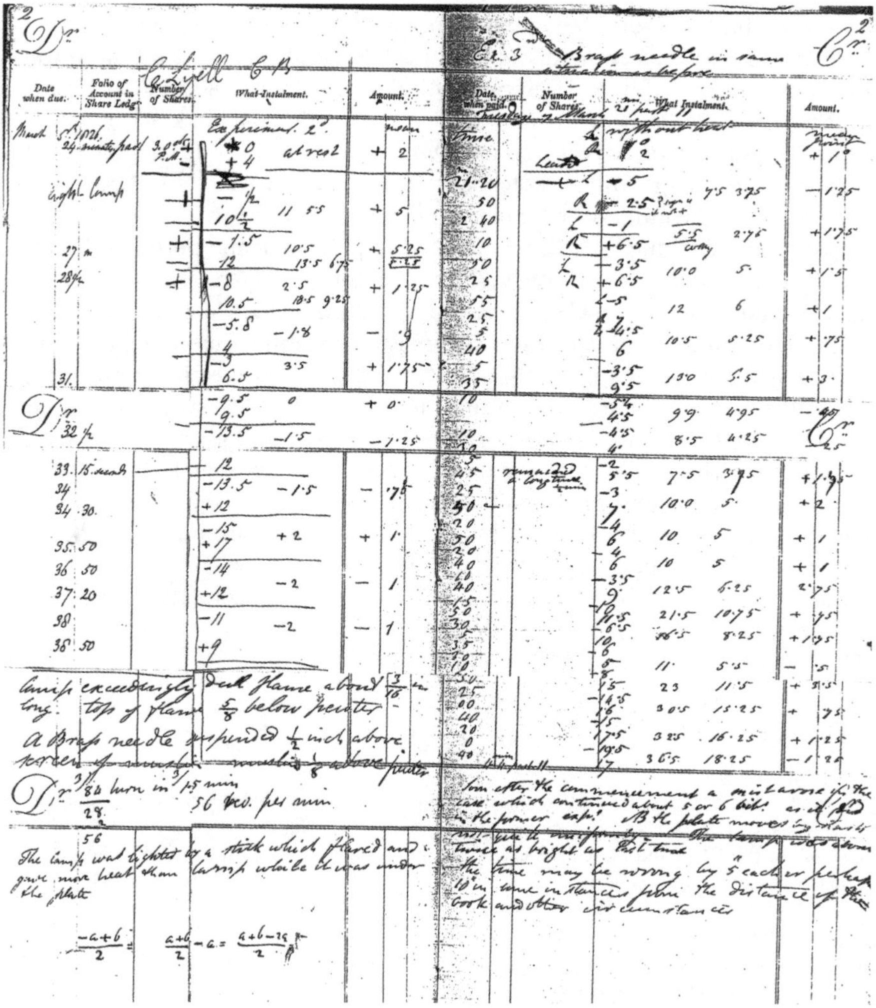

Figure 7. Babbage's notebook page 2 (Note book by Charles Babbage, *Experiments on revolving needle by electricity 5 March–11 June 1826*, Cambridge University Library MSS Add 8705.38 fols 1–10).

brewing culture as did, to a large extent, his overall attitude towards narrating by numbers. Thirdly, Joule's mode of keeping an account of laboratory experiences was not a singular practice but has to be seen as part of a broader cultural development of science which we may call the rising culture of precision.

Notebooks have a history of their own. The comparison with Babbage's and Faraday's accounts shows that in the period under investigation researchers could draw on various traditions of recording experiences, i.e. Faraday's gentlemanly literary practice of keeping a personal diary in contrast to Joule's and Babbage's numerical practices of accounting. However, the brief comparison of these notebooks is just the beginning of a historical investigation of the extent to which this core technique of the emerging analytical world around 1800, i.e. narrating by numbers, became the seed of the establishment of the notebook as the research technology of the exact sciences in the 19th century. Finally, one might argue that these two forms of writing notebooks just expressed the difference between qualitative and quantitative research. And, indeed, in previous historical scholarship on Joule this simplistic distinction had led to conflicting portraits. It is well represented in Donald Cardwell's defence of Joule as a creative, original man against a position held by George Sarton who described Joule as a metrologist; a polarisation which has its consequences in historical writings until today. But such a polarisation between apparently uncreative practices of precision measurement and a qualitative and innovative attitude leads the reader astray. Detailed studies of laboratory notebooks – treated as a research technology of science at a specific historical period – reveal that this polarisation probably originated in this very period of cultural transformation in which the new moral economy of the expert culture took command.[27] Joule's laboratory book is an important landmark in the emergence of this expert culture in which precision measurement was pursued as a means to effect even qualitative change in the early nineteenth-century exact science of heat.

NOTES

† I would like to thank the participants of this workshop and Simon Schaffer for stimulating discussions. I cite material from archives at the University of Manchester Institute of Science and Technology (UMIST) and the University of Cambridge and am grateful for permission to do so.

[1] James Prescott Joule's *Laboratory Book from 1843–1858*, University of Manchester Institute of Science and Technology (UMIST), Manchester.

[2] Michael Faraday, *Chemical Manipulation; Being Instructions to Students in Chemistry, on the Methods of Performing Experiments of Demonstration or of Research, with Accuracy and Success* (London: W. Phillips, 1827), p. 546.

[3] On the notion of literary technology see Steven Shapin and Simon Schaffer, *Leviathan and the Air-Pump. Hobbes, Boyle, and the Experimental Life* (Princeton: Princeton University Press, 1985), pp. 60ff. In a recent brief article by H.-J. Rheinberger he suggests to treat the notebook as a "Forschungswerkzeug," see Hans-Jörg Rheinberger, "Wissenschaftsgeschichte und experimentelle Praxis," Karl-Sudhoff-Gedächtnisvortrag, *Nachrichtenblatt der Deutschen Gesellschaft für Geschichte der Medizin, Naturwissenschaft und Technik e.V.* 49 (1999) 3: 198–210, 206.

[4] Marcel Mauss, "Les techniques du corps," *Journal du psychologie normale et pathologique* 32 (1935): 271–293, 278: "J'appelle technique un acte *traditionnel efficace* (et vous voyez qu'en ceci il n'est pas différent de l'acte magique, religieux, symbolique). Il faut qu'il soit *traditionnel et efficace*. Il n'y a pas de technique et pas de transmission, s'il n'y a pas de tradition."

[5] With the spread of laboratory science during the 19th century the demand for providing a systematic body of theoretical and practical knowledge by which the students of science were guided to keep trustworthy records of laboratory events increased. Unfortunately this important part of the history of the notebook as a research technology cannot be dealt with in this paper.

[6] With regard to the actual readership of this book we have to remember that the manual was held at the library of the Royal Institution in London and only a few copies existed "which are in the hands of collectors and libraries." See publisher's note in Michael Faraday, *Chemical Manipulation* (1827), op. cit. (note 2). Therefore experimentalists in Faraday's time – mostly working outside learned institutions – hardly had a chance to read his book and often pursued or developed their own craft of keeping a personal account book.

[7] Concerns like this may have led David Gooding to reflect on the limits of Faraday's literary representations of experimental practice. For his reflections on what he has called the "cognitive regress" and the attempts to repeat experiments see David Gooding, *Experiment and the Making of Meaning*. Human Agency in Scientific Observation and Experiment, (Dordrecht: Kluwer Academic, 1990). David Gooding, "Can We Tell What Really Went On?," in Frank A.J.L. James, ed., *The Development of the Laboratory: Essays on the Place of Experiment in Industrial Civilization* (London: Macmillan, 1989), pp. 63–82.

[8] The laboratory book under investigation starts June 16th, 1843 and ends with the year 1858, 435 pages (Lowery No. 5 (b)) UMIST (HST). His first laboratory book (1839–43) does not exist any more.

[9] Donald S. L. Cardwell, *James Joule. A biography* (Manchester and New York: Manchester University Press, 1989), p. 47. However, Cardwell did not try to provide further evidence for his speculation.

[10] Leon Rosenfeld, "Joule's Scientific Outlook," *Bulletin of the British Society for the History of Science* 1 (1952) 7: 169–176; Gordon Jones, "Joule's Early Researches," *Centaurus* 13 (1968) 2: 198–219; see also John Forrester's argument that Joule was reasoning about electrical phenomena in terms of a closed system in which nothing can be lost. John Forrester, "Chemistry and the Conservation of Energy: The work of James Prescott Joule," *Studies in History and Philosophy of Science* 6 (1975) 4: 273–313, 290.

[11] This approach includes the detailed reconstitution of the working gestures and course of experimental practices through performing the experiment with a full scale replica. The experiment in question is Joule's paddle wheel experiment to determine the mechanical equivalent of heat (Figure 4). See H. Otto Sibum, "Reworking the Mechanical Value of Heat: Instruments of Precision and Gestures of Accuracy in Early Victorian England," *Studies in History and Philosophy of Science* 26 (1995) 1: 73–106; as well as H. Otto Sibum, "Experimental History of Science," in Svante Lindqvist, ed., *Museums of Modern Science*, Nobel Symposium 112 (Canton, MA: Science History Publications, 2000), pp. 77–86.

[12] H. Otto Sibum, "Les Gestes de la Mesure. Joule, les pratiques de la brasserie et la science," *Annales: Histoire, Sciences Sociales* 4–5 (Juillet-Octobre, 1998): 745–774.

[13] For a study of the important collaboration between Joule and the instrument maker and natural philosopher John Benjamin Dancer see H. Otto Sibum "Shifting Scales. Microstudies in Early Victorian Britain," Paper held at the conference *Varieties of Scientific Experience*, Berlin, 1997, Max Planck Institute for the History of Science, Preprint No. 171. For the conception of space see Crosbie Smith and John Agar, *Making Space for Science. Territorial Themes in the Shaping of Knowledge* (Houndsmill: Macmillan, 1998) as well as Hans-Jörg Rheinberger, Michael Hagner and Bettina Wahrig-Schmidt, *Räume des Wissens: Repräsentation, Codierung, Spur* (Berlin: Akademie Verlag, 1997).

[14] Joseph Bateman, *The Excise Officer's Manual: Being a Practical Introduction to the Business of Charging and Collecting the Duties Under the Management of Her Majesties Commissioners of Inland Revenue*, second edition (London: William Maxwell, Bell Yard, Lincoln's Inn – Law and General Publisher, 1852), p. 259. Italics by Otto Sibum.

[15] Several historians have commented on this issue. Gordon Jones, "Joule's Early Researches," *Centaurus* 13 (1968) 2: 198–219; John Forrester, "Chemistry and the Conservation of Energy: The Work of James Prescott Joule," (1975), op. cit. (note 10). Haldane Gee has provided us with a rare account of the now missing first laboratory book. W.W. Haldane Gee, "Joule's Laws of Electric Heating", *Memoirs and Proceeding of the Manchester Literary and Philosophical Society* 69 (1924/

25): XIII – XV, XIV; for practices of mensuration see also Joseph Bateman, *The Excise Officer's Manual* (1852), op. cit. (note 14).

[16] Joule's *Laboratory Book from 1843 – 1858,* pp. 181ff, also p. 223–24, UMIST, Joule archive.

[17] In the early Victorian period brewing was still treated as integral part of chemistry.

[18] Historians of science have not studied Joule's thermometrical work closely. For this research project see the author's conference paper "Shifting Scales," op. cit. (note 13).

[19] A. P. Hatton, L. Rosenfeld, "An Analysis of Joule's Experiments on the Expansion of Air," *Centaurus* 4 (1956) 4: 311–318, 312–13.

[20] James P. Joule, "On the Mechanical Equivalent of Heat," *Philosophical Transactions of the Royal Society* 140 (1850): 61–82; alternatively in *The Scientific Papers of James Prescott Joule*, published by the Physical Society of London, vol. 1 (London: Taylor and Francis, 1884), pp. 289–328, pp. 305–306.

[21] James Prescott Joule, *On the Mechanical Equivalent of Heat*, manuscript, Archive of the Royal Society, London, PT. 37.3.

[22] Notebook by Charles Babbage, *Experiments on revolving needle by electricity 5 March – 11 June 1826*, Cambridge University Library MSS Add 8705.38 fols 1–10. For the related publication see Charles Babbage, "On Electrical and Magnetic Rotations," *Philosophical Transactions* 116 (1826): 494–528.

[23] Joule's biographer Donald Cardwell points to engineering and accountancy as his cultural resources for his scientific attitude. Donald S. L. Cardwell, *James Joule. A biography* (1989), op. cit. (note 9), p. 47. For a discussion of Babbage and the business of astronomy see William Ashworth, "The Calculating Eye: Baily, Herschel, Babbage and the Business of Astronomy," *BJHS* 27 (1994): 409–441.

[24] See in particular Ashworth, "The Calculating Eye," (1994), op. cit. (note 23), p. 409.

[25] For further details on the rise of the culture of precision see M. Norton Wise with the collaboration of Crosbie Smith, "Work and Waste: Political Economy and Natural Philosophy in Nineteenth Century Britain (I), *History of Science* 27 (1989): 262–301; "Work and Waste (II)," *ibid*, pp. 391–449; and "Work and Waste (III)," *History of Science* 28 (1990): 221–256; M. Norton Wise, ed., *The Values of Precision* (Princeton, N.J.: Princeton University Press, 1995) and T. Porter, *Trust in Numbers. The Pursuit of Objectivity in Science and Public Life* (Princeton, N.J.: Princeton University Press,1995); Jack Morell, Arnold Thackray, *Gentlemen of Science: Early Years of the British Association for the Advancement of Science* (Oxford: Clarendon Press, 1981); Crosbie Smith, *The Science of Energy: A Cultural History of Energy Physics in Victorian Britain* (Chicago and London: University of Chicago and Athlone Press, 1998); Julian Hoppit, *Risk and Failure in English Business 1700–1800* (Cambridge: Cambridge University Press, 1987); Boyd Hilton, *The Age of Atonement: The Influence of Evangelicalism on Social and Economic Thought 1785–1865* (Oxford: Oxford University Press, 1988).

[26] See the author's publications (note 11) as well as the different modes of inquiry discussed in this volume.

[27] For the controversy see Donald S. L. Cardwell, J*ames Joule, A bioprahy,* (1989), op. cit. (note 9), p. 8; George Sarton, "The discovery of the law of conservation of energy," *Isis* 13 (1929/30) 40: 18–44. On the notion of moral economy in science see Lorraine Daston, "The Moral Economy of Science," *Osiris* 10 (1995): 3–24.

ANDREA LOETTGERS*

EXPLORING CONTENTS AND BOUNDARIES OF EXPERIMENTAL PRACTICE IN LABORATORY NOTEBOOKS: SAMUEL PIERPONT LANGLEY AND THE MAPPING OF THE INFRA-RED REGION OF THE SOLAR SPECTRUM

INTRODUCTION

For historians of science with a special interest in experimental practice, laboratory notebooks provide an outstanding source. They offer the opportunity to look behind the facade given by the final representation of the published results. In the publications, the way along which the experimenter reached the results and conclusions presented are more or less invisible (Holmes 1992). If one finds descriptions of the development of the investigation, they are reduced to some substantial points in which a retrospective logic is expressed.

The inside view which one gets by examining the laboratory notebook is not unlimited, the resolution of the fine structure of experimentation differs from notebook to notebook, depends on the function of the notebook, the circumstances under which it was written and the personal style of the author. But nevertheless the notebooks lead back to the activities and developments in the laboratory, to the realm of production of scientific knowledge, which includes such different factors as experimental strategies, skills, experiences, material and personal resources, social relations, instruments, scientific facts, and theories. This broad concept of experimental practice (Pickering 1995) automatically leads out of the laboratory into the historical and disciplinary context in which the scientific research project was embedded and consequently to the limit of that what one can find about this practice in the laboratory notebooks. I will try to determine these limits more precisely and to show pathways, rooted in the notebooks, which lead out of the laboratory. A further question concerns the activities in the laboratory. How far can one follow the activities of the experimenter in, for example, performing the experiments? What can one say about skills and experiences which were necessary and emerged in the course of experimentation?

Within the limits given by the historical record I will focus on two points, first the organisation of a laboratory and second the development of scientific instruments, for which especially the laboratory notebooks under investigation

---

* ETH Zürich

Frederic L. Holmes, Jürgen Renn and Hans-Jörg Rheinberger (eds.), Reworking the Bench: Research Notebooks in the History of Science, 159–182 © *2003 Kluwer Academic Publishers. Printed in Great Britain*

provide an outstanding source. The development of scientific instruments leads into regions of experimental practice which have received little attention. Exceptions are Peter Galison's and Alexi Assmus' case study of the bubble-chamber (Gooding 1987) and Andrew Pickering's concept of *dialectic of resistance and accommodation* (Pickering 1995) in which the instruments are no longer static elements, in contrast to Bruno Latours concept of the*"actor network"* in which he reduces instruments to *"inscription-devices."* By examining the dynamics connected with the instruments and the fact that they themselves form objects of investigation (Dörries 1994), the dynamics of the entire process of investigation becomes more complex and increases and leads to a network that goes beyond that of Latour's.

The objects of the case under review are the investigations of the American astrophysicist Samuel Pierpont Langely of the infra-red region of the solar spectrum. These are extensively documented in thirty laboratory notebooks, each 300 to 800 pages in length. They offer an insight into a complex and dynamic process that was strongly shaped by the interaction of performing experiments in this region of the solar spectrum with the development of instruments and experimental methods. To extract this process out of the huge amount of material and to make it transparent, I tried to follow up different *pathways of development* which I found in the notebooks, to look at their interactions and to determine the different factors relevant for the dynamics, like the material and personal resources, experimental strategies, skills, experiences and theories of the time which enter into Langley's investigations. Because these factors themselves were not constant, they lead to additional dynamics in the research project.

Langley's experiments do not count among the "big" experiments, in the course of which new unexpected results are discovered. But one should not undervalue them. He designed a new instrument, the bolometer, a radiation measuring instrument, based on the detection of the effect of the heat of the radiation, and he was the first to make with this instrument the nearly unexplored dark longwave part of the solar spectrum accessible. So he was, so to say, a pioneer in this region. Photographic methods as they were successfully used by Henry Rowland (Hentschel 1998) in the UV- and visible part of the spectrum, were not applicable because of the lack of an infra-red sensible photoemulsion.[1]

In 1883, in the beginning of his systematic investigation of the infra-red region of the solar spectrum Langley wrote:

> Especially when we consider that these rays are invisible, and that the whole process may be compared to a patient groping in the dark, does the need of an instrument which will record the precise point where a hot or a cold line was felt become obvious. This is the object of the spectro-bolometer [...]. (Langley, 1883)

At that time, the bolometer was far away from being a precision instrument. Langley's groping in the dark was performed with an instrument which gave doubtful results. So he had to deal with the following connected unknowns: the properties of the bolometer, sensitivity and stability, and the properties of the infra-red region of the solar spectrum, number and position of the absorption lines, and the expansion of the spectrum. The calibration of the bolometer was restricted to the part of the solar spectrum already explored. Langley did show that he could determine the Fraunhofer lines in the visible part of the solar spectrum. But the trustworthiness of this instrument in this part of the spectrum could not be transferred to the infra-red region.

The research process documented by Langley's notebooks during the next 17 years displays a very complex stabilization process. This process was marked by an oscillation of the bolometer between two different functions. The bolometer in its function as an infra-red measuring instrument was used in the experiments for mapping the infra-red spectrum. It was also itself an object of investigation. In the investigations in the infra-red region, different strategies were developed and used in order to handle the uncertainty and sources of error connected with the instrument. Also the results, the number and position of absorption lines, were carefully examined in order to separate fact from artifact. As an object of investigation, the bolometer constituted a research process in which single instruments of the apparatus were improved, stabilized, tuned against each other, and in which the measuring and registration process was automatized and special programs were developed in order to isolate and eliminate sources of error.

Next to these processes the laboratory grew and was organized more and more according to the demands of the division of labour, as assistants with different experiences and skills came in to the laboratory and left their marks. All these different facets of the experimental practice are reflected in the laboratory notebooks.

In 1900 the investigations resulted in the publication of the first volume of the *Annals of the Astrophysical Observatory of the Smithsonian Institution* (Langley 1900). Here Langley published a map of the infra-red region of the solar spectrum extending from 1,1μ to 5,3μ and containing the position of 750 absorption lines.[2]

Before describing some of the pathways[3] of development I shall give a short overview of, including some biographical notes and a short summary of prior studies in the infra-red solar spectrum, in order to put Langley and his work into the historical and disciplinary context of his time. Furthermore, I shall also provide some necessary information about his radiation measuring instrument, the bolometer.

OVERVIEW

Biographical Notes

Samuel Pierpont Langley, who counts among the founders of astrophysics in the United States (Lankford 1997) and had worked as an engineer and architect in Chicago and St. Louis, started his career in science at age thirty and without any university training, as an astronomer at the Harvard Observatory serving as assistant to John Winlock.[4]

Only one year later Langley obtained a professorship at the Naval Academy, where he was expected to renew the observatory in addition to his tasks as a professor for astronomy. But in the same year he moved to Western University in Pittsburgh where he had received a professorship in physics and astronomy and took over the directorship of the Allegheny Observatory. The observatory was part of the university but at the time existed "*only in name*" (Jones, 1965). To raise money for equipment, Langley invented, following the example of the European observatories like Greenwich and Potsdam, a time-signal service system and sold the time-signals to the Pennsylvania line, a railroad company. He won the support of one of their trustees and officers, Wilham Thaw, who substantially financed a great amount of Langley's research projects.

Until about 1876 the main focus of Langley's research concerned the observation of sunspots. In 1876 Langley began with his investigations of the heat distribution in the solar spectrum in connection with the determination of the solar constant. By using spectroscopical methods he hoped to gain insight into the solar-terrestrial relation, into the physical-chemical processes of our atmosphere. Ultimately, he hoped to be able to answer meteorological questions concerning the influence of solar radiation on the earth's climate and to make weather forecasts (Langley1884). In the course of these investigations, about 1880, he designed a radiation measuring instrument, the so called bolometer. A short description of its functioning and the apparatus will be given in the next section. The opportunities the new instrument offered to him, to perform measurements in the unexplored infra-red region of the solar spectrum, led to a turn in his research program. Langley realized these opportunities on an expedition to Mount Whitney in South California. There, in the dry and clear air of the mountains, he performed measurements taking him further in the infra-red region, as everyone before him. On that occasion he also measured, for the first time, a cold band (absorption band) which he called Ω at 1,8μ.[5] About 1882 he started making systematic measurements to map this part of the solar spectrum.

In 1887, Langley was appointed first secretary of the Smithsonian Institution in Washington. The fact that he became the director of this renowned Institution shows that Langley was among the most famous and influential scientists in the US during his time. When he came to Washington no observatory existed. In 1890, a wooden building in the garden of the Smithsonian Institution was erected to serve as an astrophysical observatory until a proper place outside Washington could be found. For various reasons this did not happen in Langley's era. So they had to perform the measurements, which were highly sensitive to changes in the temperature, under very unfavourable conditions. Nevertheless, the astrophysical laboratory, which was more and more based on the principles of division of labour, grew during its time in Washington, and the research project was successfully brought to an end, as the publication of the results in 1900 shows.

Prior Studies in the Infra-Red Solar Spectrum

The history of infra-red spectroscopy started in 1800 with William Herschel's (1738–1822) experiments in which he discovered that a thermometer exposed beyond the red visible part of the solar spectrum exhibits heating effects. This surprising result raised the question whether heat and light are two distinct entities. In order to answer this question Herschel performed a series of experiments. But because of the lack of an instrument which was more sensitive then the thermometer at this time he did not come to a satisfying answer.

About 1830 a new instrument was invented by the Italian physicists Leopoldo Nobili (1784–1835) and Macedonio Melloni (1798–1854), the thermopile. This instrument, based on the thermoelectric effect discovered by Thomas Johann Seebeck (1770–1835) in 1822, did fulfil the demand of being more sensitive. Melloni himself and other physicists like Hippolyte Louis Fizeau (1819–1896) and Léon Bernand Jean Foucault (1819–1868) took up Herschel's investigations again with this new instrument. They came to the conclusion that heat and light were not distinct entities, rather that each ray has three distinct qualities – heat, light and chemical action.

For the investigation of the infra-red region of the solar spectrum, the thermopile was used in 1858 by Johann Heinrich Müller, and in 1872 by Sergei Iwanowitsch Lamansky. In spite of the increased sensitivity, the detection of the heat of the radiation was, as the following quotation from Müller shows, far behind the accuracy of the optical investigation of the solar spectrum:

> An den Nachweis wärmeloser Streifen im Spektrum, welche den Fraunhofer'schen dunklen Linien entsprechen, scheint vor der Hand noch nicht gedacht werden zu können, wie denn überhaupt die thermische Untersuchung des Spektrums noch weit von dem Grade der Genauigkeit entfernt ist, den man in der optischen Untersuchung des Spektrums erreicht hat. (Müller 1858)

By using lenses Lamansky increased the intensity of the radiation in the infra-red region and was able to detect three absorption bands in this region when he worked at the laboratory of Hermann Helmholz in Heidelberg (Figure 1). This proceeding, however, was problematic because of the absorption of the radiation by the lenses.

In addition to the detection of the heat of the radiation, the detection of the chemical effect was tried. In 1840 John Herschel (1792–1871) published a thermograph which showed for the first time unequal absorption in the infra-red region (Figure 2).

In 1842, John William Draper (1811–1882) detected three bands in this region by phosphorescence. He marked them as, α, β und γ. Also Edmond Alexandre Becquerel (1820–1891) and Antoine Henri Becquerel (1851–1908) used this method in their investigations of the solar spectrum. In 1883, H. A. Becquerel improved this method. He roughly mapped out the solar spectrum by using a Rutherford grating up to about $1{,}4\mu$. Around 1873 William de Wiveleslie Abney

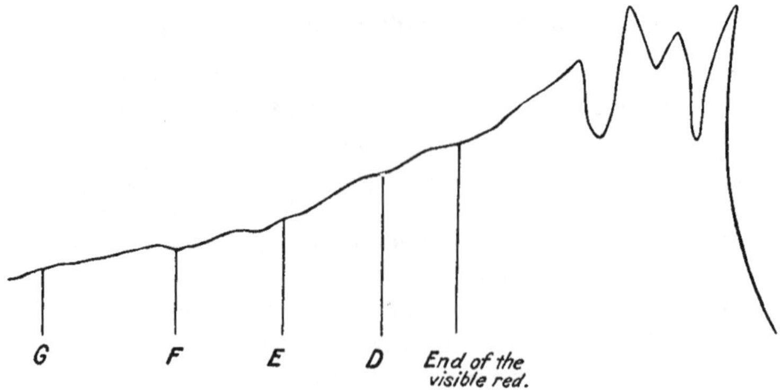

Figure 1. Three absorption bands in the infra-red spectrum detected by Lamansky in 1872 (Lamansky 1872).

Figure 2. Thermograph by John Herschel 1840 (Herschel 1840).

(1843–1920) developed a special photoemulsion by means of which he succeeded in photographing the solar spectrum up to wave-lengths of 1μ with great detail. Further efforts by Abney shifted the limit up to $1{,}8\mu$. At this point the possibilities of exploring of the infra-red region of the solar spectrum by photographing were exhausted. This limit could only be overcome by Langley's bolometer.

The Bolometer and the Spectrobolometer

The functioning of the bolometer is based on the nearly proportional variation of the electrical resistance of metals R, with temperature T.

$$R = r(1 + \alpha T) \tag{1}$$

in which r stands for the resistance at $T = 0$, R at $T > 0$ and α is a constant.

Translating this relation into a radiation measuring instrument, Langley put a very thin metal strip, approximately 1μ [6] in one arm of a Wheatstone bridge. Because of its thinness the metal strip was highly sensitive to changes in temperature and therefore of the electric resistance. The Wheatstone bridge which was commonly used in precision measurements for determining resistances of electrical conductors, consisted of four resistors and a galvanometer (see Figure 3). When the bridge was balanced, the voltage drop at all four resistors was equal and there was no current through the galvanometer. If the thin metal strip,

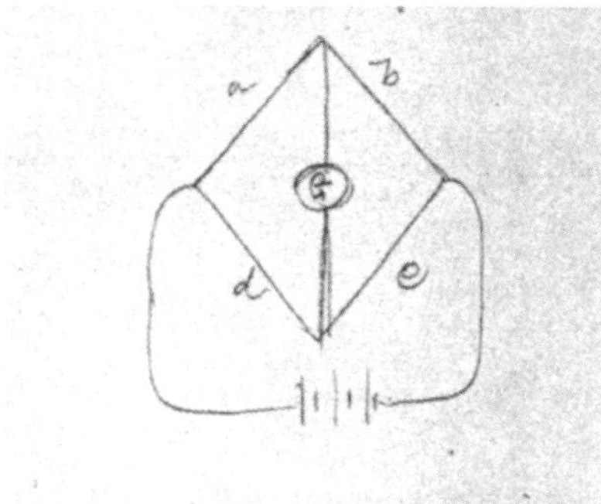

Figure 3. Sketch of a Wheatstone-Bridge connected with a battery, out of the waste-book from 1892, by Lawton Olcott Wadsworth.

the radiation receiver, was exposed to a radiation source, its temperature changed resulting in a variation of its resistance and a current was measured by a deflection of the galvanometer needle.

In the course of this transformation process (see also Figure 4), the change in temperature was made visible by the deflection of the galvanometer needle. Already the first bolometer achieved a sensitivity of $(\frac{1}{100000})$°C.

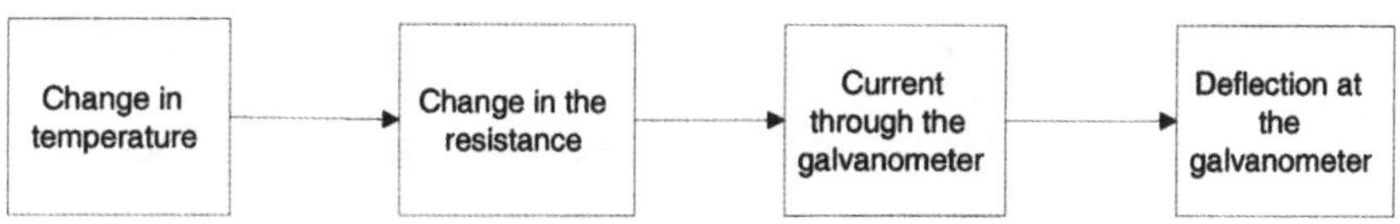

Figure 4. The transformation process by which the change in temperature at the radiation receiver became visible.

The unit made up of the bolometer and the galvanometer formed the nucleus of an apparatus which was called spectrobolometer. In principle, this apparatus differed from a spectrometer only in that, instead of a telescope or photoplate, the bolometer was used as a detection instrument. The sunlight coming from a heliostat was dispersed by a reflecting grating and later a prism. The bolometer was moved through the infra-red region of the prepared solar spectrum and changes in temperature were measured. The main question concerning the deflection of the galvanometer, which indicated the changes in temperature, was whether these deflections corresponded to absorption lines or were due to one of the diverse sources of error which were also made visible in these deflections.

LANGLEY'S LABORATORY NOTEBOOKS: THE ENERGY SPECTRUM BOOKS AND WASTE BOOKS. A SHORT DESCRIPTION

In his years at Allegheny Observatory when Langley was only assisted by Frank Very and by postdoctorals as James E. Keeler, who was in Allegheny from 1880 to

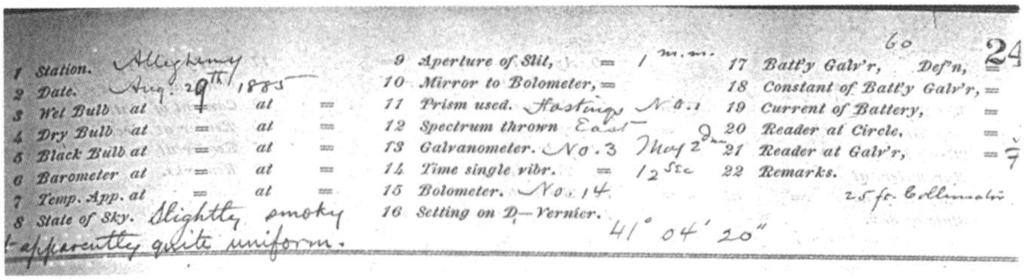
1 Station. Allegheny
2 Date. Aug. 29th 1885
3 Wet Bulb at = at =
4 Dry Bulb at = at =
5 Black Bulb at = at =
6 Barometer at = at =
7 Temp. App. at = at =
8 State of Sky. Slightly smoky apparently quite uniform.
9 Aperture of Slit, = 1 m.m.
10 Mirror to Bolometer, =
11 Prism used. Hastings No. 1
12 Spectrum thrown East
13 Galvanometer. No. 3
14 Time single vibr. = 12 sec
15 Bolometer. No. 14
16 Setting on D—Vernier. 41° 04′ 20″
17 Batt'y Galv'r, Def'n, 60
18 Constant of Batt'y Galv'r, =
19 Current of Battery, =
20 Reader at Circle, =
21 Reader at Galv'r, = 7
22 Remarks. 25 for Collimator

Figure 5. Example of the list at the beginning of the measuring records in the energy spectrum books.

1881 before he went to Herman Helmholtz in Berlin, bound notebooks were used with numbered pages and with a table of contents and graphical representation of the results at the end. These laboratory notebooks called *energy spectrum books* were specially printed for Langley's laboratory. Entries by Langley and Very were made by hand. The notebooks start with a summary of the results of the preceding year and an outline of the further aims of the investigations. At the head of every second page one finds a list (see Figure 5), where next to the station and date, external parameters that had some influence on the experiments, like the temperature of the laboratory, the pressure, the humidity and the state of the sky, were noted. Furthermore, parameters of the apparatus, the names of the reader at the dividing circle and the galvanometer were noted. At the end of the list space was left for remarks.

This list was followed by the object of the respective experiments. These ranged from:

- The determination of number and position of absorption lines in the infra-red region of the solar spectrum.
- The determination of absorption lines in the lunar spectrum.
- Analysis of the results of the measurements.
- Testing of the instruments and different parameters like the aperture of the slit, through which the sun light came in and the time of exposure of the bolometer. In order to test the sensitivity and stability of the bolometer, for example, measurements in the spectrum of a constant source of radiation were performed. Usually Leslie cubes, copper cubes filled with boiling water or another liquid like aniline, served for this purpose. Another test was the repeating of measurements in the infra-red region during which instruments like the bolometer were exchanged.
- Descriptions for arrangements of apparatus for different purposes.
- Measurements for calibrating the prism.
- Measurements to determine the absorption lines of H_2O and CO_2, which are both elements of the atmosphere. These measurements served for an assignment of the absorption lines, which were determined in the infra-red region of the solar spectrum.

After moving from Allegheny Observatory to the new astrophysical Observatory at the Smithsonian Institution, Langley established a second notebook, the so-called waste book, in addition to the energy spectrum books. This new notebook differed in its appearance from the energy spectrum books by the fact that the single pages did not have the described heads to note the external conditions and the state of the apparatus. Entries in the waste books include:

- Detailed descriptions of single instruments and suggestions for their improvements.
- Detailed analysis of different sources of error.
- Memoranda from Langley in which he commentated, for example, on suggestions for the improvements of instruments or made suggestions of his own.

This additional notebook was necessitated by the new research and working conditions in Washington. Langley's new position as director of the Smithsonian Institution, with its great range of different interests in science and art, let him very little time to be in the laboratory. He tried to coordinate the work in the laboratory by memoranda. The waste books offered a possibility for him to follow, to a certain degree, the activities in the laboratory. Furthermore, on the principle of the division organized laboratory, the waste books served in this context as a means of communication between the assistants.

DIFFERENT PATHWAYS OF DEVELOPMENT AS RECORDED IN THE NOTEBOOKS

The different objects of the energy spectrum books and waste books are connected with different pathways of development in the entire development of the experimental setup. As I will try to show, these pathways of development led into different regions of experimental practice.

Before starting to follow up extracts out of these pathways, I will introduce a classification of the different *main* pathways based on the different objects of investigation. Each of the main pathways include all the developments which were connected with the respective objects. These pathways possess different time scales and some of them were connected. The developments in these pathways depended on different factors, which were not constant and interacted with other developments. A description by main pathways of development represents, therefore, the coarse structure of the network.

The main pathways of development yielded by this classification are:

- The mapping of the infra-red region of the solar spectrum, (MIR)
- Optimization of the sensitivity of the instruments (OSI)
- Optimization of the stability of the instruments (OSTI)
- Tuning the instruments of the apparatus/Optimization of the sensitivity and stability of the apparatus (OSSA)

After these general remarks about the concept of pathways of development, I will discuss some examples of these pathways and show into which region of experimental practice they led, how these pathways led outside of the laboratory and where one encounters limits. Let us start with the pathway of development (MIR), the object of which was the determination of the number and position of the absorption lines.

Mapping the Infra-red Region MIR

As mentioned above, the greatest problem in mapping the infra-red region of the solar spectrum was that both the properties of the bolometer and the infra-red region were unknown. When the bolometer was used in the experiments as an infra-red measuring instrument, each deflection of the galvanometer needle had to be examined for being fact or artifact, and also every position of the potential absorption lines had to be carefully determined. The position of the absorption lines was problematic because of an error source which Langley called "drift," a continuous wandering of the galvanometer needle during measurements. In the drift, several different sources of errors became manifest, as, for example, variations of the external temperature, which affected the bolometer, and variations of battery current. The localization and elimination of these error sources, i.e. increasing the stability of the bolometer, was the main aim in handling the unknown properties of the bolometer and of the infra-red region of the solar spectrum. But this was a very difficult and complex process and formed a program for itself which was, as we will see, located in the instrumental region.

The first example I will give in this paragraph shows how the effects caused by the drift were to be eliminated in order to determine the exact position of the potential absorption lines. If there are real absorption lines was object of further investigation. The second example concerns the uncertainty connected with the expansion of the infra-red spectrum. At this time different inconsistent approximations made by scientists as Müller ($1{,}8\mu$) existed.

Figure 6 is a page from the energy spectrum book from 1885. The object of the measurements is noted in the first line: "Measures on Ω and ω_1 and ω_2," the determination of the position of these bands.

The first of the five columns in the measuring report, headed "Circle Reading," gives the deviation of the radiation measured on the dividing circle. The first of the following four, headed "Before," gives the galanometer deflection without exposing the bolometer to the radiation. The next one, headed with "Exposed," gives the galvanometer deflection by exposing the bolometer to the radiation and the next one headed "Return," gives the galvanometer deflection without exposing it to the radiation. In the last column, headed "Deflc.," the mean value of the differences between the first and second values and the second and third values is given. On the right side, the time of the measurement is noted, along with a remark concerning the deviation of the readings at which the measurements were performed.[7]

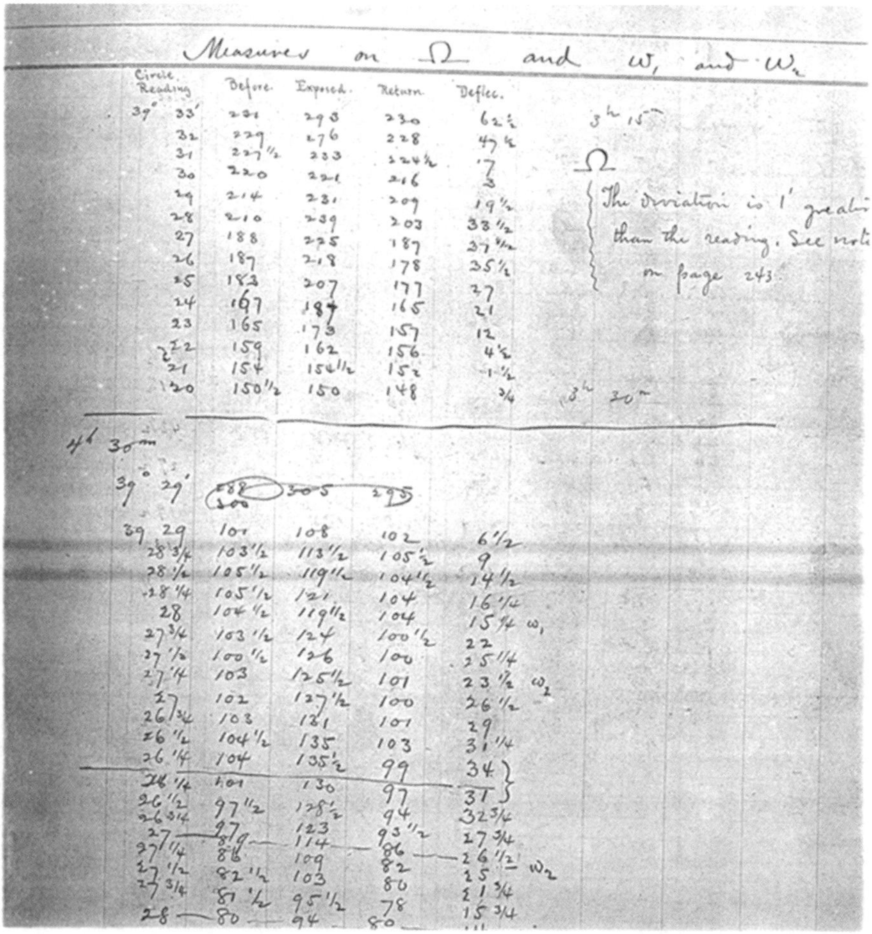

Figure 6. Extract from a measuring record in the energy spectrum book of 1885. Object of the experiment is the determination of the position of the absorption band Ω and absorption lines ω_1 and ω_2.

By taking the mean value of before-and-after readings the drift was taken into account. The zeropoint position, was determined in this way, for every single reading.

The measurements for ω_1 and ω_2 were performed, as the second record on the page shows, in the same way. Here the only difference is that they used smaller distances between the measuring points, resulting in a higher resolution.

Figure 7 shows the graphical representation of the measurements. The ordinate refers to the angles of the deviation of the radiation on the dividing circle, and the abscissa to the galvanometer deflection. At the deviation near 39.31, one sees clearly the minimum which refers to the band Ω. The remark on the right handside

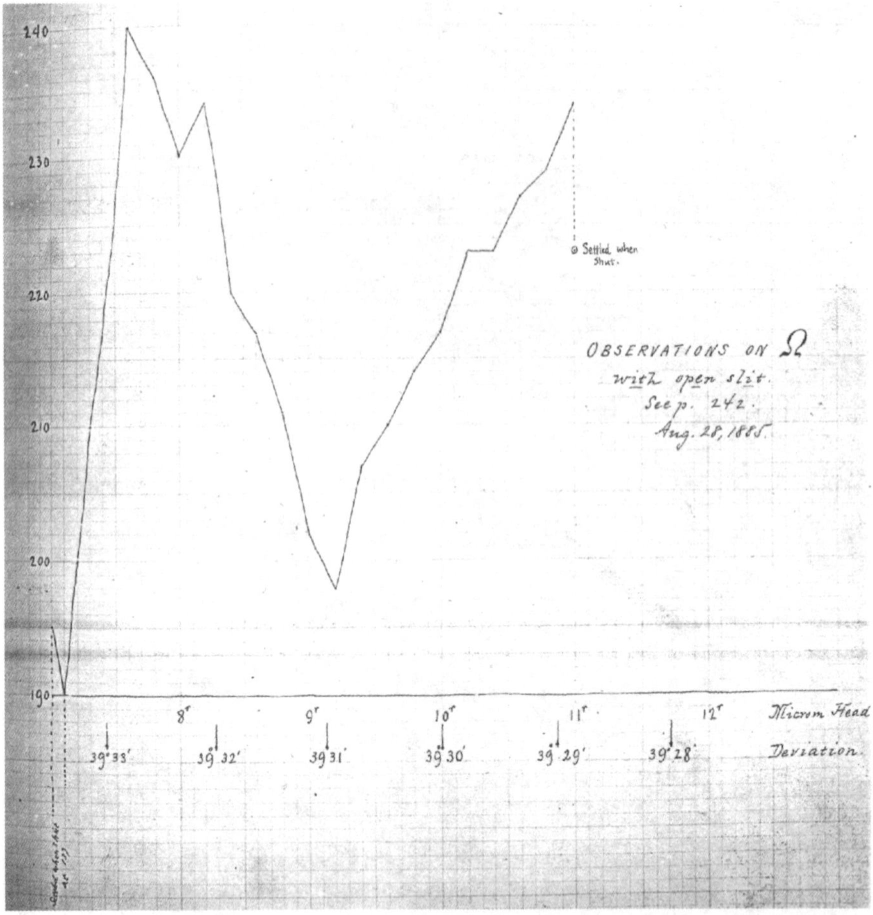

Figure 7. Graphical representation of the results of the determination of Ω

refers to the experimental method used. Instead of closing the slit through which the radiation came in between every single measurement, the measurements were performed with the slit open all the time. This was regarded as vaguer, but it was not so time consuming as closing the slit after every step and opening it again for the next measurement. The marks at the right and left side of the curve refer to the values measured without exposure before and after the measurement. With these values the contribution of the drift becomes clear.

Figure 8 is the representation of the measurements of ω_1 and μ_2. The remark on the curve: "Hunting for ω_2" shows that the indication of this band was very much smaller, and that meant that it was much harder to determine the exact position of this band and to distinguish it from artifacts.

The second example, the uncertainty of the end of the infra-red spectrum is illustrated by an extract from an analysis of a measurement in the energy spectrum

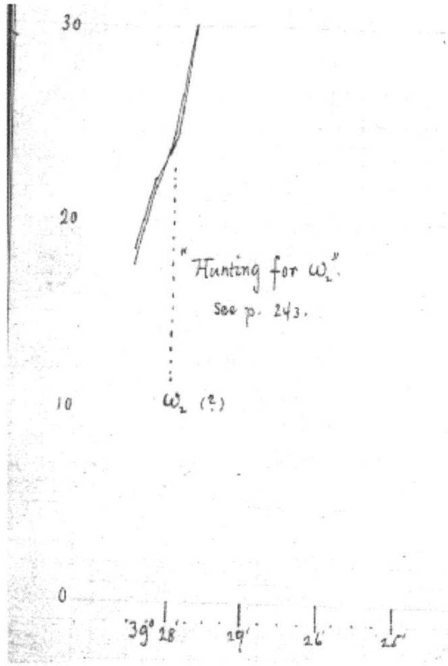

Figure 8. Graphical representaion of the results of the determination of ω_2.

book of 1882. The subject of this measurement was: "To trace the limits of the infra-red spectrum" and in the analysis of the measurements Langley wrote:

> [...] The result is an extension of the spectrum far beyond the limits at which it was supposed to end. The energy is however excessively feeble, indicating either that the original solar energy in these regions is very small, or else that the absorption is enormous. Further measurements will be necessary to decide which of these explanations is correct. The precipitous fall of the curve just beyond Dev. 44 according to all analogies drawn from other parts of the spectrum must include a new absorption band. If the curve does not rise again to its previous height we must conclude either the great extent of the cold band or the rapid diminution of the original energy.

Following this entry, Langley had reached a limit in the measurements for which, according to his discussion, the following two explanations were possible:

1. a broad cold band, which he concluded by the strategy of drawing analogies from other explored parts of the spectrum. This strategy one finds again and again in handling the new results in the unexplored region, whose trustworthiness was doubtful because of the fact, that the properties of the bolometer in the unexplored infra-red region were not determined.
   or
2. a strong decrease of the solar energy.

In order to decide which of the two explanations was correct the sensitivity of the radiation measuring instrument had necessarily to be increased further, which was also the only way to get a stronger indication for the band ω_2.

These two examples show how the unknown properties of the infra-red region of the solar spectrum are connected to the unknown properties of the bolometer. They show the uncertainty regarding the question which of the deflections actually corresponded to absorption lines, the exact position of absorption lines, if one had recorded all the absorption lines, and the total expansion of the infra-red spectrum. In order to eliminate the first and second uncertainties, the error sources had to be eliminated, i.e. the stability had to be increased. For the third and fourth uncertainties, the sensitivity of the bolometer had to be increased. Increasing the stability and sensitivity connected the pathways of development centered around the experiments for determining the number and position of the absorption lines with the pathways of development in the instrumental region, aimed at optimization of the sensitivity and stability of single instruments and the whole apparatus.

In the next section, I shall follow the connection into the instrumental region and take a closer look at the pathway of development concerned with improving the stability of the instruments of the apparatus OSTI.

THE PROGRESSION OF THE PATHWAYS OF DEVELOPMENT IN THE INSTRUMENTAL REGION AND THEIR CONNECTIONS

The Pathway of Development OSTI

One central part in the pathway of development OSTI concerned the localization, isolation and elimination of the different sources of the drift and the fluctuations which accompanied the increasing sensitivity of the instruments. These improvements revealed additional structures in the spectrum, which were hard to distinguish from the deflections connected with the absorption lines and which could not be averaged out by the strategy described in the last section. At this point one of the main characteristics of the developments in the instrumental region becomes obvious, namely, that sensitivity and stability stood in a competing relation.

The different sources of error, which were the object of the pathway under investigation, included external sources of error and sources of error which are located in the instruments themselves and in their connections. In the waste books, which were introduced after the move to the new astrophysical observatory at the Smithsonian Institution, one finds special programs concerned with the localization and isolation of the different sources of error.

Perhaps one might wonder at this point, why one does not find such programs in the energy spectrum books from the time at the Allegheny Observatory? There are two main reasons for this circumstance which were connected with the different organization of the laboratory in Washington, and with Langley's plan to automatize the measuring and registration process. The fact that the organization in Washington was based much more on the principles of division of labour, led to the above mentioned fact, that the waste books formed a means for communication between the assistants and Langley. The automatization plan for the measuring and registration process led to an intensified search for sources of error because, as the memorandum from Langley from the 1892 waste book shows, the elimination of the drift, was a requisite for the automatization. I will cite an extract out of this memorandum, because it also provides insight into the communication between Langley and his assistants.

> My idea (if drift could be eliminated) would be to have a vertical strip of sensitive paper rolled perfectly upward by a clockwork in the focus of the galvanometer mirror. The sides of this paper are marked in degrees and minutes, corresponding to the divisions of the spectrobolometer circle whose arm is moved by the same clockwork (through electric or other intermediary) so that when the circle is moved through n minutes of arc, the paper is moved upward linearly by a quantity corresponding to the same angular measure. A light is reflected from the mirror on to the paper on which are traced the movements of the mirror due to the varying heat of the spectrum and to passing inequalities of sky transmission.

In the same year one also finds an entry of one of Langley's assistants, Frank Lawton Olcott Wadsworth, who had just joined the laboratory and who had assisted Albert A. Michelson with his experiments for the determination of the meter performed in Paris at the *Bureau International des Poids et Mesures* from 1890 to 1892. The entry starts with the words:

> When the galvanometer was connected with bolometer it deflected badly so that it was impossible to get any comparison readings. Must study to eliminate drift in present galvanometer.

This statement is followed by an analysis of the different causes of the drift which became visible by the deflections of the galvanometer. I will cite some passages out of it in order to give a further impression of the form and function of the waste book, and also into the problems which were connected with the drift.

> The magnetic drift of the galvanometer is undoubtedly very small.
> It is caused:
> 1) by daily changes in the value of *H*
> 2) by changes in temperature, in the masses of iron in the laboratory
> 3) by the moving of small masses of iron, such as keys etc.
> All these disturbing effects can be almost completely neutralized by enclosing the galvanometer in a heavy sheet iron case of suitable thickness placing weak controlling magnets inside the case.
> All the effects due to these disturbing causes are in general small compared to the other drift effects which are due to thermal effects.

> Any change of temperature in the bridge relatively to the two arms of the bolometer will cause drift in two ways.
> First, because the two balancing arms of the bridge are never affected equally by changes of temperature because of slight differences in the wire character [...] Working with a very fine bridge enclosed in a double box with all four arms of the bridge inside for purpose of testing coils I have seen a very large deflection produced simply by opening a window of the room and allowing the air (a comparatively feeble breeze) to blow in.
> Second, changes are also caused by the fact that the two arms of the bolometer are not accurately balanced against each other and in fact cannot be without the use of an external resistance. [...]
> Bad contacts in any part of the circuit
> This seems to be by far the most fruitful source of trouble and the one most difficult to eliminate.
> 1) At begin every possible contact which is not absolutely necessary should be dispensed [...]
> 2) Use preferable mercury contacts such as used on all low resistance bridges work testing standard ohm coils [see Glazebrooks papers] etc.
> 3) Pin contacts at bolometers are especially troublesome [...].

Wadsworth listed in this entry different sources of the drift and different parts of the apparatus and instruments which were affected. If one takes into account that at the beginning, in Allegheny, only the external variations in temperature and fluctuations in the battery current counted as sources of error, this entry shows that the experience and results gained in the experiments and in dealing and working with the instruments had led to a deeper insight into the instruments, their functioning, and their connections. The instruments and the apparatus were far from being black boxes.

This very complex part of the experimental practice, the dealing and working with instruments, took place in different places and is often neglected by historians of science. Only those which were performed in the laboratory by, for example Wadsworth, are documented in the notebooks. At this point, one reaches a limit of the notebooks with respect to the experimental practice. One can follow the more and more differentiated search for sources of error but one cannot follow up in the waste books the pathways which led from this point out of the laboratory. But what one finds are hints, like Wadsworths reference to Richard Glazebrook through which, with the help of additional sources, one can follow pathways in our example from Washington to England, to the *Standards Committee of the British Association* at the Cavendish Laboratory in Cambridge, whose director was Glazebrook. About the tasks of this Institution, Glazebrook wrote: "[...] that I as secretary of the B. A. Standards Committee, should undertake to test and issue certificates for resistance coils [...]." In order to understand the connection to this laboratory, one has to know that similar to the measurements in the infra-red region of the solar spectrum for the precise determination of the electrical resistances of coils, very highly sensitive and stable galvanometers were needed. With this knowledge, one gets an additional viewpoint for reconstructing the activities in the laboratory. It hence becomes clearer from where Langley and his team got some ideas for changes and improvements in their instruments.

Other hints led into the shops of instrument makers: in the present case to the shop of the instrument maker William Grunow in New York, who manufactured the first bolometers for Langley and worked on these instruments until about 1896. Fortunately, a great part of the correspondence between Grunow and Langley has been preserved. This correspondence gives sense of the great fragility of the instrument and the difficulties associated with this fact. Fixing the very thin metal strips required a lot of patience and skill. This becomes obvious in a letter Grunow wrote to Langley in 1893:

> [...] I feel sorry to perceive my inability to follow up the making of bolometers, on account of the circumstances of my situation, the bad effect on my health (eyes and nerves) caused by the anxiety which the making of bolometers always creates on me, and that I should give up the making of them, rather then continue without being able to improve or perfect them, [...].

This correspondence also depicts Grunow's situation as an instrument maker in the U.S. at this time. Due to the fact that American scientists gained access to the European market through some business-minded instrument makers, those like Grunow, who had only a small business and limited capacity, lost orders and often had to give up their business.

This example shows, that in the present case one can get information about the little examined role of the instrument maker in such a research project, and, furthermore, about the context in which he and his work were embedded.

One of the major improvements Grunow made to stabilize the bolometer, was to provide the instrument with a solid jacket to protect it from external changes of temperature and avoid internal air currents. Neither disturbance was eliminated by this change. In 1892, this construction was extended by a constant water flow through the case, following an idea of the predecessor of Wadsworth, William Hallock, and which was also implemented by Grunow. Figure 9 shows such a water-cooled bolometer. Hallock commented in the waste book after the first tests of his construction with the enthusiastic words: "I killed the drift." Later experiments and the use of the device in the experiments in the infra-red region showed, however, that this was an illusion.

Rather then going in further technical details here, let me just mention that for the optical part of the apparatus, i.e. prism, lenses and mirrors the situation was similar to that of the bolometer. These parts were manufactured by John A. Brashear (1840–1920) who had his shop near the Allegheny observatory, and not by chance, because Langley helped him to build up his business. Also here the correspondence is preserved. Apart from the different objects Brashear and Grunow worked on, Brashear's story was much more a story of success. His

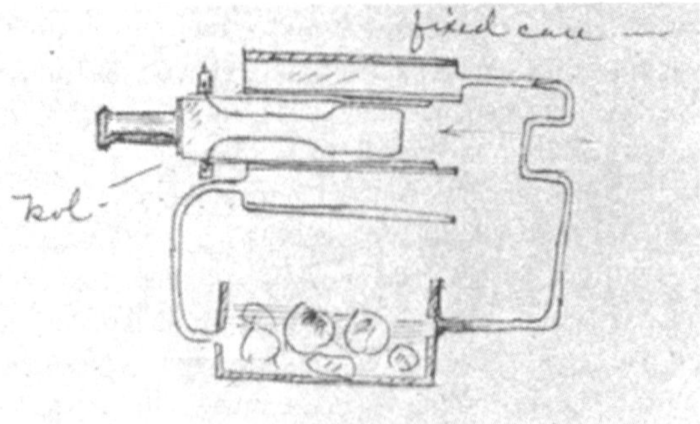

Figure 9. Sketch of a water-cooled bolometer. The sketch shows a bolometer ("*bol.*") enclosed by a *fixed case* connected with a water bath.

business grew, while Grunow's lost more and more orders. Other instruments incorporated in the apparatus, such as, the galvanometer, were modified and improved in the laboratory by the assistants.

All instruments – the bolometer, the optical instruments as well as the galvanometer – were in respect to their stability tested, after the changes in special experiments which were performed for this purpose. For example, the galvanometer was connected alone in an electric circuit and the vibrations of the galvanometer needle, due to the different causes were registered. Figure 10 shows some examples of such records. These additional very fine deflections made it very difficult to distinguish between deflections due to absorption lines and those which were accidental. Beyond the analysis of the stability of the galvanometer, these curves were also used in the analysis of the results where they served as reference curves. They were put underneath the intensity curve, the so-called bolograph, the output of the automatic registration procedure, and the structures of the two curves were compared.

Now I will come to the pathway of development OSI which was concerned with the improvement of the sensitivity of the instruments.

The Pathway of Development OSI

As discussed for the case of the stabilization of the instruments and illustrated by the experiments to explore the limits of the infra-red spectrum, it was mostly the results of the experiments in the infra-red region that provided the source for the improvement of the sensitivity. A further parallel to OSTI was that the improvements led into the shops of instrument makers and into other regions of research. And furthermore, the improved instruments were tested in special experiments. One example for such a test was to create the experiments, in the spectrum of a constant source of radiation, like the Leslie cube. Figure 11 shows a sketch of this arrangement from the energy spectrum book of 1885.

One sees the bolometer in front of such a Leslie cube filled with boiling water.

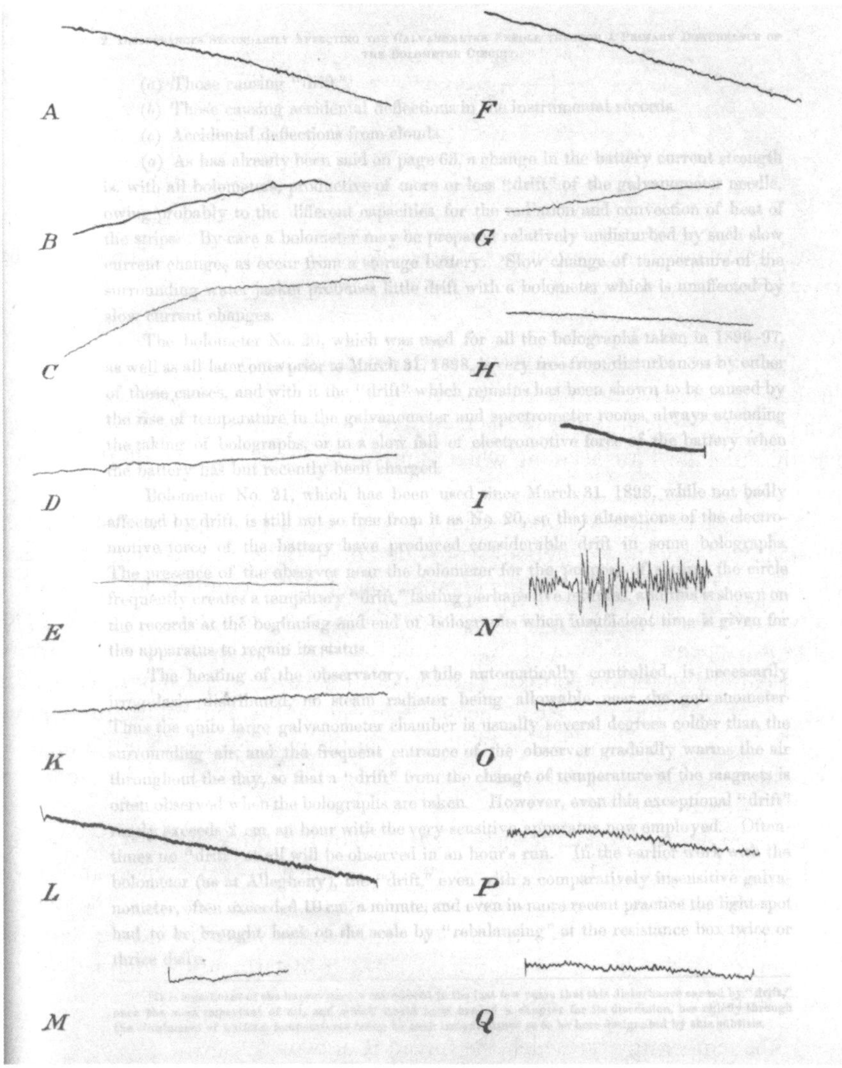

Figure 10. Galvanometer and battery records, 1895 to 1898.

The bolometer was connected with a galvanometer. The deflection on the galvanometer was noted and compared with the deflections of other bolometers produced under the same conditions. This sounds very simple and reliable but was not at all. Difficulties arose mainly for two reasons. First, as stressed in the last section, the galvanometer and bolometer were highly sensitive instruments with respect to external disturbances. So it was nearly impossible to reproduce the same conditions. Second, the sensitivity of the bolometer was due to the fact that the deflection of the galvanometer which was at the end of the chain of the

Figure 11. Bolometer in front of Leslie cube filled with boiling water. This arrangement served for the determination of the sensitivity of bolometers.

transformation process, made the difference in temperature visible, defined through the sensitivity of the galvanometer. The galvanometer was also improved with respect to its sensitivity in the course of the investigation, so that one always had this to take into account when comparing the sensitivity of the different bolometers.

In the last example of this section, I will take a look at the question what information laboratory notebooks provide about skills and the activities in the laboratory. This brings me to the comparison of the laboratory notebook with a keyhole. By looking through a keyhole one observes something that takes place just as it happens. The form of the keyhole determines what one can see. Here we have a parallel to the notebooks. The form of the keyhole is comparable with the notes of the experimenter. Here one also sees only a fragment, but here we have a fragment from something which took place in the past. So we do not look at the experimenter over his shoulder by reading his notebooks, we cannot follow his activities step by step.

An example of this circumstance is the following entry in the waste book of 1891, in which one of Langley's assistants recorded his experiences in drawing quartz threads. These threads were used to suspend the galvanometer needles and mirror in the Thomson mirror galvanometer, which they used in their experiments. For him, the following considerations seemed to determine the diameter of the quartz thread:

> 1. the temperature at which the thread is drawn,
> 2. the speed of the arrow,
> 3. the diameter of the little sphere formed in the melted quartz rod.
>
> To get the finest possible fibre the quartz rod should not be over 0.5 mm in diameter, the temperature should be raised to the highest possible point. The exact instant for discharging the arrow is when the rod is about to drop in two by its own weight. The speed of the arrow should be as great as possible with rubber bands for the motion power. Of these three considerations I should say the last (3) is of most consequence.

This cryptic description is only understandable for someone who is familiar with this process of drawing quartz threads. There is no description in the waste book of how to do it, because it was not necessary, the other assistants already knew how. The assistant did sum up his experiences with the process and did put it forward to the others in order to improve the drawing of quartz threads. A historian of science who reads this description more then 100 years later with the aim to reconstruct and understand the experimental practice in the laboratory, and who is, furthermore, unfamiliar with such techniques, will first need to find more information, from other sources. A hint in the waste book led to V. Boys. In 1882 he invented the method for drawing quartz threads. A short description of this method which one finds in the literature makes the entry in the waste book understandable: "Clamp the one end of a quartz rod and attach the other one to the bolt of a crossbow which is launched when the rod softens in a gas flame." This almost adventurous description seems to clarify what the assistant wrote, and one gets an idea of how skilled the assistant had to be. But something is missing, and this is a deeper insight into the practice. It is not possible on the basis of the laboratory notebooks to reconstruct the assistant's drawing of the threads, his first tasks, the process in acquiring the necessary skills and the gathering of experience. All these aspects are hidden from sight by the view through the keyhole.

When the tests confirmed that the changes in the instruments increased sensitivity, respectively the stability of the apparatus, in the next step the instrument was tuned to the other instruments of the apparatus.

The Pathway of Development OSSA

Both pathways of development, the one which included the optimization of the stability (OSTI) as well as the one which included the optimization of the sensitivity of the instruments (OSI), were connected with a pathway of development aiming at the optimization of the stabilization and sensitivity of the entire apparatus by tuning its different instruments. This pathway of development splits, as the laboratory notebooks, specially the waste books show, into two interacting pathways. In the first one, because of the fact, that the changes of the bolometer and galvanometer with respect to their stability and sensitivity were performed in different places by different people, the more or less independent developments of these two instruments, came together. In the second one, the other instruments were tuned to the stability and sensitivity of the detection unit of bolometer and galvanometer.

In the case of the tuning of the bolometer and the galvanometer, it became again obvious that the sensitivity of the bolometer could only be defined in terms of the range of the sensitivity of the galvanometer. If changes on the bolometer didn't meet the aspired aim, an increase in the sensitivity, this could mean two things: first the changes didn't lead to an increase in the sensitivity or second the sensitivity of the galvanometer was too low to reflect the improvement. On this point one was all the time confronted with high uncertainty.

Objects of the second pathway were, for example, to tune the width of the slit and prism on the sensitivity of the detection unit. The measuring records show a decreasing of the width of the slits with increasing sensitivity. This resulted in a higher resolution in the spectra. More absorption lines became visible. In the case of the prism, the search for materials with a high transparency in the long wave part of the spectrum started. The notebooks show that Langley started to use rock salt prism, which had the disadvantage that they are very sensitive to moisture. They had to be repolished at regular intervals and their optical constants had to be redetermined. This again led to uncertainty, because in the analysis of the results these circumstances – the changing of the optical constants – had to be taken into account, but there was no reliable method available.

CONCLUSION

The case study of Langley's experiments shows that laboratory notebooks offer the opportunity of gaining an insight into experimental practice. By working out the different pathways of development one gets a very complex and dynamic network in which the production of scientific knowledge took place. This network did not include all the different aspects and factors on which the research process depended, but does go far behind what one finds in publications. This network is not restricted to the laboratory, it led us out of the laboratory to the workshops of the instrument maker, into other laboratories and into the historical and disciplinary context.

In the case study under investigation an important, often neglected part of the experimental practice, the meaning of the development of instruments became clear. In the process the instruments themselves became objects of an investigation, which was strongly connected to the experiments performed in order to map the unexplored infra-red region of the solar spectrum. They were changed and improved depending on the experiments for which they were designed. All this took place in a complex and dynamic stabilization process, in it various interconnected pathways of development build up a multi-layered network of forces and constraints.

ACKNOWLEDGMENTS

Research on this project was funded by the Deutsche Forschungsgemeinschaft. I wish to thank the Institute for the History of Science at Göttingen University, the Smithsonian Institution Washington, and the Max Planck Institute for the History of Science in Berlin for providing a stimulating and supportive working atmosphere. It was Klaus Hentschel who made this research possible by his support, advice, and suggestions. I also wish to thank Falk Ries and the members of the history of science group at Oldenburg University and Tilman Sauer for many fruitful and stimulating discussions, Frederic Holmes for valuable comments and criticism on a draft of this paper, and Hans-Jörg Rheinberger for the possibility to work in his department.

NOTES

[1] Furthermore Langley's measurements of the distribution of heat in the spectrum of an artificial source of radiation, which as we will see, he used for the calibration of the bolometers, created an important contribution in the research of the black body radiation formula (Kangro 1970).
[2] The map of the infra-red region is accessible as pl. XX at http://adsbit.harvard.edu/books/saoann/.
[3] These are part of the results of an extensive case study which will be published separately.
[4] The fact of the lack of an university training may seem very unusual but it wasn't. John Lankford shows in his book *American Astronomy, Co-unity, Careers and Powers, 1859–1940* (Lankford 1997), that the lack of an university training did not prevent from becoming a professor for astronomy before World War II in the US.
[5] The position of the last visible Fraunhofer line is at 0,76μ.
[6] $1\mu = \frac{1}{1000}$ mm.
[7] A remark on page 243 of this notebook shows, that the deviation of 1' refers to the setting in the morning. Usually they used Fraunhofer lines as calibration points and checked them several times during the measurements.

REFERENCES

Dörries, Matthias (1994), "Balances, spectroscopes and the reflexive nature of experiments," *Studies in History and Philosophy of Science* 25.

Gooding, David, ed. (1987), *The Uses of Experiments: Studies in the Natural Sciences* (Cambridge: Cambridge University Press).

Glazebrook, T. R. (1910), "The Rayleigh Period," in *A history of the Cavendish Laboratory* (New York, Bombay, and Calcutta: Longmans Green, and Co).

Hentschel, Klaus (1998), *Zum Zusammenspiel von Instrument, Experiment und Theorie* (Hamburg: Verlag Dr. Kovac).

Herschel, John (1840), "On the chemical action of the rays of the solar spectrum on preparations of silver and other substances, both metallic and non-metallic," *Philosophical Transactions* 130, p.1.

Holmes, Frederic L. (1992), "Do we understand historically how experimental knowledge is acquired?," *History of Science* 30.

Jones, Bessie Zaban (1965), *The Lighthouse of the Skies* (Smithsonian Institution: Washington, D.C.).

Kangro, Hans (1970), *Vorgeschichte des Planckschen Strahlungsgesetzes: Messungen und Theorien der spektralen Energieverteilung bis zur Begründung der Quantenhypothese* (Wiesbaden: Steiner).

Lamansky, Sergie (1827), "Untersuchungen über das Wärmespectrum des Sonnen- und Kalklichtes," *Annalen der Physik* 146: 200–233.

Langley, Samuel (1883), "The Selective Absorption of Solar Energy," *American Journal of Science* 25: 169–183.

Langley, Samuel (1884), *Researches on solar heat and its absorption by the earth's atmosphere. A report of the Mont Whitney expedition* (Washington: Government Printing Service).

Langley, Samuel (1900), *Annals of the Astrophysical Observatory of the Smithsonian Institution*, vol. 1 (Washington: Goverment Printing Office).

Lankford, John (1997), *American Astronomy* (Chicago: University of Chicago Press).

Latour, Bruno, and S. Woolger (1986), *Laboratory Life: The Construction of Scientific Facts* (Princeton: Princton University Press).

Latour, Bruno (1987), *Science in Action: How to Follow Scientists and Engineers through Society* (Cambridge: Cambridge University Press).

Müller, Johann H. J. (1858), "Untersuchungen über die thermischen Wirkungen des Sonnenspektrums," *Annalen der Physik* 11: 337–359.

Pickering, Andrew (1984), *Constructing Quarks: A Sociological History of Particle Physics* (Chicago: Univesity of Chicago Press).

Pickering, Andrew (1989), "Editing and Epistemology: Three Accounts of the Discovery of the Weak Neutral Current," in L. Hargens, R. A. Jones, and A. Pickering, eds., *Knowledge and Society: Studies in the Natural Sciences* (Cambridge: Cambridge University Press), pp. 275–97, 989.

Pickering, Andrew (1995), *The mangle of practice: time, agency, and science* (Chicago: University of Chicago Press).

CHRISTOPH HOFFMANN*

THE POCKET SCHEDULE
Note-taking as a Research Technique: Ernst Mach's Ballistic-Photographic Experiments[1]

I

In the middle of the 1880s Ernst Mach (1838–1916), then director of the Physical Institute of the University of Prague, and Peter Salcher (1848–1928), professor of mechanics at the Austrian Naval Academy in Fiume (Rijeka), started a research-project on the phenomena in the air caused by a rifle-projectile moving at supersonic speed. The most spectacular result of the experiments was a set of schlieren-photographies, which depicted the bullet in flight surrounded by a system of shock-waves at its tip and body. Less spectacular but more important in the long run was Mach's and Salcher's observation that the speed of sound is a fundamental threshold for all dynamical processes in gases, first published 1887 in their paper presented to the Imperial Academy of Science in Vienna. Mach-number, Mach-angle, Mach-cone and some more terms still reminds us of the great importance this discovery had for the development of gas dynamics and experimental ballistics in the 20th century.[2]

The subject of the following is neither Ernst Mach's place in the history of physics, nor Peter Salcher's lack of place.[3] Instead Mach's research-notebooks will become the leading part of the story. Small non-human actors as they are, notebooks normally are regarded as external memories, which can be read as mere passive reflections of experimental operations or cognitive processes. At least with respect to Mach this simple understanding of the process of noting does not completely correspond to its function in his research. In fact Mach's notebooks not only hold together the several dimensions of everyday-work in a physics institute. But note-taking was also an important tool in the emergence of an epistemological framework in his research. As Mach's notes on the ballistic-photographic experiments will show, the process of noting itself possesses a performative power. Therefore I would propose that a reader of today should not firstly perceive these writings as the (missing) link between experiments and publications or theoretical concepts and cognitive processes, but as a technique of science, which has to be analysed in its effects on the knowledge-production in

* Max Planck Institute for the History of Science, Berlin.

Frederic L. Holmes, Jürgen Renn and Hans-Jörg Rheinberger (eds.), Reworking the Bench: Research Notebooks in the History of Science, 183–202 © *2003 Kluwer Academic Publishers. Printed in Great Britain*

the same way as – for example – the effects of photography or the differential calculus.

II

When Ernst Mach took up the Chair of Experimental Physics at the University of Prague in the spring of 1867, he found that his working conditions had thoroughly improved.[4] Unlike in his first post as "Privatdozent" at the University of Vienna or later as Professor of Mathematics and Physics at the University of Graz, he now had a budget for equipment, however small, at his disposal. Still more important, he could count on the support of an assistant, a mechanic, and soon afterwards a second assistant for his experiments. For the time being, his research facilities were somewhat less favourable. Before the Physics Institute moved on Mach's initiative into a new Natural Science Institute building at the start of the 1880s, two rooms in a university accommodation block had to suffice for experiments. At least from then onwards costlier physical experiments, other than the cheap physiological self-experiments to which Mach had seen himself limited, became possible.

From the point of view of his experiments – not to mention his other, in particular, theoretical work – Mach's decades in Prague until his move to Vienna in 1895 were thus extremely productive.[5] It was also a period in which Mach saw his working time divided between diverse tasks – however plain and simple the enterprise he was directing may in fact have been. Lectures had to be held, courses had to be planned, apparatus and materials had to be purchased, new funds had to be obtained, he had to participate in faculty business. Two periods as Rector, in 1879 and 1883, demanded all his attention, he had to engage in negotiations with the Vienna Academy and the ministry, and finally articles for the liberal press in Prague and Vienna were also in demand. These obligations were not without consequences for Mach's way of working: research, above all, meant under these circumstances primarily directing experiments, and allowing planned experiments to be carried out by doctoral students or co-workers. It could certainly no longer mean carrying out all laboratory work himself, following all projects down to the last detail with his own eyes. A glance at Mach's bibliography confirms it: almost without exception all experimental works from the Prague period are co-authored. But the "institutional revolution"[6] in the natural sciences of the outgoing nineteenth century led not only to new ways of organising and publishing. In Mach's case, this sort of science also initiated a particular way of administration and of coping, which kept together the different activities and events which occur in the running of a physics institute. This schedule's physical manifestation was, as a rule, some 14 cm x 20 cm, between one hundred and two hundred pages in length, and fitted neatly into the pocket of the gown of a professor of the Austro-Hungarian Empire: Mach's notebooks.

The catalogue of Mach's papers, which are kept in the archive of the German Museum in Munich, specifies a total of fifty-three of these slim books, covering a forty-year research period. The chronology shows that the note-taking begins in

Prague, with Mach's new sphere of activity. The first of the notebooks is dated around the beginning of the 1870s, exactly forty more follow until the move to Vienna, and the last is begun in 1910. At this time the by now famous scholar had, because of a stroke, for many years only been able to write with his left hand. This misfortune seems finally to have brought the note-taking to an end; at any rate, from 1910 until his death in 1916, Mach preferred to write with a typewriter. However, thirty years earlier, when the pocket schedule was at full strength, things looked rather different. Then the books accompanied their master to the most diverse places: to the laboratory, on travel, and in the study. There was space in them for juxtaposing reported data and observations, throwing together arrangements for experiments with unsolved questions, commingling beginnings of letters and bibliographical references, or recording a book borrowed by a student. This was also the obvious place for the reminder: "buy notebook!"[7]

How Mach's research comes to life again on the pages of his notebooks is shown on a glance at the little volume begun on "11 Jan 1887," today number 25 in the archive. At the top of one page a train connection between Prague and Laibach (Ljubljana) is noted (Figure 1).[8] Underneath is the detail of the arrangements for an experiment, accompanied by notes and commentary. Finally there is a calculation, which takes up the rest of the page. Each of these notes indicates different aspects of scientific activity: organisation, experimentation, and data-processing follow one another randomly and without gaps. But it is not only the manner of the notes which is incoherent. They also point to different research projects, or at least into different areas of one research enterprise. The train connection appears in the context of a journey Mach made to Fiume in April 1887. There he would visit Peter Salcher, his co-worker in the just finished ballistic-photographic experiments. However, as the records during his stay reveal, Mach's interest was not in projectiles flying in the air, but schlieren, flows, and waves in water. These sorts of things are already mentioned some thirty pages before, only there they are not treated – as we will see later – as the events of experiments, but as the epistemological framework in which the research on projectiles receives a theoretical context.[9]

Whilst the note of a train connection lets one read of the separating off of a new field of experimentation, the sketch below it belongs to an experiment which appears in the most diverse places throughout this notebook. To the extent that the notes can be interpreted, Mach intended to investigate the effects of a gas flow which is led through a narrow channel. According to Mach's sketch, the upper border of the channel forms a metal "double roof [Doppeltes Dach]," in which an exactly-calculated amount of water is enclosed. A thermometer is carefully placed in the channel itself. The sketches and notes thus lead to measurements of the mechanical equivalent of heat, as this had been carried out since the mid-nineteenth century. They show, in the midst of original experimental work, the presence of Mach's ongoing interest in the theory of heat and thermodynamics in general. In the calculations which follow, the writer returns to the projectile

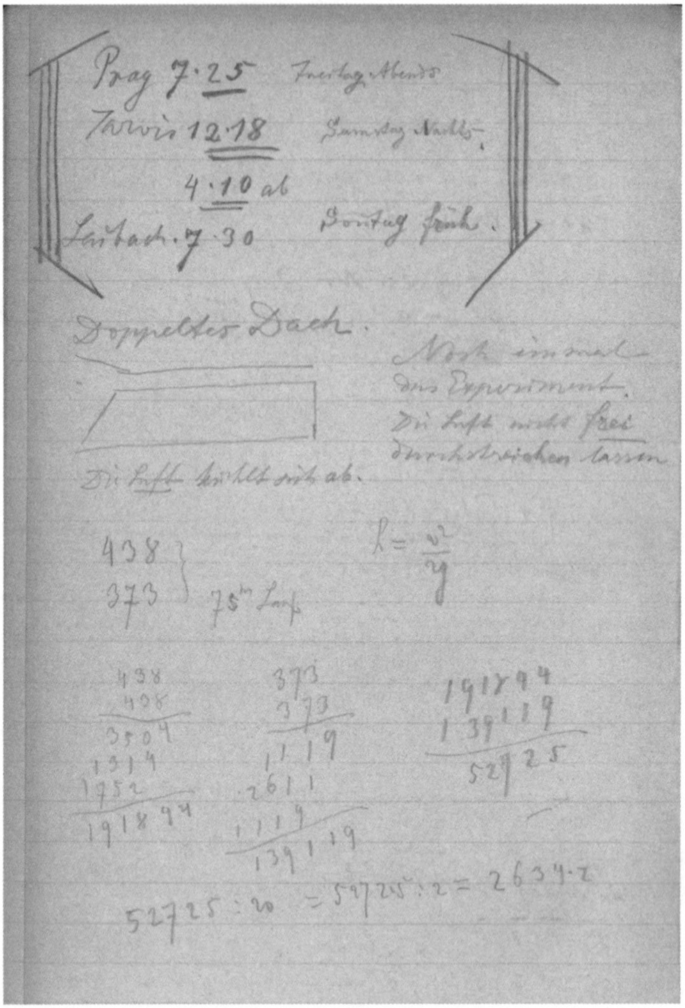

Figure 1. Ernst Mach, Notebook 25, p. 124 *(Archive of the Deutsches Museum, Munich)*.

experiments. One aspect of them occupies him, one which in the experiments themselves seemed to be only a side issue: the question of the resistance encountered by the projectile during its flight. The data employed are taken from a letter of Salcher's. On the other hand, the formulae used are from contemporary ballistics textbooks. With the origin of the notes, the relationships into which Mach's investigation enters is revealed: the representation and theory of the shock-waves around projectiles is bound to empirical work done on the firing-range.

If one wants to name the rule according to which this succession of words, drawings, and numbers coheres on a single page, one can only say that they have been brought together there one after another. Whether this occurred over one day or over several days must remain undecided, since the time-span which the notebook entries cover is totally unclear. They attest to the coexistence of the different dimensions and projects of research, or, opening at another page, the coexistence of research, teaching, and the politics of science. If one follows the linkages with other projects and spaces of knowledge[10] to which the notes point, and if one looks into their origins, similarities, and temporal distribution across the pages, then the proximity of the entries makes them readable as simultaneous events in Mach's work-life. For example projects which appear as a simple succession or as totally unconnected when looked at from the outside would now perhaps reveal new, unexpected references. That exhausts only one aspect of this pocket schedule, however. Two covers, usually cardboard, hold together an academic life. But even more than that, on the pages between these cardboard covers, interconnections and constellations emerge in which experiments appear part of an agenda, and results appear in context. One gets an impression of this function of the notebooks when one makes the form of the notes, their clashes, and the nature of their arrangement the centre of one's attention. A good example is provided by the ballistic-photographic experiments, which made Mach's name in the twentieth century.

III

Looking back at the starting point of this work, Mach remarked:

> In 1881, on hearing in Paris the lecture of the Belgian artillerist Melsens, who hazarded the conjecture that projectiles travelling at a high rate of speed carry masses of compressed air before them which are instrumental in producing in bodies struck by the projectiles certain well-known facts of the nature of explosions, the desire arose in me of experimentally testing his conjecture and of rendering the phenomenon, if it really existed, perceptible. The desire was the stronger as I could say that all the means for realising it existed, and I had in part already used and tested them for other purposes.[11]

It is true, the "set-up" of the resulting experiments consisted in tried and trusted elements: Leyden jars, spark discharges, the schlieren process for optically representing a medium's density differences, and projectiles had already, in another arrangement, been the basis for Mach's older work on shock-waves. A camera was the only new addition, which, in place of the human eye, would register the schlieren picture.[12] But contrary to the impression bequeathed us in Mach's description, the examination of Melsens's hypothesis appears anything but trivial. The first tests in 1884 and 1885 did result in photographs of the flying bullet, but there was no mass of air to be discovered at its tip.

The reason why quickly becomes evident: the bullets of the pistol used for the Prague experiments only reach a relatively slow speed; one needs modern weapons which reach a higher speed. With things standing this way, Mach asked Salcher at the end of January 1886 to repeat the tests with rifle bullets, which fly distinctly

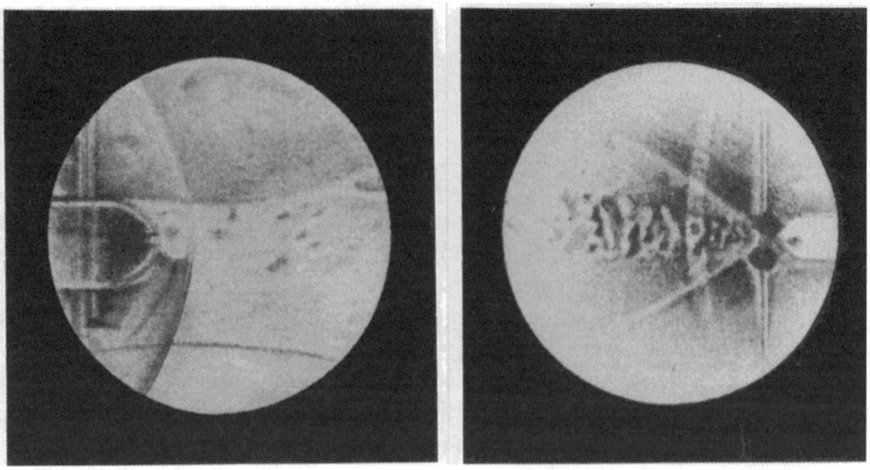

Figure 2. Left: head wave in form of hyperbola. Right: wave strips at the base of the projectile and the structure of the turbulent wake downstream in the path of fire. Photolithography, enlargement, originally circa 2 cm × 2 cm (Ernst Mach and Peter Salcher, "Photographische Fixirung der durch Projectile in der Luft eingeleiteten Vorgänge," *Sitzungsberichte der Mathematisch-Naturwissenschaftlichen Klasse der Kaiserlichen Akademie der Wissenschaften* 95, 2 (1887), plate, figs. 2 and 6).

faster than the speed of sound. An attempt, which up till then had been without a positive result, was now apparently leading to a happy end. Mach was already able by the end of May 1886 to report to the Viennese Academy of Science on the first of Salcher's photographs of the sought-after "compressed mass of air."[13] Experimental work admittedly remained to be done. Still the photographs showed "some details, the secure interpretation of which must await further tests."[14] Their explanation was to lead to a completely different object of research than the one that Mach and Salcher supposed. In the introduction of Mach's and Salcher's report from 1887 "masses of air" were still only mentioned in citations from earlier publications. What appeared was from now called a "kind of stationary sound-wave" [15] or "stationary flow" [16], and a whole system of such appearances around bullets flying faster than the speed of sound emerges (Figure 2).

This result was not only unforeseen, it was in the strictest sense the "unprecedented event" of the experiment itself.[17] Only with a progressively better arrangement of the experiment and with the switch to a higher projectile speed did the nature of this apparent "mass of air" acquire a characteristic image: "The whole process appears to have many similarities with the movement of waves created by a moving ship," remarked Salcher in December 1886.[18] And only after this connection had happened, the phenomena could be treated as cases of new knowledge on the border between acoustics and fluid mechanics which, up till then, had been developed more or less wholly as a theory of hydrodynamics, i.e. a theory of the flow of liquids.

An expansion in correspondence helped to maintain this experimentation. Salcher sent Mach nearly forty letters within twelve months, and not many less were sent in the opposite direction.[19] But above all several dozen photographic plates, this intercourse's real conditions of possibility, went from Fiume to Prague. For only fixed and transportable pictures of these most fleeting appearances permitted the separation in time and space of experiments in Fiume and discussions of the results in Prague. Such "immutable mobiles"[20] keep science going, and change the course of science itself: they allow divisions of labour over large distances, and dissolve the chronotopical connection between experiment, discussion, and model building. As a result the evaluation of observations and data can be dealt with at any time, so long as discrete processes and torn apart events are stuck together elsewhere.

In early February 1887 Mach once again received mail from Fiume: the almost thirty plates are for the time being the last that Salcher will make up. The site of the tests, an old shed in the grounds of the Marine Academy, has been pulled down a few days earlier.[21] As a consequence the days of shooting and photography are over, and the moment reached for publishing a major treatise. The photographs had barely arrived when Mach wrote to Salcher: "I would like to check all the plates, prepare a draft, and send you the plates along with the draft."[22] Mach's notebook number 25, mentioned above, formed the medium for this work. Shortly before the middle of the volume there appears a torrent of notes, running through all the puzzles, references, questions, and images relating to the projectile experiments which, either in letters or photographs "held at rest"[23], had accumulated in the course of the last twelve months in Prague. Before considering the elaboration of the projectile tests at a certain point in the notebooks, a few remarks will be made about possible readings of scientific notebooks.

IV

Mach's notebooks differ specifically in appearance from other, related forms of note-taking. They differ from laboratory journals in that only the day that they are begun is, as a rule, noted. The entry procedure appears, in contrast to a protocol, to know no norms. Mach also gave the notebooks no titles, and it cannot be said that the notebooks are devoted to any one theme. Finally there is no coherence dominating the standard of the notes: from the sketches of an experiment's arrangement to the remark "clean windows"[24], everything can in principle get into them. Thanks to their inconsistency Mach's notebooks resist traditional categories of description. They were neither purely sites for reflection nor purely observation journals. They can be read neither as mere records of a "res extensa" nor as the merely written signs of an internally-reasoning mind. Accounting for these circumstances means appreciating the peculiarity of Mach's notebooks – their specific form and function – and above all it means understanding what can be discovered or regained from them. This is a question which not only concerns the possibilities for knowledge which exist in Mach's notebooks, but which fundamentally touches on the case studies of research-practice and theory-development in the natural sciences to which the history and theory of science have devoted themselves for almost the last twenty years.

From the start, these studies have granted notebooks in their various forms from laboratory journals to diaries, an elevated status as historical material.[25] One expects that their contents will reveal what processes precede the acquiring of a stock of knowledge and what contingencies and distortions are concealed within these processes. Thus Frederic L. Holmes, in his studies of the quantification of chemistry in Antoine-Laurent Lavoisier's experimental-cabinet, reads the available laboratory journals as the history of an "investigative pathway" which, in contrast to the published version, unfurls and possesses its own dynamic.[26] On the other hand, for Jürgen Renn and Tilman Sauer, in their readings of Albert Einstein's famous "Zürich Notebook" from the years 1912–1913, it is a matter of appropriately understanding a cognitive paradox. The fact that in Einstein's reflections in his notebook the field equations for gravitation are discarded which in 1915 will prove to be the way to a solution, loses much of its incomprehensibility when one learns to follow the notes from the standpoint of Einstein's "heuristic principles and their effect on his research."[27]

These different forms of access undoubtedly correspond to the form of each document. A day by day or at least regularly kept laboratory journal already suggests in its chronology the steps of a process. In contrast, a strict mathematical representation already generates the appearance of a cogent intellectual strategy. At the same time the style of the notes provides the grounds for the very different perceptions of what exactly can be read in them. If Holmes puts forward Lavoisier's records as traces of experimental operations, so Renn and Sauer discover Einstein's records as signs of a regular thought-process. These positions are in one respect not so far from one another as it seems: namely, in both cases figures and numbers are made into evidence of totally different events and processes. For experiment records are not identical with a research process – not to speak of the experiments themselves – nor are calculations identical with the axiomatics in which they seem to have originated. What one sees when one opens Einstein's volume is rather that: "Due to the character of a research notebook, most of the calculations are extremely sketchy, display false starts, and come with almost no explanatory text."[28] In order to make this confused picture analysable in the desired way, it does not suffice to go through the characters and figures. Neither does it suffice to return to the theories and models hidden in them which give them meaning. In order to make up the unwritten steps of these notes and calculations, Einstein himself, his "intellectual biography," must be drawn in.[29] The calculations in no way speak for themselves. To the contrary: the necessity of contextualising them makes one emphatically understand that Einstein's notes are not simply evidence of the thought-processes for which they stand in. This applies no less to readings which want to pursue experimental sequence in the notebooks.

Obstacles to the repetition of historical experiments suggest that considerable parts of experimental acts and events do not find expression in laboratory journals, protocols, or sketches of instruments and "set-ups." H. Otto Sibum's reconstruction of James Joule's experiments on the mechanical equivalent of heat, for example, show how much "gestural knowledge," according to which instruments and apparatuses have been assembled and used, is intrinsic to the procedure.[30] Joule's notebooks are silent about this knowledge: the data, calculations, and notes on the calibration of the thermometer scale to be used for measurement only acquire a significance on consideration of the materiality and the treatment from which they originated.[31] Written knowledge, therefore, shows no less contingency in what is handed down – and what is not. One can try, by having recourse to the material conditions of an experiment, the personal capabilities of an experimenter, the structure of an axiomatics, or an individual style of reasoning, to bridge these gaps and uncertainties; but this would exceed what is on the page. This is a legitimate historical procedure, it is certainly also one which treats the distinctive nature of these documents only as an obstacle to knowledge. To come back to Mach's notebooks: if one treats their pages not as evidence for something *outside* them, then what typically takes place *in them* emerges. One thus gets a glimpse of the particular productivity of noting, which should not be completely (mis)taken for cognitive occurrences or sequences of experimental events. This will now be more closely set out by examining a key point in the modelling of the projectile experiments.

V

The stream of records from the start of 1887 in which Mach describes the projectile experiments consists in almost equal parts of drawings, notes, and calculations which in their local context refer to one another, but whose sequence offers a more meandering sight. If one can observe no genetic progress in this jumble of records, one can however observe a unifying theme. In Mach's words: "Relationship of stationary flow to sound propagation."[32] It is no coincidence that this relationship forms the centre of Mach's notes. In discussing this relationship, the different results and observations from the projectile experiments are brought into a unified representation and, in relation to observations and theoretical conceptions, are integrated in a model in a considerably developed area of physics. But this relationship was by no means established. On the contrary, Mach's achievement had been to make it thinkable. The beginning of this process was the above-mentioned comparison between the appearances on the projectile photographs and the wave-features produced by a moving ship. This image only became strategically effective when, shortly afterwards, Salcher brought in the English engineer William Froude's investigations into the relationship between hull form and the resistance of ships. Froude's towing experiments in an experimental tank not only resulted in the sketches of the wave-features, to which Salcher referred in his comparison, they also gave way to the description of the so-called "wave making resistance" which a submerged solid body experiences in a fluid.

In respect to the "theory of stream-lines", Froude analyses the fluid motion through the ship's hull in paths or lines of particle flow.[33] From this point of view it makes no difference whether one imagines the solid body of the ship to be moving through a stationary fluid or whether one imagines a fluid moving against the stationary solid body of the ship. Decisive is only the fact that, when the fluid (water) passes the body, depending on the cross-section of the hull, the direction of the stream-lines change, local differences in the velocity of the flow appear, and correspondingly the pressure acting on the particles increases or diminishes. In the case of a ship moving in water these local differences in direction, velocity, and pressure of the particles produce the wave-mountains and valleys, which form the wave-features perceptible on the water's surface.

Froude's work on the "wave-making resistance" and his theoretical description of the observed wave-features had immediate effect on Mach's and Salcher's research on projectiles in winter 1886–1887. Attempts to represent the appearances directly on photographs as streamlines of a gas flow did not indeed lead to a result, but this failure did not hinder the reworking of the appearances in a hydrodynamical framework, only now events were played out in a series of graphical "hybrids" in Mach's notebooks instead of in photographs or experiments.[34] One can analyse these hybrids in terms of their purpose; correspondingly one will principally take into account the result that appears in them. But one can also analyse them in terms of their characteristics and hence obtain a glimpse of the dynamic of the note-taking itself. In what follows, the second approach will be primarily taken up.

The first of these hybrids stands right at the start of notebook 25. A little sketch appears between notes on Leyden jars and experiments with electric sparks (Figure 3).[35]

Anyone who is familiar with Mach's and Salcher's experiments and especially with their famous photographs sees on first glance at the drawing a projectile reduced to its basic elements, whose tip is enveloped by a stationary wave-cone. Whether, at the time this sketch was made, Mach saw in it what is immediately apparent today, is doubtful. But it is possible to state without speculation what Mach certainly did see in the drawing, for the sketch is not alone on the page. Next to it a rather erratic sentence is scribbled: "Like a bridge pier, travelling [Wie ein Brückenpfeiler, der fortschreitet]." If one says that the words reproduce what was before Mach's eyes, that would only be a half truth. Mach sees in the drawing something more; he sees something comparable to a bridge pier, but which is only present here in the comparative index, as a "like." It is no secret: this second thing is in fact composed of the projectile and the appearance which can be seen around its tip on the photographs. One will therefore be able to say: what can be seen on the photographs seems like a travelling bridge-pier.

Things are naturally more complicated. The sentence which apparently comprises the significance of the hybrid can only be expressed in the act of mediating between drawing and commentary. Its only counterpart on paper is that "like." On the other hand, even there no travelling bridge pier is to be seen. What

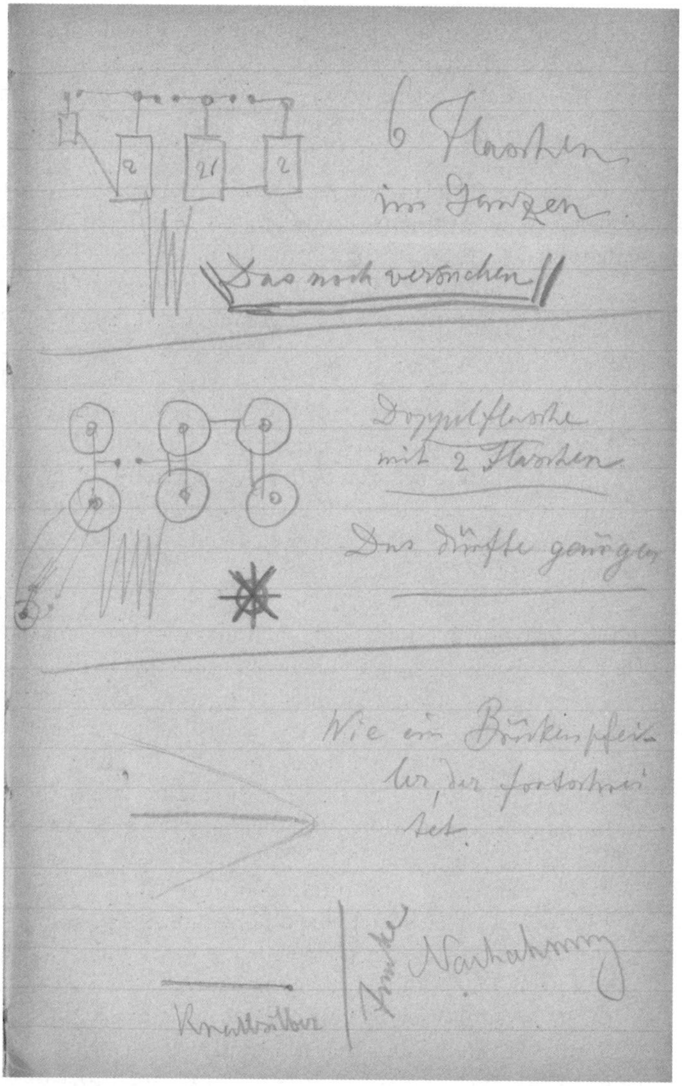

Figure 3. Ernst Mach, Notebook 25, p. 9 *(Archive of the Deutsches Museum, Munich)*.

can be seen is an overturned conical object and a horizontal line which suggest something for which there is no "outside" reference. Under any circumstances, one will look for travelling bridge piers in vain. Strictly speaking, the thing, and the meaning for which it is supposed to stand, is only visible in between the sketch and commentary. The hybrid therefore seems less like a representation of an outer object of nature or an inner mental object. Rather it functions as primarily as a mediation between heterogeneous objects. With that, however, the question of what a projectile has in common with a bridge pier is not finished with.

Actually, the hybrid's potential depends fundamentally on not speaking here of just any bridge pier, but of a bridge pier that is understood as being in a state of motion. Such a thing contradicts everyday experience, but in the context of a certain epistemology it possesses a perfectly clear counterpart. From the standpoint of "stream-line theory" a bridge pier which is moving forward is equivalent to a bridge pier against which a medium is moving: both cases represent – other things being equal – variants of one process. A difference exists between them only with regard to the fact that normally in theoretical discussions the second case is presupposed. By setting in motion the bridge-pier, now the model-case "stationary body in a moving fluid" approximates to Mach's experimental object "moving projectile in the air." At the same time, the phenomena which appear on the photographs are submerged in the wave-features around the bridge-pier. What looks to one eye like waves gathering around a projectile, looks to the other like the flow of water around a still bridge pier. The economy of this hybrid, however, marks it as oscillating between both cases; it exactly corresponds neither to the one nor to the other. Air and water, acoustics and hydrodynamics, waves, flows and streams are drawn together in a sketch and a commentary and, in this object, they become interchangeable.

Several weeks and dozens of pages later, as Mach set about examining and explaining all the remaining questions, data, and uncertainties from the projectile experiments in his notebook, the relation between projectiles in the air and bodies submerged in water appeared quite self-evident. No "like" any longer bore witness to a process of mediation, no more unmoving bodies had to be laboriously set in motion in order to reconcile the model-case of hydrodynamics with wave-like phenomena in the air: now it was the ship's body and no longer the bridge pier which became the notes' referent. More precisely, one can no longer talk at all of a referent. At this stage of "model-building'[36], projectiles and ship's hulls are *one* thing: "In the wake, warmed air and rotating air," noted Mach once under the sketch of two wave cones, which encompass a horizontal.[37] The seamlessness with which the commentary binds together phenomena in water and air tallies with the majority of the pages, in that without a dividing line, notes and sketches from the projectile experiments are followed by laws of hydrodynamics and the special case of stationary flow. Mixed in among them is a third type of sketch which belongs neither to the one field nor to the other, but to that of the theory of electricity.

At the Paris Electrotechnical Congress in 1881 Mach not only came into contact with the ideas of the ballistics expert Melsens, he also admired the galvanic patterns presented to the public by the physicist Adrien Guébhard (Figure 4).

In this type of experiment – first reported by Leopoldo Nobili in the 1820s – two wires tied to the negative and positive poles of a galvanic battery are laid on a silver-plated copper plate which is submerged in a metal saline solution. In Guébhard's opinion, the coloured rings which stand out on the copper plate concentrically around both poles represent lines of equal voltage.[38] However, in an article published in 1882, Mach commented that the rings represent lines of equal current intensity, though these admittedly seem very similar in their form to the

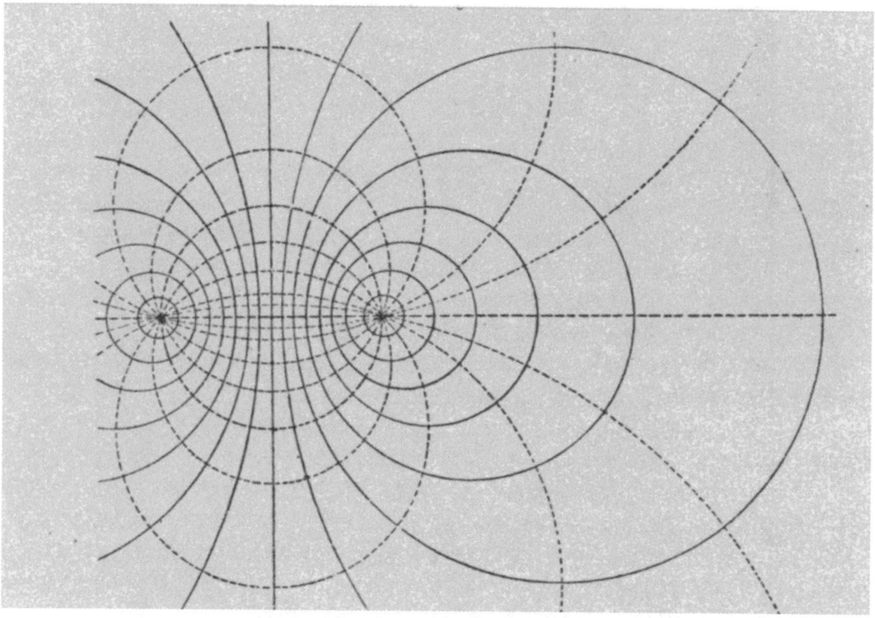

Figure 4. Diagram of the lines around two poles of equal size and different charge (Adrien Guébhard, "Figuration electrochimique des lignes equipotentielles sur des portions quelconques du plan," *Journal de physique théorique et appliquée*, second series, 1 (1882), fig. 2).

expected lines of equal voltage.[39] Five years later, such galvanic patterns again appear among Mach's sketches for the projectile experiments. At first the drawings show the familiar system of rings around a negative and a positive pole. The next sketches, several pages later, then show the outline of a projectile surrounded at the head and tail by concentric rings. "Galv.[anic] current experiment" it says, on the right hand side.[40] This shift ends for the time being in a third sketch which reaches at once beyond the "current experiment.'

When seen together with the words underneath it, the object in the centre of the page turns out to be extremely complex (Figure 5).[41] The remark, "with the middle of the battery [Mit der Mitte der Batterie]," initially reveals the jumble of lines, signs, and digits as the sketch of an attempt – mentioned in the paper of Mach and Salcher – to represent approximately, "according to the Nobili-Guébhard method," curves of equal density and pressure in the air around the projectile.[42] According to the description there, in the middle one sees a projectile-shaped, non-conducting cylinder, to the top and tail of which – marked plus and minus – the wires coming from the positive and negative poles of the galvanic battery are drawn. In front of and behind the cylinder, two metal sheets – marked "I" and "II" – are placed, whose shape resembles the wave-strips visible on the photograph at the head and behind of the projectile. In order to assist the memory of the experimenters, there is a note nearby to the effect that these metal sheets

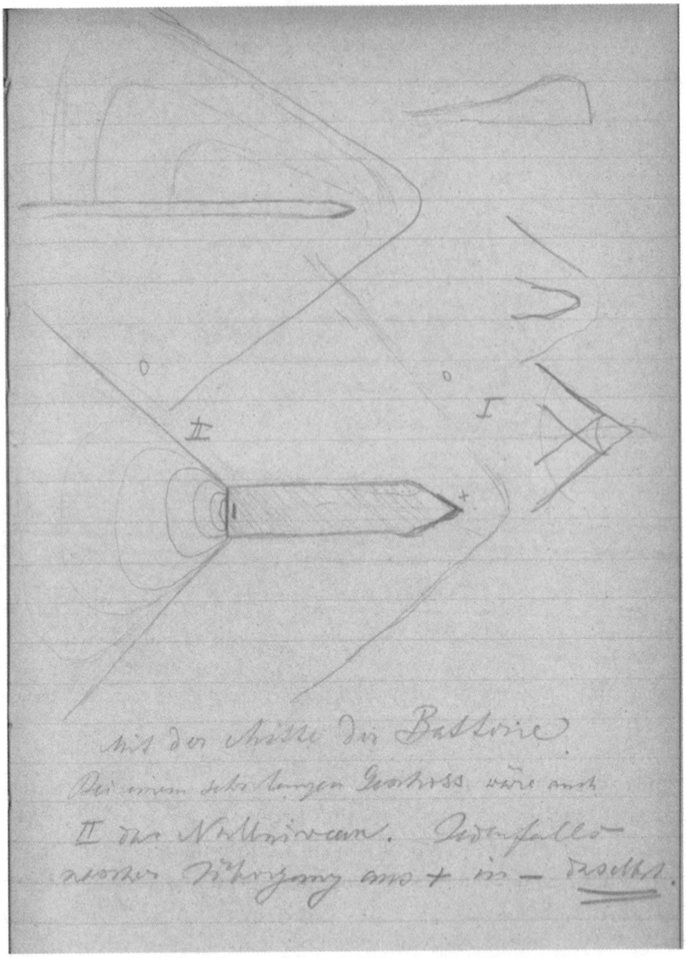

Figure 5. Ernst Mach, Notebook 25, p. 101 *(Archive of the Deutsches Museum, Munich).*

must likewise be connected to the negative pole, which at that time was in the "middle of the battery." Finally, behind the cylinder there appears a system of rings, familiar from the galvanic patterns.

To a certain extent this reading of the sketch is, however, misleading, for the notes move in another direction. Considering the following sentences a second meaning comes to light. Mach notes:

> With a very long missile II would also be zero level. At any rate rapid transition there from + to – [Bei einem sehr langen Geschoss wäre auch II das Nullniveau. Jedenfalls rascher Übergang aus + in – daselbst].

Rather than the arrangement for an experiment, the object now represents an expression of the pressure conditions in the air around the flying projectile.

According to Mach's words, for projectiles of different lengths, different pressure conditions would come to light against the projectile's tail at point "II": with the missiles used by him, at that point the compression of the air at the tip of the projectile – marked with a plus – is followed by a rarefaction; with a "very long missile," on the other hand, the air pressure would show the normal "level" as in the undisturbed air in front of the projectile at point "I." The various associations generated by these words cannot be differentiated in the sketch itself. Furthermore, only the epistemologically unclear use of plus and minus signs permits the transition from the arrangement of the galvanic pattern to the representation of pressure conditions in the air at specific points on the body of the projectile. Even more conspicuous than the extent to which physical processes ordinarily treated separately are graphically interwoven, is the absence of any reflection whatsoever on this activity. On the contrary, an electric current which flows in the distance between the two wire-endings, and a stationary flow, which represents a dynamic state of equilibrium – be it in the air around a projectile or in the water around a ship's hull – are held together in one figure without any sign of mediation and reflection on the epistemological differences between them. Instead of laboriously claiming the unity of the processes, this unity is put into practice in the note-taking.

Analogy-building, as it takes place here, to speak in cognitive concepts, is again and again used by Mach in his theoretical writings as the dominant principle of any progress in knowledge.[43] In his lecture "On the Principle of Comparison in Physics" (1894), Mach refers to the exemplary case of precisely that "peculiar relationship of similarity" in which, seven years earlier, he had brought electric currents and gas-flows into his drawings. The "analogy," the paragraph ends, is "an effective means of mastering heterogeneous fields of facts in unitary comprehension."[44] A look at Mach's notebooks can show how this "means" converts into practice, how analogies happen under the regime of note-taking even before they are marked *as* analogies.

Transitions and shifts between heterogeneous processes do not always converge so seamlessly as in the hybrid under discussion. For the most part data, notes, or sketches, non-uniform as they are, stand together without being made to fit together. But both possibilities represent complementary modes of the function of Mach's notes. Regarded operationally, temporally, and formally, we are dealing here with transitory objects. Read together and, in being read together momentarily bound together indistinguishably, the heterogeneous elements once again fall apart when the modelling of phenomena comes to an end. In this process of analogy-noting Mach achieved something which he never succeeded in achieving on the "royal road" of theoretical physics. When one goes through the pages on which a theoretical framework for the projectile experiments takes shape, it is striking that mathematical representations of the air-phenomena around the projectile are almost completely absent. The few that are started with reference to Euler's equations of fluid motion soon break off after several rows, and are finally dropped. Equally unsuccessful follow-up attempts are strewn in a loose bundle of sheets, kept separetely in Mach's papers under the title "Sound-waves as Stationary Flows."[45]

Mach's dealings with the problem of formulating the phenomena mathematically, allows one once again to understand that his notebooks are not the place where research was originally conducted or recorded. They are much more primarily the place where things being conducted elsewhere are extracted, negotiated and brought to articulation, that is, take on meaning. Hence the meander in which motions of liquids, electric currents and phenomena in the air are interwoven, after a good twenty pages dissolves into the discrete elements of a treatise – from methods to the discussion of individual phenomena.[46] What has before been imperceptibly held together is again distinguished neatly and tidily here. In Roman-numbered sections I to III, what were previously components of one hybrid or transitory object now form the discrete terms of a written argument (Figure 6).

Whilst the outlined concept is separated off and furthered in a different place on different paper, new activities and processes acquire space in the notebook. What has thus far been collected in it is now either past – partly still without results, partly crossed out and finished with – or will become future – partly as outstanding work, partly as publications which will set in train new experiments, letters, sketches, and writings. In April 1887 the mathematics and natural sciences class of the Vienna Academy made the receipt of a paper with the title "Photographische Fixirung der durch Projectile in der Luft eingeleiteten Vorgänge" by Ernst Mach and Peter Salcher public.[47] The seemingly self-evident transition from the presentation of air-phenomena around projectiles to a now "long-known phenomenon" in the water around ship's hulls and bridge piers,[48] which occurs in this paper, shows no longer any sign of the mediating operations, negotiations and alterations with which this connection had once been integrated in Mach's notebook into the research project.

VI

At the end of the nineteenth century Johann Gustav Droysen distinguished all the historian's material in "remnants," which come down to the presence without intention, and "sources," which are intentionally shaped for presenting an account of events.[49] According to this classic division notebooks fall into a curious grey zone. Remnants with respect to their use, their explicit purpose as a medium of preservation nevertheless brings them very close to sources. This lack of clarity is a result not of a particular unruliness of this material, but of a lack of clarity in the historical method itself. Every historical material – remants as well as sources – is always already shaped, that is conditioned by the circumstances of its genesis. Whoever opens a notebook is, in doing so, first of all confronted with the process of noting which from time to time unwieldily covers, shortens, transforms the noted events. Criticism and an intensification of material allow this obstacle to be overcome. Yet as this happens, it simultaneously clouds one's vision to the fact that the process of noting is still a significant factor which inscribes itself in the

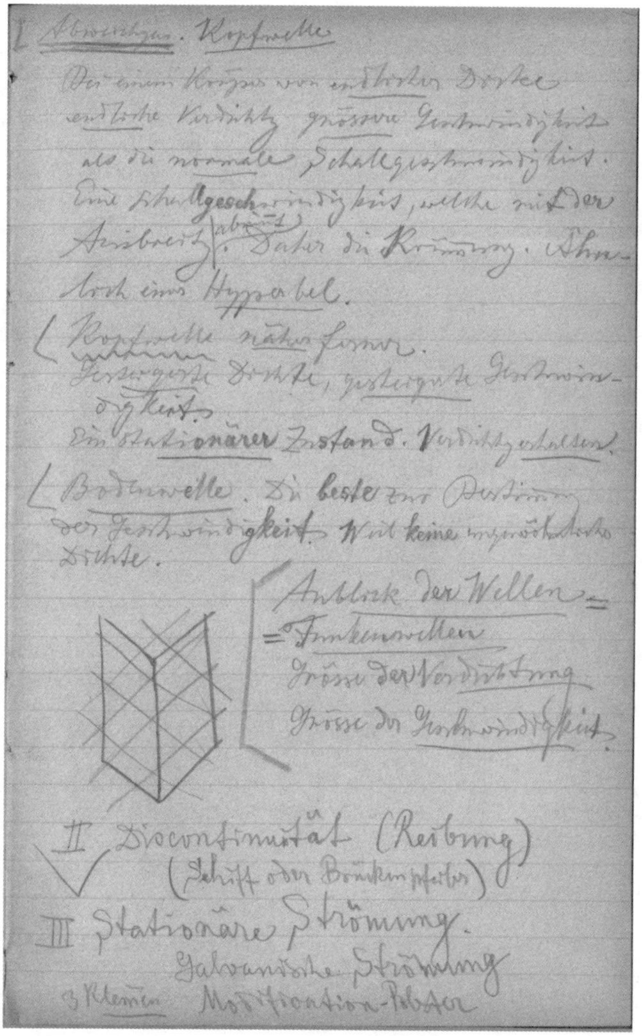

Figure 6. Ernst Mach, Notebook 25, p. 113 *(Archive of the Deutsches Museum, Munich)*.

past – which makes history. In Mach's research and scientific practice the notebook undoubtedly possesses a logistic as well as an epistemological function. The process of noting is – in a sense related to that which Michel Foucault claimed for the discourse – a violence "we do to things."[50] By this violence the business of science is made possible. Through this violence it receives its unity and its coherence. On opening Mach's notebooks, the poetics of a note-taking regime reveals itself, as Paul Valéry put it in the *Cahiers*:

Je n'arrive pas à ce que j'écris, mais j'écris ce qui conduit – ou?[51]

NOTES

[1] The text was translated from the German by Dan Stone. I am especially grateful to Dr. Wilhelm Füßl, head of the Archive of the Deutsches Museum in Munich, who generously allowed me to reproduce several pages of Ernst Mach's notebook 25. For copies of Mach's letters to Salcher I would like to thank Doris Brandner M.A. (Vienna), Dipl.-Ing. Dr. Günter Salcher (Hermagor) and Dr. Ing. Peter Krehl (Freiburg i. Br.). The work on Mach's notebooks was begun during my stay at the DFG post-graduate seminar "Representation-Rhetoric-Knowledge" at the Europa University Viadrina in Frankfurt/Oder, and completed under a grant from the Max-Planck Institute for the History of Science in Berlin, Department III. A section from Mach's Notebook 25, discussed here, has been edited as a facsimile with comments in Christoph Hoffmann and Peter Berz, eds., *Über Schall. Ernst Machs und Peter Salchers Geschoßfotografien* (Göttingen: Wallstein Verlag, 2001), pp. 49–141. I am especially indebted to Peter for many discussions and comments and to Caroline Welsh, who kindly helped me revising the translation.

[2] See Wolfgang F. Merzkirch, "Mach's Contribution to the Development of Gas Dynamics," in Robert S. Cohen and Raymond J. Seeger, eds., *Ernst Mach. Physicist and Philosopher* (= Boston Studies in the Philosophy of Science VI), (Dordrecht: D. Reidel Publishing Company, 1970): 42–59; Hans Reichenbach, "Contributions of Ernst Mach to Fluid Mechanics," *Annual Review of Fluid Mechanics*, 15 (1983): 1–28; Rudolf Dvorák, "Contribution of Ernst Mach to Gas Dynamics," in Václav Prosser and Jaroslav Folta, eds., *Ernst Mach and the Development of Physics* (Prague: Karolinum, 1991): 217–236; G. A. Tokaty, *A History and Philosophy of Fluid Mechanics* (New York: Dover Publications, 1994), pp. 193–195; and John D. Anderson, *A History of Aerodynamics and It's Impact on Flying Machines* (Cambridge, New York: Cambridge University Press, 1997), pp. 375–376.

[3] For a short biographical sketch of Salcher see Heinrich Bayer von Bayersburg, *Österreichs Admirale und bedeutende Persönlichkeiten der k. u. k. Kriegsmarine 1867–1918* (Vienna: Bergland Verlag, 1962), pp. 158–160.

[4] For Mach's Prague period see John T. Blackmore, *Ernst Mach. His Work, Life, and Influence* (Berkeley, Los Angeles, London: University of California Press, 1972), pp. 38–46, and especially: Dieter Hoffmann, "Ernst Mach in Prag," in Dieter Hoffmann and Hubert Laitko, eds., *Ernst Mach. Studien und Dokumente zu Leben und Werk* (Berlin: Deutscher Verlag der Wissenschaften, 1991): 141–178.

[5] See Dieter Hoffmann, Ondrej Pöß and Ján Chrapan, "Ernst Mach as an Experimenter," in Václav Prosser and Jaroslav Folta, eds., *Ernst Mach and the Development of Physics* (Prague: Karolinum, 1991): 325–341.

[6] See David Cahan, "The Institutional Revolution in German Physics, 1865–1914," *Historical Studies in the Physical Sciences* 15 (1984): 1–65.

[7] Notebook 2, dated "1871," *Archive of the Deutsches Museum Munich*, Ernst Mach papers, NL 174/0506, p. 160.

[8] Notebook 25, dated "11 Jan 1887," *Archive of the Deutsches Museum Munich*, Ernst Mach papers, NL 174/0529, p. 124.

[9] Notebook 25, p. 125–129.

[10] On spaces of knowledge see Hans-Jörg Rheinberger, Michael Hagner and Bettina Wahrig-Schmidt, eds., *Räume des Wissens. Repräsentation, Codierung, Spur* (Berlin: Akademie Verlag, 1997).

[11] Ernst Mach, "On some Phenomena Attending the Flight of Projectiles" (1897), in *Popular Scientific Lectures*, transl. by Thomas J. McCormack, fifth edition (La Salle: The Open Court Publishing Company, 1943): 309–337, p. 310.

[12] For Mach's material ressources and the course of experiments in 1886/87 see Christoph Hoffmann, "Mach-Werke. Die "Photographische Fixirung der durch Projectile in der Luft eingeleiteten Vorgänge" (1887) von Ernst Mach und Peter Salcher," *Fotogeschichte* 16 (1996) 60: 1–18.

[13] *Anzeiger der Kaiserlichen Akademie der Wissenschaften, Mathematisch-Naturwissenschaftliche Klasse* 23 (1886), p. 136.
[14] Ibid.
[15] Ernst Mach and Peter Salcher, "Photographische Fixirung der durch Projectile in der Luft eingeleiteten Vorgänge," *Sitzungsberichte der Mathematisch-Naturwissenschaftlichen Klasse der Kaiserlichen Akademie der Wissenschaften* 95,2 (1887): 764–781, p. 771.
[16] Ibid, p. 776.
[17] See Hans-Jörg Rheinberger, "Experimental Systems and Epistemic Things," in *Toward a History of Epistemic Things. Synthesizing Proteins in the Test Tube* (Stanford: Stanford University Press, 1997): 24–37, p. 31.
[18] Peter Salcher to Ernst Mach, Fiume, 10. Dezember 1886, *Archive of the Deutsches Museum Munich*, Ernst Mach papers, NL 174/2745.
[19] The letters from Salcher to Mach are included in the papers of Ernst Mach in the archive of the Deutsches Museum in Munich. The extant letters from Mach to Salcher are in hands of the Salcher family in Austria.
[20] See Bruno Latour, "Drawing things together," in Michael Lynch and Steve Woolgar, eds., *Representation in Scientific Practice* (Cambridge, MA, London: The MIT Press, 1990): 19–68.
[21] See Peter Salcher to Ernst Mach, Fiume, 2. and 5. Februar 1887, *Archive of the Deutsches Museum Munich*, Ernst Mach papers, NL 174/2755–2756.
[22] Ernst Mach to Peter Salcher, Prag, 7. Februar 1887. Property of the Salcher family.
[23] Mach, "On some Phenomena attending the Flight of Projectiles," p. 311.
[24] Notebook 25, p. 145.
[25] See Frederic L. Holmes, "The Fine Structure of Scientific Creativity," *History of Science* 19 (1981): 60–70, p. 65.
[26] Frederic L. Holmes, "Scientific Writing and Scientific Discovery," *Isis*, 78 (1987): 220–235, pp. 234–235. For further details see: Frederic L. Holmes, *Antoine Lavoisier: the next crucial year; or the sources of his quantitative method in chemistry* (Princeton: Princeton University Press, 1998).
[27] Jürgen Renn and Tilman Sauer, "Heuristics and Mathematical Representation in Einstein's Search for a Gravitational Field Equation," in Hubert Goenner, Jürgen Renn, Jim Ritter and Tilman Sauer, eds., *The Expanding World of General Relativity* (Boston, Basel, Berlin: Birkhäuser 1999): 87–125, p. 95.
[28] Ibid., p. 90.
[29] Ibid., p. 122.
[30] See Heinz Otto Sibum, "Reworking the Mechanical Value of Heat: Instruments of Precision and Gestures of Accuracy in Early Victorian England," *Studies in History and Philosophy of Science* 26 (1995): 73–106.
[31] Ibid., p. 78.
[32] Notebook 25, p. 92.
[33] See William Froude, "The Fundamental Principles of the Resistance of Ships" (1876), *Notices of the Proceedings of the Meetings of the Members of the Royal Institution of Great Britain* 8 (1875–1878): 188–213. For the relation of Froude's to Mach's and Salcher's experiments see Christoph Hoffmann, "Das Projektil im Wasser. Zur Konjunktur eines Zwischendings," in Christoph Hoffmann and Peter Berz, eds., *Über Schall. Ernst Machs und Peter Salchers Geschoßfotografien* (Göttingen: Wallstein Verlag, 2001), pp. 259–288.
[34] On "hybrids" see Latour, "Drawing things together," p. 29, and Rheinberger, *Toward a History of Epistemic Things*, pp. 135–136.
[35] Notebook 25, p. 9.
[36] For the notion of "model building" see Timothy Lenoir, "Practice, Reason, Context: The Dialogue between Theory and Experiment," *Science in Context* 2 (1988): 3–21, p. 5.
[37] Notebook 25, p. 94.
[38] See Adrien Guébhard, "Figuration Electrochimique des Lignes Equipotentielles sur des Portions Quelconques du Plan," *Journal de Physique Théorique et Appliquée*, Second Series 1 (1882): 205–222.

[39] See Ernst Mach, "Über Herrn A. Guébhard's Darstellung der Äquipotentialcurven," *Sitzungsberichte der Mathematisch-Naturwissenschaftlichen Klasse der Kaiserlichen Akademie der Wissenschaften* 86,2 (1882): 8–14.
[40] Notebook 25, p. 97.
[41] Ibid, p. 101.
[42] See Mach and Salcher, "Photographische Fixirung der durch Projectile in der Luft eingeleiteten Vorgänge," p. 778.
[43] Ernst Mach, "On the Principle of Comparison in Physics" (1894), in *Popular Scientific Lectures*, transl. by Thomas J. McCormack, fifth edition (La Salle: The Open Court Publishing Company, 1943): 236–258, pp. 249–250.
[44] Ibid, p. 250.
[45] Archive of the Deutsches Museum Munich, Ernst Mach papers, NL 174/0447.
[46] Notebook 25, pp. 110–113.
[47] *Anzeiger der Kaiserlichen Akademie der Wissenschaften, Mathematisch-Naturwissenschaftliche Klasse* 24 (1887): 102–103.
[48] See Mach and Salcher, "Photographische Fixirung der durch Projectile in der Luft eingeleiteten Vorgänge," p. 775.
[49] See Johann Gustav Droysen, "Grundriß der Historik" (1882), in *Historik. Rekonstruktion der ersten vollständigen Fassung der Vorlesungen (1857) Grundriß der Historik in der ersten handschriftlichen (1857/58) und in der letzten gedruckten Fassung (1882)*, Textausgabe von Peter Leyh (Stuttgart: frommann-holzboog, 1977): 426–427.
[50] Michel Foucault, "The Discourse on Language" (L'Ordre du discours, 1970), in *The Archaeology of Knowledge and the Discourse on Language* (New York: Pantheon Books, 1982): 215–237, p. 229.
[51] Paul Valéry, *Cahiers*, Édition établie, présentée et annotée par Judith Robinson (Paris: Gallimard, 1973), vol. 1, p. 7.

DANIEL P. TODES*

FROM LONE INVESTIGATOR TO LABORATORY CHIEF: IVAN PAVLOV'S RESEARCH NOTEBOOKS AS A REFLECTION OF HIS MANAGERIAL AND INTERPRETIVE STYLE

When I began archival research on Ivan Pavlov's life and work I looked forward eagerly to emulating such scholars as Larry Holmes and Gerald Geison who have used laboratory notebooks so fruitfully in their analysis of the fine grain of experimental science. That, alas, was not to be. Pavlov's personal papers in the St. Petersburg branch of the Archive of the Russian Academy of Sciences do include scattered research notebooks – and these shed important light upon his experimental procedures – but, for a variety of reasons, they do not enable the type of analysis that I originally had in mind.

Several of these reasons reflect the mundane contingencies in any archival research – for example, the selective preservation of any individual's papers – but one is of more general interest: The changing nature of Pavlov's research notebooks reflects his transformation from a lone investigator to the chief of a large laboratory enterprise. In this short paper, then, I will offer some preliminary reflections about these notebooks and the changing social-cognitive dynamics of Pavlov's research, and will conclude with a few thoughts about the particular challenges posed to historians of science by the emergence of large laboratories.

PAVLOV'S SCIENTIFIC STYLE AND A PERIODIZATION OF HIS SCIENTIFIC CAREER

Ivan Pavlov (1849–1936) enjoyed an exceptionally long and productive scientific career. In the 1870s and 1880s, he conducted research on both the physiology of digestion and the physiology of circulation (completing a doctoral dissertation in 1883 on the nerves of the heart). From about 1889–1903 he worked exclusively on digestive physiology, receiving a Nobel Prize for that work in 1904. By that year, he had launched the investigations of what he termed "higher nervous activity" that would occupy him for the next three decades until his death at age 86.

We might also periodize Pavlov's career according to the circumstances of his laboratory work: From the mid–1870s through 1890, he was a "workshop

---

* Johns Hopkins University, Baltimore, MD

Frederic L. Holmes, Jürgen Renn and Hans-Jörg Rheinberger (eds.), Reworking the Bench: Research Notebooks in the History of Science, 203–220 © *2003 Kluwer Academic Publishers. Printed in Great Britain*

physiologist," a lone investigator who conducted his experiments in somebody else's laboratory, utilizing crude and often homemade equipment. He occasionally enjoyed the services of an assistant, but Pavlov himself chose his subjects, planned the course of experimental trials, conducted experiments, and interpreted and wrote up the results.

This changed radically in 1891, when Pavlov became chief of the Physiology Division of the newly-created Imperial Institute of Experimental Medicine in St. Petersburg.[1] He thus became a "factory physiologist," the head of a large laboratory enterprise in which he tightly coordinated the labors of assistants, attendants, and about twelve coworkers per year in pursuit of his own investigative goals. Pavlov used these coworkers as extensions of his sensory reach – as "skilled hands" (as he once put it) attached to his own head. These coworkers (or, *praktikanty*) were usually physicians seeking a quick doctorate: they arrived at the laboratory with little or no background in physiology and with a maximum of two years to choose their thesis topic, conduct the necessary research, write their dissertation, and defend it.

Praktikanty were incorporated into a laboratory system designed to maximize production while giving the chief control over the interpretive moments in research. Pavlov assigned them topics along clearly delineated lines of investigation that flowed from his longstanding scientific vision, and provided each with a suitable, surgically-altered dog for their experiments and an assistant to incorporate them into laboratory culture. That culture combined *glasnost* with a hierarchical structure featuring the chief's immense institutional and intellectual authority. Through conferences at the bench, laboratory-wide discussions, and thorough editing of all written products, Pavlov shaped the experimental results obtained by his coworkers into the unified and powerful vision of digestive physiology that he presented in his synthetic *Lectures on the Work of the Main Digestive Glands* (1897).

This managerial system remained basically unchanged in later decades, but the years 1921–1936 nevertheless constitute a third period in the history of Pavlov's laboratory circumstances. Having abandoned thoughts of emigration and reached an accomodation with the Bolsheviks, Pavlov became the beneficiary of essentially unlimited state funding – and his laboratory enterprise expanded qualitatively. Aside from his main laboratory at the Institute of Experimental Medicine (which was expanded and renovated), he also directed experimenters at his refurbished laboratories at the Academy of Sciences and the Military-Medical Academy; and, beginning in the late 1920s, at his own science city in Koltushi, just outside of Leningrad. Changes in the workforce and other circumstances related to Soviet power elicited some changes in the laboratory's social-cognitive dynamics – but these constituted minor variations upon familiar themes.

One unchanging feature of Pavlov's scientific style is also important for understanding his managerial practices: His approach to physiology revolved around the tension between two main goals: 1) To investigate the normal, intact animal, and 2) To obtain entirely consistent, determined results, which, ideally, were expressed quantitatively (in the form of secretory curves for digestion or salivary patterns for higher nervous activity). Pavlov understood the animal organism as a machine – but as a very complex machine in which, therefore, uncontrolled variables abounded in even the most carefully designed experiment.

For him, this explained why the very same experiment generated somewhat different results with two different animals, and even with the same animal on two different days. As a firm believer in determinist physiology in the spirit of Claude Bernard, Pavlov needed to explain these differences in terms of uncontrolled variables (which might reside, for example, in the psyche of the experimental animal or in still-undiscovered nervous processes). There was, then, an unavoidable interpretive moment at the heart of every experiment – and Pavlov's managerial system was designed to provide him with both the quantitative data generated by experiments and the ability to make these interpretive judgments himself.

THE CHANGING STYLE OF PAVLOV'S NOTEBOOKS

Pavlov's personal papers at the Archive of the Russian Academy of Sciences contain six of his research notebooks. This collection is clearly incomplete and does not permit a fine-grained analysis of his most important scientific achievements. They do, however, provide insights into his interpretive and managerial style at important junctures in his research – and, taken together, they clearly reflect the changing circumstances of his scientific practice.

The first notebook dates from the late 1880s, when Pavlov worked as a lone investigator.[2] This large thick notebook was clearly not chosen for portability, and it contains, in his own hand, experimental protocols, scattered interpretive comments, numerous doodles, and the initial drafts of some articles. (Figure 1) This notebook reflects his practice as a workshop scientist, at which time he alone (with the occasional aid of an assistant) planned, prepared, conducted, interpreted, and wrote up his conclusions from experiments.

Proceeding chronologically, the next notebook in Pavlov's archive is a pocket-sized calendar book from 1894. By this time, he was the chief of the large laboratory in the Physiology Division of the Imperial Institute of Experimental Medicine, where he supervised the work of about twelve coworkers per year. The year 1894 proved pivotal in his laboratory's research on digestion: The Pavlov isolated stomach was, for the first time, successfully created in a dog, allowing Pavlov and his coworkers to measure the secretory responses of an intact animal to various foods. Enthusing about the initial experimental results, Pavlov confided to his wife that "An enlivening of our projects is inevitable – success after success, not only new but downright beautiful."[3] In part as a result of these experiments, Pavlov conceived in this same year the metaphor that guided subsequent research and framed his *Lectures on the Work of the Main Digestive Glands*: the digestive system was "a complex chemical factory."

The calendar book for 1894, however, contains no reference to these

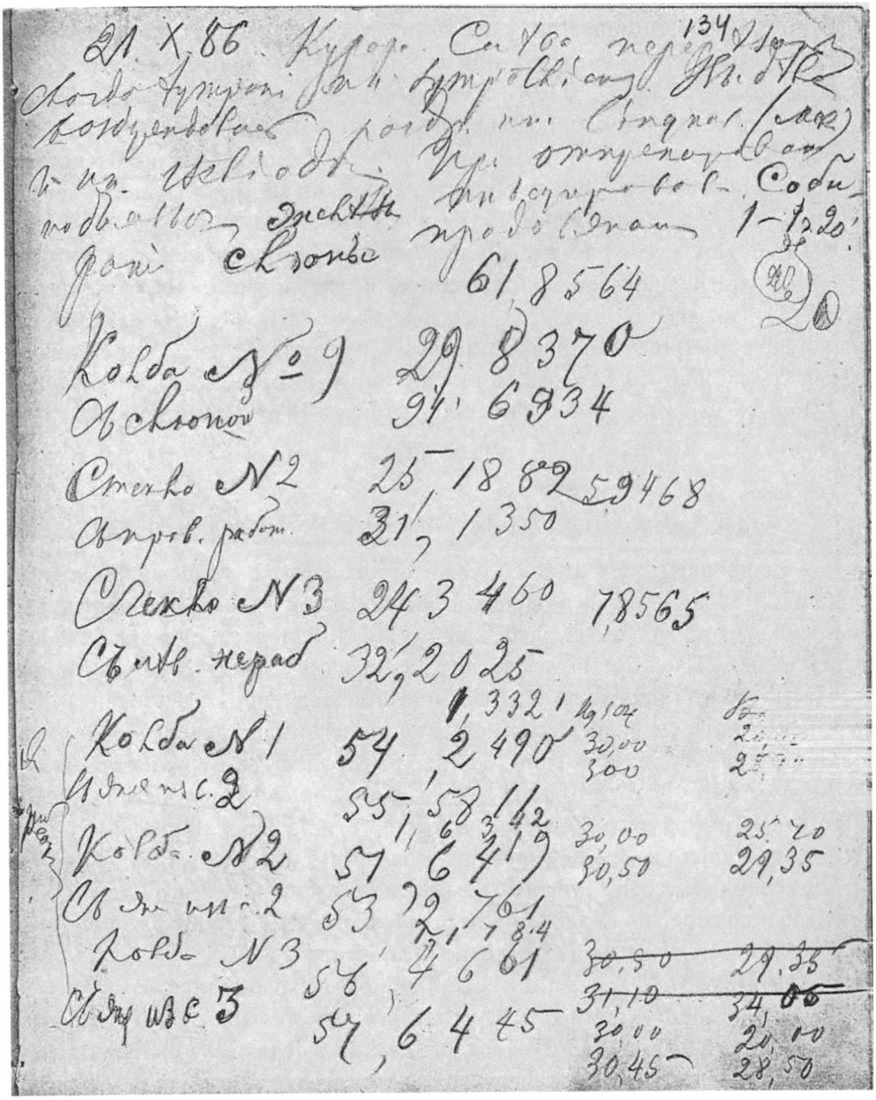

Figures 1a, 1b. *Figure 1* shows two typical pages from this notebook, with the results of experiments upon a curarized dog.

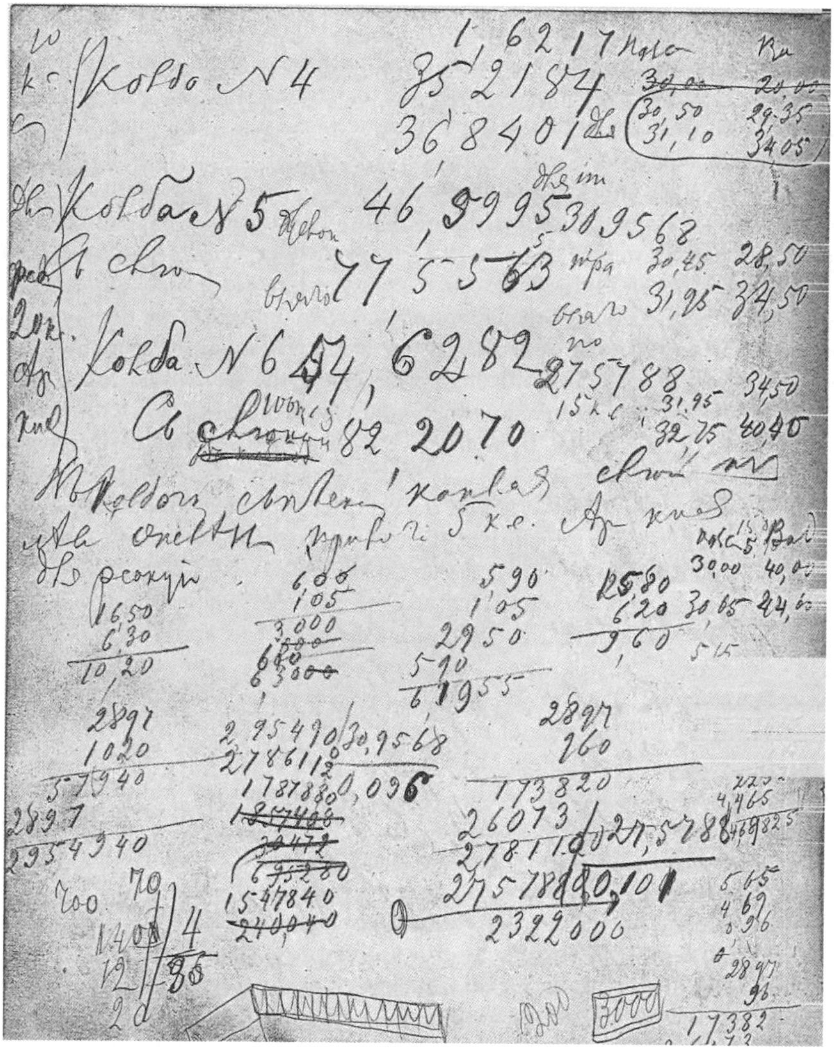

developments. Pavlov used it in the same way that many of us use such notebooks today: for example, to record appointments, write reminders and addresses, or make an occasional note about a book or article we might want to read.

Where, then, did Pavlov record the experimental data that he had earlier inscribed in his laboratory notebook? Upon becoming chief of his large laboratory operation in 1891, he instituted strict and standardized procedures according to which his coworkers recorded data in separate notebooks arranged by dog. Coworkers provided a simple description of procedures and the dog's demeanor during experiments, and recorded the rows of quantitative data that were a hallmark of Pavlovian physiology. Pavlov forbade them to write their own interpretive thoughts in these protocol books, which remained in the laboratory so the chief had constant access to them. These notebooks, then, recorded the labors of the chief's "skilled hands," preserving the results for his own interpretive efforts. Unfortunately, with one important exception (see below) these notebooks have disappeared.[4]

As for his own thoughts, Pavlov prided himself on his prodigious memory – and so did not, apparently, feel the need to jot them down. He supervised his coworkers closely, moving constantly from bench to bench, and so was fully engaged in the several lines of investigation being pursued in the laboratory. The social-cognitive structure of the laboratory enabled him to control the interpretive moments arising in the experiments of his coworkers: the chief aired his interpretations in meetings with his coworkers at the bench and in occasional gatherings of the laboratory collective, and, through his careful editing of all publications, these interpretations shaped the doctoral theses, reports, and articles generated by his laboratory.

Pavlov subsequently converted his pocket calendar into a laboratory notebook of sorts. His archive contains five of these small, portable notebooks – each from the later years when Pavlov managed an expanding laboratory operation engaged in research on higher nervous activity. Two of these calendar books cover the years 1909–1918, the other three date from the late 1920s and 1930s.[5]

As their form testifies, these are the notebooks of a laboratory chief. As in 1894, Pavlov used them to jot down addresses, reminders, and even political comments; but he filled most of their pages with his reactions to his laboratory's research. Reading these notebooks, one imagines him carrying it with him from bench to bench, and then returning to his study in the laboratory to gather his thoughts and make sense of what he's seen. Sometimes he pulls out his notebook during a moment of inspiration; sometimes, he takes it out at home (where he regularly spent the post-dinner hours at work in his study) to record his thoughts about the general course of research, to remind himself to instruct a coworker to perform a specific experiment, to group experimental results according to an emerging hypothesis, to outline a synthetic article, or to vent his strong feelings about the life of his country. (Figure 2 shows several pages from these notebooks.)

The extant notebooks are too few, and too sparsely distributed over the years, for us to trace in detail their evolution from everyday calendar books to the hybrid records of later years. Clearly, however, Pavlov lost confidence in his ability to track and direct laboratory research by relying upon memory alone. For one thing, the size of his laboratory operation was constantly increasing – growing qualitatively in the wake of his Nobel Prize of 1904 and, even more so, after he came to terms with the Bolsheviks in 1920–1921 and became the beneficiary of virtually unlimited state funding.[6] For another, the interpretive issues in research on higher nervous activity proved much more complex, and much less easily resolved, than those in his earlier research on digestive physiology. Finally, of course, Pavlov was ageing – and he may have lost some confidence in his memory.

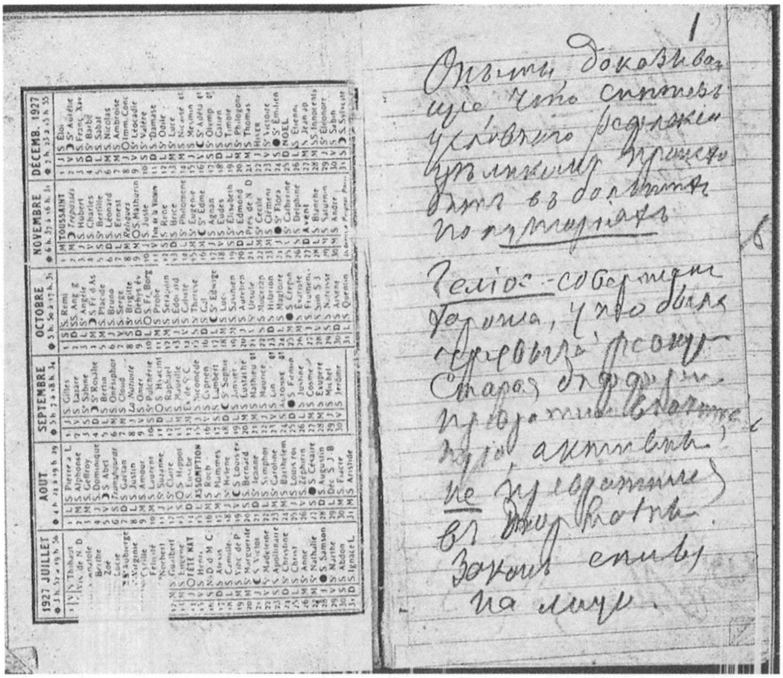

Figures 2a, 2b, 2c. Several pages from Pavlov's notebook for 1927–29. They begin with a list of experiments proving a particular proposition, followed by thoughts about follow-up experiments, reflections upon one experiment trial, and some tentative theorizing.

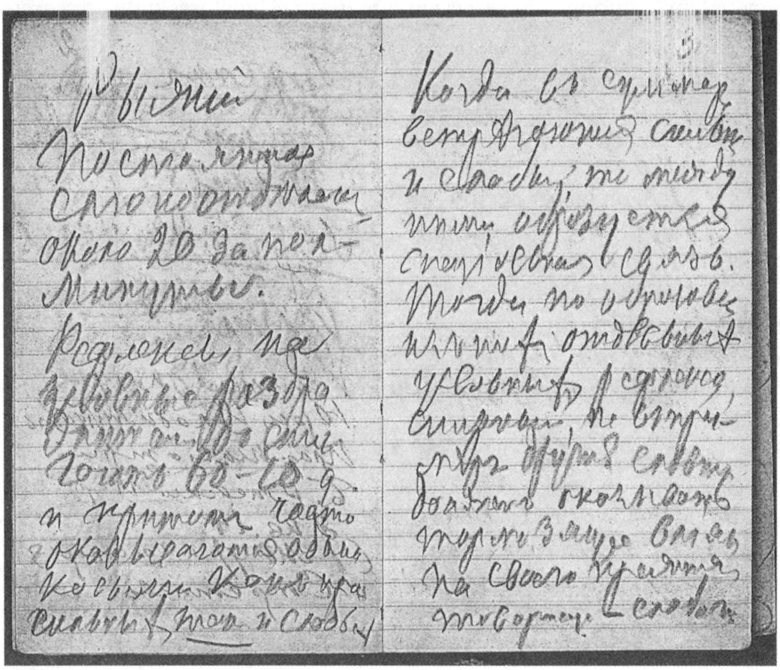

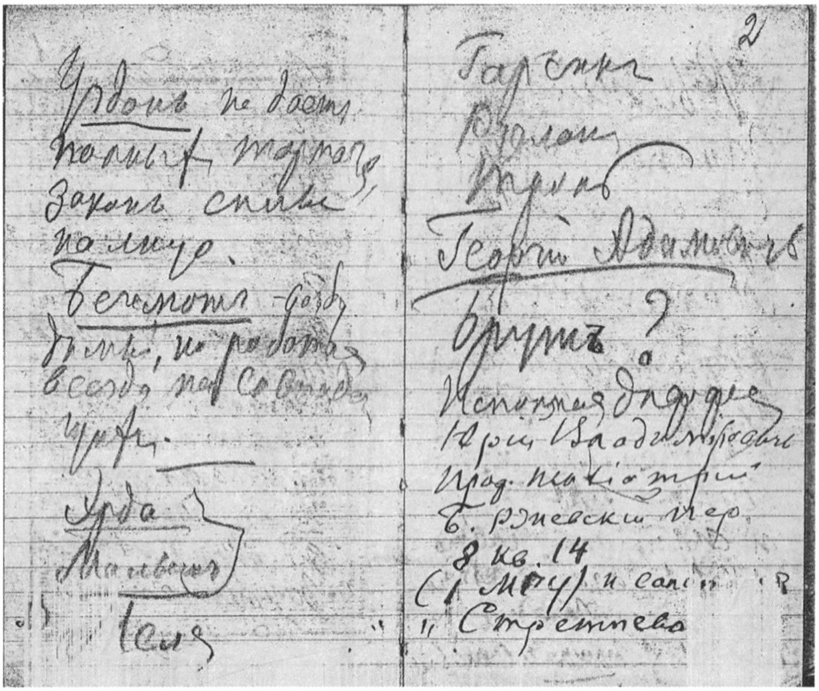

In any case, these notebooks offer a revealing look at Pavlov's interpretive and managerial style during some especially interesting junctures in his research.

REFLECTIONS OF A FACTORY PHYSIOLOGIST: THE NOTEBOOK FOR 1911–1918

One small brown notebook dates from an especially critical and fruitful period in Pavlov's research on higher nervous activity. The entries are mostly from the years 1911–1913, when Pavlov was attempting to develop a systematic classification of higher nervous processes and their relationship to one another, to understand the nature of inhibition, and to place his research within a broader framework. The entries became infrequent in mid–1914, when, with the outbreak of World War I, his coworkers departed for the front, bringing his research almost entirely to a halt; and thereafter are reduced to scattered, mostly personal notations through 1918.

We are concerned here with the form and general contours of Pavlov's notebook entries, and not with the content of his scientific research. Even for this limited purpose, however, it is useful to keep in mind two general issues that arose from the pivotal experiments of M.N. Erofeeva. Erofeeva – a member of the first cohort of female investigators who entered Pavlov's laboratory after the revolution of 1905 – succeeded in obtaining what she and Pavlov considered a "food reaction" to an electrical shock. That is, after she had repeatedly administered a mild shock to the dog and then fed it, the dog came to respond to the shock, not with its earlier "defensive" reaction, but rather by salivating. According to laboratory doctrine, the shock – which previously had elicited an unconditional defensive response – had now become a conditional stimulus associated with food. This, for Pavlov, illustrated the great flexibility of conditional reflexes, which thus provided an animal with flexible responses to a changing environment. Erofeeva then gradually increased the strength of the shock – and she (and Pavlov) claimed that the food response was maintained even at levels that caused the dog observable injury. She then conducted similar trials with acid. Pavlov swiftly incorporated her results into a general view of higher nervous activity that explained, among other things, the smiles on the faces of Christian martyrs on the cross.[7] These experiments also stimulated his interest in the nervous mechanisms responsible for these results, and in the relationship of excitation, inhibition, and other higher nervous processes.

A second important result of Erofeeva's experiments was that, shortly after salivating in response to a mild electrical shock, her dog frequently fell asleep on the stand. For both coworker and chief, this "reflex of sleep" raised questions about the relationship between inhibition and sleep, and this, in turn, brought them to the issue of the relationship of various nervous processes.

These, then, were the basic experiments and issues that Pavlov was pondering in the years covered by this notebook. His comments fall roughly into four categories: 1) Observations about individual experiments, sometimes with ideas for further experiments and tentative conclusions, 2) Grouping of the experiments of various coworkers as Pavlov sought to generalize, 3) Thoughts about the general enterprise and philosophical musings, and 4) Miscellany. I provide below a few examples of each.

Observations About Single Experiments

The first three of these are responses to Erofeeva's experiments:

> "With an acidic exciter, as with an electrical one, the food effect grows with repetition. See how the conditional inhibitor maintains itself during the first and subsequent exposures to acid."

> "Inhibition passes into sleep more sharply and is more frequently observed when acidic reflexes are inhibited than when food [reflexes] are."

> "How does the exinguishing inhibitor act upon the electrical conditional exciter? How, in a dog with an electrical and acidic exciter, does the natural acidic exciter act? (Both questions for Maria Nikolaevna [Erofeeva])."

> "How to understand the sharp distinction between secretory and motor responses in Vol'bort's [dog]?"

"Must think about the sharp negative response to conditional inhibitors and especially about its onset in Fridman (Chebotareva's [dog]) . . . ."

"Investigate anew and systematically (paying attention to the new data about the spreading of inhibition) how extinction, conditional inhibition and differentiated inhibition of one conditional reflex act upon the conditional reflex of another group: that is, of the food [group] upon the nonfood [group], and vice-versa."

"How will reflexes of time change under the influence of exciter substances: caffeine and so forth?"

"An interesting episode with Kal'm, that impudent and aggressive dog."

Grouping and Generalizing

"Don't very strong exciters act in a special manner, eliciting the complex of phenomena that are subjectively termed fright, for example in Beliakov's [dog] Dogonia . . . and in Vol'bort's [dog] "Trubach" . . . and in Grom – they salivated immediately or subsequently from a pair of conditional inhibitors (the first and second); as if by the destruction of the inhibitory functions of the [cerebral] hemispheres, there arises a condition analogous to the condition of the spine when poisoned by strychnine."

"We must distinguish between three groups of inhibition: internal, external, and sleep. It is interesting that sleep inhibition is identical with external [inhibition] and the disinhibition of internal [inhibition], but, of course, is entirely different, so to speak, by its nature. External inhibition – active reflexes, sleep – passive reflexes."

"External inhibition can be understood as follows: the excitation of the central reflex attracts to itself the nervous energy of another center, and so in this latter center excitability declines. If this is so, then from this perspective one can also understand the formation of a conditional reflex, that is, the existing active center attracts to itself any excitation entering the large [cerebral] hemispheres. That this is actually the case is also obvious from the formation of the food electrical conditional reflex, that is, the excitation, usually going to the center for defensive movements, is directed in large part to the food center, as the more vitally powerful [one]." [Here Pavlov is attempting to explain Erofeeva's results with electrical shocks, and, he relies, as he did frequently, upon an implicit metaphor drawn from physics (the metaphor of attraction).]

[In the following note, Pavlov muses about the "aggressive" dog, Kal'm, who uncharacteristically fell asleep on the stand:] "Inhibitedness, there arose with this a sleepy state entirely unusual for this dog . . . . Did this latter not occur in this manner?: The irradiation of excitation from the aggressive center reached the food [center], destroys there the inhibitory side of the conditional inhibitor, freeing its excitatory, always concealed component – and, conjoining now with the exciting conditional reflex, it gives a summed effect. Consequently [its effect is] increased in comparison with the effect of the conditional reflex alone. This explanation [accords with] the fact of 'disinhibition' of the conditional inhibitors in Vol'bort's [dog]. . . ."

"What cases do we have of the useful inter-relation or summation of various types of internal inhibition [?]" [Pavlov then considers a few possibilities.]

Several of the citations above reveal Pavlov's tendency to think anthropomorphically about the responses of experimental animals. (As I argue elsewhere, this anthropomorphic moment played an important and usually hidden role in Pavlov's interpretation of experimental data.[8]) This is clearly evident in the following entries:

"In many cases [in trials with the dog Norka] a conditional salivary reflex is observed, but the motor reaction of chewing food disappears. This clearly occurred with the development of a hypnotic state. Does this not represent an analogy with that phase of hypnosis in man, when there is an entirely developed cataleptic state, but the person preserves a normal consciousness and conducts conversations [?]"

"When playing *gorodki* [Pavlov's favorite sport, which resembles skittles] one sometimes senses clearly the distance and the muscular tension that is necessary for a good throw, and from this a clear certainty in the aim. At other times, one throws as if by chance, into empty space, without any confidence. This is either the suppression through inhibition of one sphere of the motor analyser or of the paths linking visual excitation with this sphere."

Thoughts About the Enterprise and Philosophical Musings

"There is probably no sharp boundary between voluntary and involuntary acts. One can probably make salivation voluntary both in relation to excitation and inhibition." [For Pavlov, this important conclusion flowed from Erofeeva's experiments with electrical shock.]

"The activity of the higher nervous system is admittedly very complex, but this also has its advantageous side. In the spine we are dealing with completed activity, established relationships, and so we don't see and don't know how this is accomplished, by what process, as a result of which qualities of the nervous mass this occurred. In the higher brain we have before us the process of the formation of reflexes in all its most complex relations – and, consequently, in the course of this process there is revealed before us all the basic qualities of nervous activity, all the preliminary and intermediate stages, and so the internal mechanism of nervous activity."

> "One can say that the central nervous system and specifically its higher division is a special organ of development – and this is why nervous particularities and pathological deviations are so easily transmitted by heredity."

> "We consider all so-called psychic activity to be a function of the brain mass, of a defined mechanism, that is, of an object conceived spatially. But how can one place in this mechanism an activity that is conceived psychologically, that is, non-spatially [?]"
> [This note is but one of several archival indications of Pavlov's uncertainty about his grand quest to use investigations of conditional reflexes to explain mental and emotional states. This private uncertainty contrasted sharply with his confident, sometimes even aggressive, public posture.]

> "I do not know what exactly we have done, in what way we have broken through, but it is clear to me that there now exists a union of thought, a mixing and unification of the ideas of all participants in the intellectual work [of the laboratory]"

> "Must study the history of the understanding of the word 'reflex'."

The following three entries concerning psychology appear consecutively:

> "I am certain that with the recognition of the true [place?] of physiology, in the direction that is now actualized in the doctrine of conditional reflexes, with the activity of the human mind in this direction, there will come a period of extraordinary discoveries, no less stunning than those in the sphere of inorganic phenomena."

> "The future belongs to the physiology of the brain, and not to psychology; in any case, physiology should precede psychology. Psychological thinking is not now economical in the sense of a strict scientific investigation."

> "I understand and sense the power of the psychologists' mind, working in this difficult uneconomical situation."

Miscellany

> "Some thoughts and dreams about the current war: And the example of Germany and England in this war shows that the idea of a world government is not a true resolution of the land question, but rather a human weakness, originating, so to speak, from the inertia of human nature."

> "Speak with A. V. [Timofeev] about a psychiatric case for collaborative work."
> [Deprived of his coworkers by the war, and so unable to pursue his laboratory research aggressively, Pavlov indulged a longstanding interest in psychiatry by beginning regular outings to the Alexander III Charity Home for the Mentally Ill, where his friend Timofeev was director.]

THE NOTEBOOK FOR 1927–1929

The remaining notebooks in Pavlov's archive date from the late 1920s and 1930s, and have the same form as that from the 1910s.

The notebook from 1927–1929 is especially interesting for several reasons. In the previous year Pavlov had completed his first – and, as it turned out, his only – monograph on higher nervous activity, *Lectures on the Work of the Large Hemispheres of the Brain*. In 1927, he continued to direct experiments to clarify what he regarded as this book's tentative analysis of several key issues. The chief intended to substantially revise the second edition of *Lectures*, and the 1927 notebook records his continuing attempt to develop a systematic understanding of some basic issues. [9]

The year 1927 also witnessed the rapid development of two directions in laboratory research: the analysis and classification of "nervous types" and the attempt to develop an "experimental pathology of higher nervous activity" that would be relevant to psychiatric practice. Pavlov's approach to these two lines of investigation was shaped fundamentally by a conception of higher nervous activity that revolved around the interaction of excitation and inhibition – a dichotomy that he associated metaphorically with freedom (excitation) and discipline (inhibition).[10]

Pavlov's thoughts about these developments are reflected in his pocket notebook for 1927–1929. A few examples suffice as illustration:

Grouping and Generalizing

> "Experiments proving that the synthesis of a conditional reflex occurs entirely in the large [cerebral] hemispheres: Gelios [a dog] – very good, he has a wonderful reaction; the former differentiation has turned into a positive, active [one], not an inhibitory one. The law of strength is obvious [here]." [Pavlov lists other dogs that seem to prove the point, then notes one that apparently doesn't – and suggests a reason:] "Brut? Incomplete differentiation."
>
> "One of the strongest impressions and most frequently encountered facts in the study of conditional reflexes is the ... summation of stimuli. In the presentation of conditional reflexes, this should occupy one of the most conspicuous places, should be presented clearly, one of the most important acquisitions of the objective study of complex nervous activity. The same is true of the fact that everything that falls upon the receptive surfaces of the dog acts upon the course of nervous activity. This comes especially to the fore during some sensory processes such as, for example, internal inhibition in the form of delay."

Pavlov's responses to individual dogs now do not merely include anthropomorphic elements, but reflect his attempt to use his understanding of higher nervous activity to bridge the gap between the laboratory and psychiatry.

"Skipin's [dog] Ryzhii is clearly a hysteric, in which the cortex is constantly inhibited."

"Umnitsa's state recalls that of a sleepy infant. Are they not one and the same?"

Miscellany

"Fate. The social struggle with it. Practice. Eugenics."
[Pavlov soon petitioned the government for funds to build his Institute of the Experimental Genetics of Higher Nervous Activity in Koltushi, just outside of Leningrad. Here he began preparations for experiments on the relative weight and interaction of hereditary and environmental influences on the nervous systems of dogs, which he hoped would allow him "to acquire, by means of selection by breeders, the most perfected nervous system."[11]]

"My God! When will our life finally be rid of the wild idea that those who created cultured Russia should somehow disappear, be eliminated?" [This comment reflects Pavlov's unalloyed opposition to Bolshevik policies from 1917 through the late 1920s, when his attitude changed somewhat.[12]]

THE NOTEBOOKS OF THE ORCHESTRA'S "CONDUCTOR" AND HIS "DEAR SOLOIST"

One especially intriguing entry in the notebook of 1927–1929 reads: "Castrate dogs of every type." This records Pavlov's acquiescence to the longstanding entreaties of Maria Kapitonovna Petrova. Petrova had completed her doctoral thesis in Pavlov's laboratory on the eve of World War I, and had remained there ever since. She and the chief had become lovers in 1914, and their affair continued for two decades until Pavlov's death in 1936. Petrova also proved an industrious and committed coworker who exercised an important influence on the direction of laboratory investigations. Committed to forging a connection between laboratory investigations and the clinic, she contributed much to the investigation of nervous types, and, as another leading coworker, Lev Orbeli, later testified, she was fundamentally responsible for the laboratory's turn toward the "pathology of higher nervous activity."[13]

To this end, Pavlov castrated his first dog, Joy, in May 1928. Many others – of "every [nervous] type" – followed in subsequent years. Chief and coworker together examined the affect of the operation upon the dogs' nervous responses and treated each with caffeine (to enhance excitation) and/or bromides (to enhance inhibition). After each operation, Pavlov wrote something suitable in the laboratory notebook kept by Petrova for each dog. For example: "Joy, it would simply be a dream if you permitted us to resolve this difficult task" and (in an entry with an agreeable poetic rhythm in the original Russian) "Don't pout any more, my dear, please conduct yourself as before."

Petrova's laboratory notebooks are preserved in her personal papers at the St. Petersburg branch of the Archive of the Russian Academy of Sciences, and these provide an opportunity to coordinate Pavlov's notebook entries with detailed experimental protocols – and thus to more closely study the play between concrete experiments and the interpretive moments in this research.

The chief's many references to Petrova's research in his own notebooks, and his occasional drawings and comments in the notebooks of the "dear soloist" in his laboratory "orchestra," attest to his close attention to both investigator and investigation.[14]

CONCLUDING REFLECTIONS

Pavlov's research notebooks reflect the changing circumstances of his scientific practice: his transformation from lone investigator to laboratory chief, and subtler changes in his large laboratory enterprise over his four decades as director. The different moments in the experimental process that, in the 1880s, transpired within one and the same person were, in later years, distributed among a large workforce in a dynamic, social-cognitive division of labor.

Pavlov's transformation, of course, mirrored that of scientific production as a whole. During the mid- and late- nineteenth century, in a development paralleling that which began much earlier in goods production, the scientific workshop increasingly gave way to large, sometimes factory-style laboratories. Justus von Liebig and Felix Hoppe-Seyler in chemistry, Carl Ludwig and Michael Foster in physiology, Robert Koch and Louis Pasteur in bacteriology, and Paul Ehrlich in immunology all presided over distinctively social enterprises involving substantial capital investment, a specially-designed workplace, a relatively large workforce, a developed division of labor, and a productive process that involved managerial decisions. Writing in 1896, William Henry Welch observed that the emergence of the "well-equipped and properly organized modern laboratory" had "completely revolutionised during the past half-century the material conditions under which scientific work is prosecuted."[15]

Like scientific workshops, these large laboratories differed much one from the other – for example, in the nature of their workforce, the managerial style of their chief, the particular demands of their scientific subjects and institutional contexts, and, consequently, in their social-cognitive dynamics. I have characterized Pavlov's operation as a "physiology factory," but size did not itself a factory make: For example, Pavlov's Russian rival, V. M. Bekhterev, also commanded a large laboratory enterprise, but this operation lacked the strict and closely supervised system through which Pavlov bound his coworkers tightly to his own investigatory ambitions and interpretive decisions.

As a number of historians have pointed out, these large laboratories must be approached as sites of social production. Too often, however (and this reflects the curious state of our discipline today) this notion of science as production has been used to devalue the importance of ideas – and even of scientists as active agents –

as if ideas (and individuals) were not an important force of production. This tendency, of course, reflects in part the changing sociology of our discipline, in which individual scholars, by their background and training, often feel much more comfortable, capable, and/or interested in addressing either the "social aspect" or the "internal aspect" of science.

Pavlov's laboratory notebooks remind us of the organic union between the two. These notebooks were shaped by the social conditions of laboratory production – and must be read accordingly. They reflect the chief's particular managerial and interpretive style, and point the historian toward other materials necessary to placing the notebooks within the laboratory's system of production. In Pavlov's case, with most of the coworkers' notebooks missing, the other moments in laboratory production must be addressed through attention to the coworkers' doctoral dissertations, presentations, and articles. In the single happy case of his research with Maria Petrova, we have at our disposal a useful record of both the original experimental protocols and the chief's fairly immediate interpretive responses.

Since the social-cognitive division of labor varied markedly among large large laboratories, the nature of the laboratory notebooks used within them no doubt varied much as well. Judging from Gerald Geison's research, for example, it is clear that, although Louis Pasteur also directed a large laboratory workforce, his managerial style differed markedly from Pavlov's. For one thing, Pasteur was secretive about his hypotheses and experiments, while Pavlov's laboratory thrived upon open discussion and *glasnost*. One consequence of the differing division of labor in the two laboratories was, as Geison's research demonstrates, that the fine grain of Pasteur's laboratory investigations can be traced through his notebooks alone. Paul Ehrlich's managerial style, on the other hand, apparently more closely resembled Pavlov's – and his archive at the Rockefeller Institute even includes the scraps of paper upon which he wrote instructions to his coworkers.[16]

It would, I think, be very fruitful to compare the social-cognitive dynamics in a number of these large laboratories, using laboratory notebooks as a point of entry. Such a task, however, seems beyond the resources of any lone investigator, requiring historians to pool their talents in a coordinated effort.

NOTES

[1] The discussion of Pavlov's laboratory that follows is drawn from Daniel P. Todes, "Pavlov's Physiology Factory," *Isis* 88, 2 (1997): 205–246; and from my book, *Pavlov's Physiology Factory: Experiment, Interpretation, Laboratory Enterprise* (Baltimore and London: The Johns Hopkins University Press, 2002.)

[2] All of Pavlov's notebooks discussed in this article are held by the Sankt-Peterburgskii filial Arkhiva Rossiiskoi Akademii Nauk (The St. Petersburg branch of the Archive of the Russian Academy of Sciences) [Hereafter, SPF ARAN] in *fond* 259 *opis'* 1 *delo* 59.

[3] Copy of letter from Ivan Pavlov to Serafima Karchevskaia, June 3 [1894], SPF ARAN *fond* 259 *opis'* 7 *delo* 1300/2.

[4] I have heard numerous explanations for their disappearance. Some may well remain in private

hands, but it seems likely that the great bulk of them were burned for fuel during the siege of Leningrad in 1941–1944.

[5] Pavlov's archive also contains a number of loose sheets, in his hand and that of his coworkers, that apparently originated in discussions at the bench. Some of these contain experimental data; in one, Pavlov sketched the work assignments and schedule for experiments in his Towers of Silence. See, for example, SPF ARAN *fond* 259 *opis'* 1 *ed. kh.* 44.

[6] See Daniel P. Todes, "Pavlov and the Bolsheviks," *History and Philosophy of the Life Sciences* 17 (1995): 379–418.

[7] See M. N. Erofeeva, *Razdrazhenie kozhi faradicheskim tokom, kak uslovnyi vozbuditel' sliunnykh zhelez* (St. Petersburg: Military-Medical Academy doctoral dissertation series, 1912). Pavlov drew extensively upon Erofeeva's experiments in his public speeches of 1918, entitled "On the Mind in General," "On the Russian Mind," and "On the Foundations of Culture of Animals and Man." These are preserved in SPB ARAN *fond* 259 *opis'* 1a *dela* 3–5. These speeches and their significance for Pavlov's approach to higher nervous activity are the subject of my short article, "Ivan Pavlov on "Reflexes, Revolution, and Russia," in the March issue of www.praxispost.com. I am currently working on a translation of these speeches, with an analysis of their significance for understanding the development of Pavlov's scientific work.

[8] This is a central theme of the work in progress mentioned in note 7.

[9] This book, *Lektsii o rabote bol'shikh polusharii golovnogo mozga*, was translated immediately into English by G. V. Anrep as *Conditioned Reflexes: An Investigation of the Physiological Activity of the Cerebral Cortex*. A second Russian edition appeared in 1927, and in his preface Pavlov noted that he had intended to amend it in view of "the continued activity of my laboratory," but that illness had prevented him from doing so. In 1935, when a third edition appeared, Pavlov again declined to rewrite it: he presented his 1926 work as a classic – as the "fundamental and first systematic presentation of our facts" – that was now long out of date. Pavlov did not enjoy writing and, especially, rewriting. He had similarly declined to amend his *Lectures on the Work of the Main Digestive Glands* (1897) when it was republished fifteen years later, although , as he conceded in his preface to the new edition, some of its basic conclusions had been fundamentally revised by subsequent scientific developments.

[10] This is a central theme of the work in progress mentioned in note 7 above.

[11] From Pavlov's letter to the Council of Peoples' Commissars, 2 August 1932, in Gosudarstvennyi Arkhiv Rossiiskoi Federatsii *fond* 5446 *opis'* 13 *delo* 2012.

[12] See Todes, "Pavlov and the Bolsheviks."

[13] Orbeli's assessment of Petrova's scientific work is preserved in her personal archive in SPF ARAN *fond* 767 *opis'* 3 *delo* 5.

[14] This characterization of Petrova is from Pavlov's inscription in her gift copy of the 1925 edition of his *Dvadtsatiletnii opyt ob'ektivnogo izucheniia vysshei nervnoi deiatel'nosti (povedeniia) zhivotnykh*. This first book about conditional reflexes was not a systematic monograph, but rather a collection of Pavlov's articles on this subject. This and other inscribed copies of Pavlov's works are preserved in SPF ARAN *fond* 767 *opis'* 1 *delo* 110.

[15] William H. Welch, "The Evolution of Modern Scientific Laboratories," *LIV* (May 28, 1896): 87–90; on p. 88.

[16] See Arthur M. Silverstein, *Paul Ehrlich's Receptor Immunology: The Magnificent Obsession* (San Diego: Academic Press, 2002), pp. 145–146.

HANS-JÖRG RHEINBERGER*

CARL CORRENS' EXPERIMENTS WITH PISUM, 1896–1899[1]

INTRODUCTION

The circumstances under which classical genetics was established at the turn from the nineteenth to the twentieth century have become an integral part of the standard narrative on the history of genetics. Yet despite considerable scholarly efforts, it has remained one of the puzzles in the history of science at large why it took more than 30 years until the seminal work of Gregor Mendel was received and taken up broadly.[2] Moreover, it has also remained a matter of debate how exactly the so-called 'rediscovery' of Mendel's laws came about around 1900.[3] In addition to the conventions of scientific writing, priority disputes often obscure the way in which a particular finding or conclusion has been reached. A belated – and triple, if not multiple – 'rediscovery' such as in the case of Mendel's rules can be assumed to have required an even more sophisticated mise en scène on the part of the participants.[4]

In such a situation, unpublished research records can be invaluable tools to arrive at a thicker and more satisfying picture of the order of historical events. This paper makes extended use of the research protocols covering Carl Correns' hybridization experiments with *Pisum sativum*, and to a certain extent of those with *Zea mays*. The resulting reconstruction sketches the picture of a scientist following a particular research question, struggling with his experimental material, and slowly building up an epistemic regimen in which questions and observations could acquire relevance that did not strike Correns when he first took note of them. The process here told is, in a sense, the paragon of a story where precise origins and decisive turning points, those epitaphs of scientific accomplishment, are not easily located if they exist at all in the process of science in the making. The microhistorical gaze through the magnifying glasses of research notes reveals the kind of delays that appear to be constitutive for empirically driven thinking. The research notes of Correns help make that point. In addition, they display some of the intricacies and material peculiarities which characterize the experimental process of hybridization and the particular type of inferences it allows to make. I will come back to this point toward the end of the paper.

The protocols are preserved at the Archive of the Max Planck Society in Berlin-

---

* Max Planck Institute for the History of Science, Berlin

Frederic L. Holmes, Jürgen Renn and Hans-Jörg Rheinberger (eds.), Reworking the Bench: Research Notebooks in the History of Science, 221–252

Dahlem.[5] They must have traveled with Correns all the way from Tübingen through Leipzig, Münster, and finally to Berlin where they happened to survive World War II, although, according to Correns' biographer Emmy Stein, the bigger part of Correns' archive was destroyed early in 1945.[6] The bulk of the still existing protocols contains lists of results on loose sheets roughly ordered according to the experimental objects and the years in which the experiments were carried out. Unfortunately not all pages are dated, and Stein, who had worked on them during the last years of the war, certainly did not preserve the original order everywhere. This chronological uncertainty now causes, as we will see, considerable interpretational difficulties.

Along with Hugo de Vries[7] and Erich von Tschermak-Seysenegg,[8] Carl Correns is one of the three botanists generally credited with the rediscovery of Mendel.[9] Robert Olby remarks:

> Of the so-called three rediscoverers and the several other plant breeders who published Mendelian ratios in and around 1900 there are good grounds for believing that only one established the connexion independent of reading Mendel's paper – Carl Correns.[10]

A few years ago, I presented some evidence that the story is more complicated.[11] In the following, I will try to reconstruct in some detail Correns' path to the segregation law, which in his noted publication of 1900 he stated as follows:

> The hybrid produces sexual nuclei which bring together the dispositions [die Anlagen] of the parents in all possible combinations, with the exception of those of the *same* character pair. Every combination occurs approximately equally often.[12]

A SHORT BIOGRAPHICAL NOTE

Carl Erich Correns (1864–1933) was born in Munich, where he studied with Carl Wilhelm von Nägeli (1817–1891). He completed his dissertation "On thickening by intussusception in the membranes of several algae" in 1889.[13] The following years from 1889 to 1892 he traveled as an assistant to Gottlieb Haberlandt (1854–1945) in Graz, Simon Schwendener (1829–1919) in Berlin and Wilhelm Pfeffer (1845–1920) in Leipzig. He obtained his venia legendi at the University of Tübingen in 1892. Hermann Vöchting (1847–1917) accepted his work on the physiology of sensitivity in higher plants depending on free oxygen, for his Habilitation.[14] Correns spent the next ten years in Tübingen, where he performed his crucial breeding experiments on peas and corn in the small botanical garden of the university town. In addition, he cultivated some of his experimental crops in the plantations of commercial gardeners around Tübingen.[15] Correns continued to conduct extensive breeding experiments throughout his later career, which led him as an associate professor to Leipzig in 1902 and in 1909 to Münster as a professor and director of the botanical garden. In 1913, Correns was named director of the newly founded Kaiser Wilhelm Institute for Biology in Berlin-Dahlem, where he continued to work until his death in 1933.

THE EARLY TÜBINGEN CONTEXT

Correns' decisive work with *Pisum* falls in his years as Privatdozent in Tübingen between 1892 and 1902. An overview of the whole collection of notes left from the decade between 1890 and 1900 shows that Correns was busy with a plethora of different observations and experiments during that time. His interests appear to have been equally distributed between morphology, physiology, reproduction, systematics, and floristic observation. He continued work on the morphology and growth of the cell membrane which had occupied him already in his doctoral dissertation.[16] As a follow-up to his Habilitation, he pursued the physiology of movement in plants, especially the physiology of tendrils.[17] He started an extended research on the vegetative reproduction of mosses that was to result in a book which he completed in 1899.[18] He undertook systematic studies of algae, especially the Oscillatoria.[19] And last but not least, Correns also collected and published observations about alpine flora.[20]

Indeed, Correns had been trained in physiology, morphology, and systematics. He started to do breeding experiments only around 1894. In a typewritten, undated short autobiography, Correns remarks:

> During the ten years which I spent as a Privatdozent in Tübingen, my works and publications mostly followed along paths already taken. ... Besides that, as soon as I had the possibility to use a botanical garden through my Habilitation, I started, as an aside, to do various experiments which we would now qualify as 'genetic' and which had been going through my head already earlier.[21]

At the beginning, the crossing experiments appear to have had little direct impact on his overall still physiologically oriented activity, at least as far as their extent and systematicity is concerned. They obviously originated as research trials connected to problems of developmental physiology. Among them was the formation of adventitious embryos in *Hosta* (Funkia), a problem Correns tried to solve in vain through hybridization. In his autobiographical sketch, he himself characterized his early breeding experiments as "Allotria" – that is, rather unsystematic trials – in the context of his other ongoing projects during his time at Tübingen.[22] As we will see, however, over the years they gained considerable momentum, and after he had finished his book on the vegetative reproduction of mosses in the Fall of 1899, which had nothing to do with questions of heredity, these questions came to dominate all of Correns' work for the rest of his long career.

XENIA

Correns' interest in maize had arisen through contemplating Darwin's observations on xenia. Xenia are characters of the pollen-giving plant that become visible directly on reproductive parts of the mother plant, especially the seeds and fruits, and not only in the individuals raised from them in the next generation. When Correns took up this work in 1894, no generally accepted explanation for the phenomenon was yet available. When he finished it six years later, the principal solution to the xenia phenomenon – resulting from a process of double fertilization in Angiosperms – had already been found and published by Sergej Navashin in Kiev and by Léon Guignard in Paris.[23] Correns made the first crosses with *Zea mays* in 1894. Corn became his main experimental plant to explore the xenia question. From the literature, Correns had gathered hints for xenia in other plants as well. Therefore, in the course of the next few years, he added *Lilium, Matthiola*, and *Pisum* to his experimental objects.

In a folder with notes taken between 1896 and 1900 there is a sheet titled "Experiments of Gaertner with *Pisum*" and dated "15/IV."[24] It refers to Karl Friedrich Gärtner's "Versuche und Beobachtungen über die Bastarderzeugung im Pflanzenreich."[25] The date of the note is not complete, no year is given. There are, however, good reasons to assume that it dates from the beginning of Correns' work on peas – probably 1896 – since the summarizing remarks make it evident that Correns was occupied with the xenia question in peas, as can be seen from the following quote:

> To conclude – as G. himself also stresses, a very striking influence on the fertilized seed (not on the fertilized fruit!) Very remarkable that usually *all* seeds of one pod, eventually all pods of one cross (c. 5) did show this influence! Does the coloration of the seed rely on that of the cotyledons? Then the result would be explainable, understandable as an *intermediary formation*![26]

Another page on "Pisum," undated, summarizes a paper of Wilhelm Rimpau and retains his observation on "reversions": "Rimpau easily succeeded in performing several crosses, all of them showed a great inclination to reversions, even after 6 years several not constant."[27]

The biggest surprise in this folder is a note dated "16. IV. 96" and taken on "Mendel (66)" (Figure 1).[28] Its full text reads:

> '16. IV. 96
>
> *Mendel* (66) distinguishes:
>
> dominant and recessive characters. For our case is

	dominant:	recessive:
– form of seed	round	wrinkled
– seed coat: ("albumen")	grey to brown	*white*

– cotyledons:	yellow	pale-yellow, *green*
– pod:	inflated	constricted
– :	green (unripe)	yellow (ripe)

The dominant and recessive characters are expressed already in the first generation in such a way that the former are present in 3, the latter in 1 individual, respectively.

| The hybrid form of seed shape and cotyledons develops immediately and directly through fertilization |

therefore

Cot. yellow ♀ + green ♂ = yellow ♂ + green ♀ = 3/4 yellow + 1/4 green
Form. round ♀ + wrinkled ♂ = wrinkled ♀ + round ♂ = 3/4 round + 1/4 wrinkled

The seed coat, the form and the color of the pods are not *changed.*

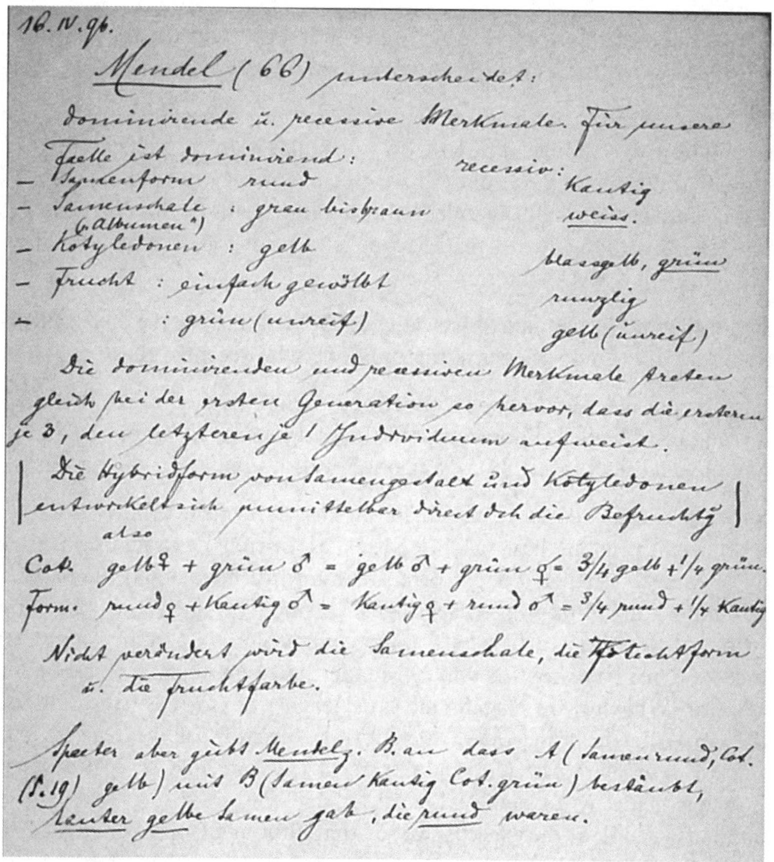
16.IV.96.

Mendel (66) unterscheidet:

dominirende u. recessive Merkmale. Für unsere Fälle ist dominirend: recessiv:
– Samenform rund Kantig
– Samenschale (u. Albumen*) : grau bis braun weiss.
– Kotyledonen : gelb blassgelb, grün
– Frucht : einfach gewölbt runzlig
– : grün (unreif) gelb (unreif)

Die dominirenden und recessiven Merkmale treten gleich bei der ersten Generation so hervor, dass die ersteren je 3, den letzteren je 1 Individuum aufweist.

| Die Hybridform von Samengestalt und Kotyledonen entwickelt sich unmittelbar direct durch die Befruchtg |

also

Cot. gelb ♀ + grün ♂ = gelb ♂ + grün ♀ = 3/4 gelb + 1/4 grün.
Form. rund ♀ + Kantig ♂ = Kantig ♀ + rund ♂ = 3/4 rund + 1/4 Kantig

Nicht verändert wird die Samenschale, die Fruchtform u. die Fruchtfarbe.

Später aber giebt Mendel z. B. an dass A (Samen rund, Cot. (S.19) gelb) mit B (Samen kantig Cot. grün) bestäubt, lauter gelbe Samen gab, die rund waren.

Figure 1. Correns' excerpt of Mendel's paper. Archive for the History of the Max Planck Society, III. Abt., Rep. 17, Nr. 115, folder "Pisum-Kreuzungen 1896–1900."

> But later *Mendel* notes, e.g., that A (seed round, cot. (p. 19) yellow) pollinated with B (seed wrinkled cot. green), *exclusively* yielded *yellow* seeds which were *round.*'

The conclusion is inescapable that despite his later recollections, Correns must have read Mendel's paper in April 1896. In the autobiographical sketch already mentioned, Correns stated:

> It was in a sleepless night in November [1899], toward morning, when the explanation for the observations on Pisum and Zea suddenly dawned upon me. But it was only when I was in the process of preparing the publication that I screened the literature systematically. Now I realized, through Focke's Pflanzenmischlinge [1881], that Mendel had found and published all this already thirty-five years ago. ... If I had found the explanation earlier, I would have published a preliminary note, despite the ongoing work on my book on mosses. For the significance of the results was quite clear to me at once.[29]

A similar version of the story is to be found in a letter that Correns wrote on 23 January 1925 on request and with an undertone of indignance to H. F. Roberts:

> The date of the day upon which, in the autumn (October) of 1899, I found the explanation, I no longer know; I do not make note of such matters. I only know that it came to me at once "like a flash," as I lay toward morning awake in bed, and let the results again run through my head. Even as little do I know now the date upon which I read Mendel's memoir for the first time; it was at all events a few weeks later.[30]

This standard story has remained basically unchallenged since the days of Roberts. To my knowledge, Onno Meijer is the only one who recently guessed

> that Correns did read Mendel's paper before that inspired night, that he failed to recognize Mendel's importance on the first reading of his paper, and that it may have worked in his mind subconsciously to reappear "like a flash."[31]

Let us see what evidence we can adduce for this view of the affair.

To begin with, let us have a closer look at Correns' extract from his early reading of Mendel's paper. First, Correns cannot just have noted the paper from Focke's book on "Pflanzen-Mischlinge,"[32] since he mentions the original pagination of Mendel's article. It is not unrealistic to assume that Correns possessed and read the reprint which Mendel had sent to his correspondent in Munich, Carl Wilhelm von Nägeli, and later teacher of Correns. It is of all events also the copy that is preserved in the reprint collection of the Kaiser Wilhelm Institute for Biology.[33] The Correns papers in Dahlem hold several other notes and drawings made by his mentor Nägeli.

Second, if we look at the excerpt, we see that although Correns noted Mendel's offspring ratio of 3:1 with respect to dominant and recessive characters, he appears to have concentrated his immediate attention on those characters of the seeds which could be related to the phenomenon of xenia that he was also pursuing in maize. This is suggested by his emphasis on the following sentence which he marked with a vertical line both to the left and to the right: "The hybrid form of seed shape and cotyledons develops immediately and directly through fertilization." This must have sounded like an interesting observation with respect to the problem of xenia.

The third point to make is that Correns in all probability was misled by Mendel's terminology and thus saw Mendel contradicting himself. Mendel speaks of the second generation of crosses as the "first generation of hybrids." Correns' summary reads: "The dominant and recessive characters are expressed already in the first generation in such a way that the former are present in 3, the latter in 1 individual, respectively." But then at the end of his note he points to an apparent contradiction: "But later *Mendel* notes, e.g., that A (seed round, cot. (p. 19) yellow) pollinated with B (seed wrinkled cot. green) *exclusively* yielded *yellow* seeds which were *round*." We can take this as a sign of a rather cursory reading.

The fourth observation points in the same direction. The aforementioned statement is also interesting in that the symbols A and B are not rendered by Correns in the way Mendel uses them in his article. Mendel uses the capital letters (A and B) for the *dominant characters*, whereas in his memo, Correns designates the two *plant varieties* to be crossed as A and B, respectively.

From the record of notes it is fairly safe to conclude that Correns had already decided to do experiments with peas before he read Gärtner, Rimpau and Mendel, since in April 1896, he had already purchased and chosen the six varieties with which to start. Accordingly, in the note on Mendel, he refers to "our cases" and then goes on listing five of Mendel's seven differing characters. The three characters he chose to follow were the form of the seed, the color of the seed coat, and the color of the cotyledons. A fourth character, the color of the seed, was a combination of the color of the seed coat and the color of the cotyledons. All this is in line with the purpose of following the possible fate of xenia. On April 23, 1896, Correns sowed six varieties of peas in six pots each with three peas of each variety (Figure 2).[34]

At this point, the difficult part of the reconstruction starts. There is no a priori reason to assume that Correns was bluntly lying when he wrote to Roberts 25 years later, and when he reported in his – undated – autobiography (also from the 1920s) that he had read Mendel only late in 1899, that is, after the completion of the fourth generation of his pea experiments and *after* the "explanation for the observations ... dawned upon [him]." Alternatively we can assume, following the guess of Meijer and in accordance with our analysis of the note on Mendel, that Correns screened Mendel's experiments from a quite different perspective when he read the paper in the Spring of 1896 for the first time. He might have made that note, put it aside, and even possibly have forgotten about it until it surfaced again a few years later and found its way into the experimental conjuncture and constellation that Correns had created in the mean time for himself.

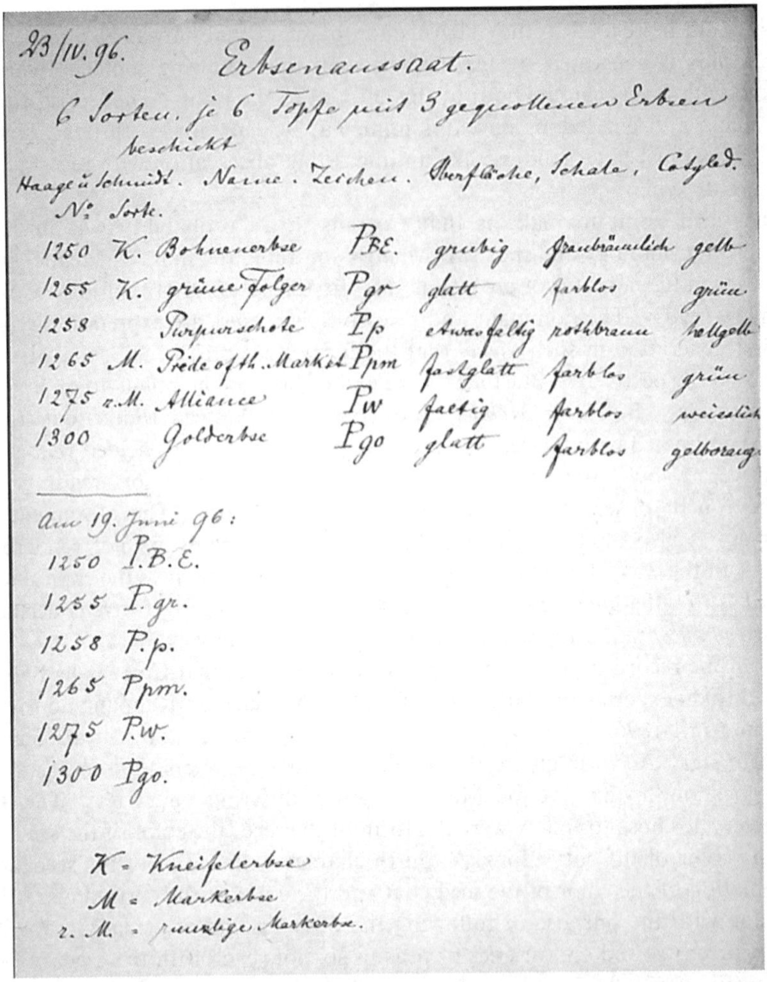

23/IV. 96. Erbsenaussaat.

6 Sorten, je 6 Töpfe mit 3 gequollenen Erbsen beschickt

Haage u. Schmidt. No	Sorte. Name.	Zeichen.	Oberfläche,	Schale,	Cotyled.
1250	K. Bohnenerbse	PBE.	grubig	graubräunlich	gelb
1255	K. grüne Folger	Pgr	glatt	farblos	grün
1258	Purpurschote	Pp	etwas faltig	rothbraun	hellgelb
1265	M. Pride of the Market	Ppm	fastglatt	farblos	grün
1275	r.M. Alliance	Pw	faltig	farblos	weisslich
1300	Golderbse	Pgo	glatt	farblos	gelborange

Am 19. Juni 96:

1250 P.B.E.
1255 P. gr.
1258 P. p.
1265 Ppm.
1275 P.w.
1300 Pgo.

K = Kneifelerbse
M = Markerbse
r. M. = runzlige Markerbse.

Figure 2. "Erbsenaussaat." Archive for the History of the Max Planck Society, III. Abt., Rep. 17, Nr. 115, folder "Pisum-Kreuzungen, Resultate 96."

THE PISUM PROTOCOLS OF 1896

What can the protocols tell us which have been preserved? First we note that Correns did not just select a particularly clearcut pair of characters as might be expected for a 'demonstrative' experiment aimed at showing whether Mendel's numbers held. What he did instead was to cross five out of the six varieties sowed in April with each other in reciprocal crosses. He made the first artificial crosses on June 19, 1896, when the plants were flowering. At the end of July and the beginning of August he harvested the first seeds. The results are listed in the following table: "Pollination and Return 96" (Figure 3).[35]

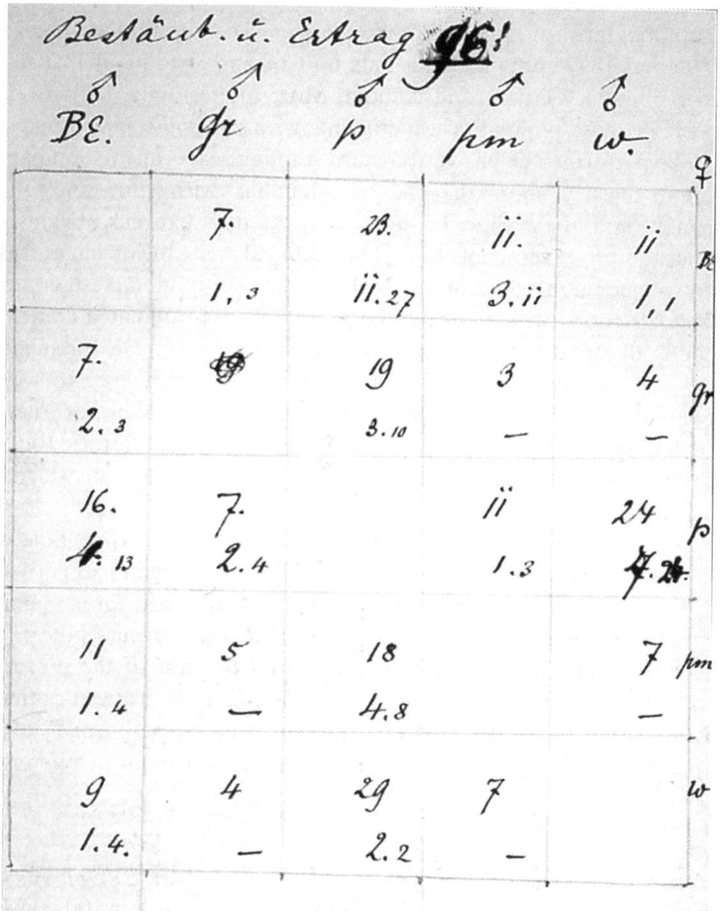

Figure 3. "Bestäub. u. Ertrag 96." Archive for the History of the Max Planck Society, III. Abt., Rep. 17, Nr. 115, folder "Pisum-Kreuzungen, Resultate 96."

The first number to the lower right of the squares represents the number of pods, the second the total number of seeds in them, whereas the number in the center of the squares represents the number of artificial pollinations. To take an example, the reciprocal crosses of P_{gr} ("grüne, späte Erfurter Folgererbse," in short "Grüne Folger") with P_p ("purpurviolettschotige Kneifelerbse," in short "Purpurschote") gave 4 and 10 seeds respectively, altogether 14 seeds resulting from 26 (7 plus 19) artificial pollinations! It is very clear from these numbers that the crosses must have been performed without the statistical counts in mind, as we might have expected if Correns had followed Mendel's precepts from the beginning.

That situation contrasts with experiments Correns performed in order to determine, during that same year 1896, the number of pollen grains necessary to achieve optimal fertilization of *Mirabilis* plants.[36] The unpublished notes from the years 1896 and 1897 contain again hints on Correns' reading at that time which included Kölreuter, Gärtner and Naudin. More interestingly, they reveal that in the context of these pollination experiments with *Mirabilis Jalapa* and *Mirabilis longifolia* he performed extensive statistical 'thought experiments' on paper.[37] His deliberations clearly show that he was familiar with this kind of treating experimental data at the time. In the case of the peas Correns obviously neither planned large scale experiments nor intended to calculate mean values. Instead, he wrote down meticulous descriptions of the seed color upon harvest, compared the artificial pollinations with a substantial sample of self-pollinated controls, noted conspicuous observations, and in some cases, even took measurements of the diameters and weights of a selection of seeds in addition. These latter characters had apparently nothing to do with the characters chosen for the crossing experiments. On the hybrid P_{pm+gr} (Pride of the Market fertilized with pollen from Grüne Folger) Correns noted laconically, thus recalling his starting point: "All Xen failed."[38]

It is excessively difficult and for the purposes of this paper unnecessary to follow in detail all the *Pisum* crosses, let alone those running in parallel with maize, through the consecutive breeding generations. Instead I will focus mainly on the one pair already mentioned: Grüne Folger (P_{gr}) and Purpurschote (P_p), while selectively extracting pertinent information from the rest of the protocols. The results obtained with this pair – yellow being dominant over green as the color of the embryo within the seed – are also represented in "Experiment I" of Correns' first paper on *Pisum* which we therefore can take as a point of reference in the course of our analysis (Figure 4).[39]

Versuch I.

Bastard zwischen der „grünen, späten Erfurter Folgererbse" mit grünem Keim und der „purpurviolettschotigen Kneifelerbse" mit gelbem Keim¹).

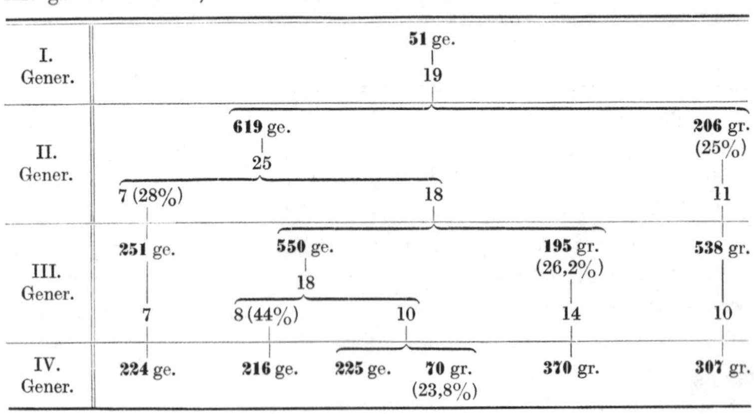

Figure 4. "Experiment I." (Correns 1900a), p. 162.

With respect to the crosses of the first generation, it should be kept in mind for the following analysis that for the time being Correns noted "yellow" as the color of the seeds for the hybrid P_{gr+p} (Grüne Folger fertilized with pollen from Purpurschote), whereas for the reciprocal hybrid P_{p+gr} (Purpurschote fertilized with pollen from Grüne Folger) he noted that the pods contained "purer green" or "conspicuously more greenish" seeds than the self-pollinated mother plant.[40]

THE 1897 PROTOCOLS

Let us now turn to the experiments performed in the next vegetation period starting in the Spring of 1897. Again we find an overview of first generation crosses (Figure 5).[41]

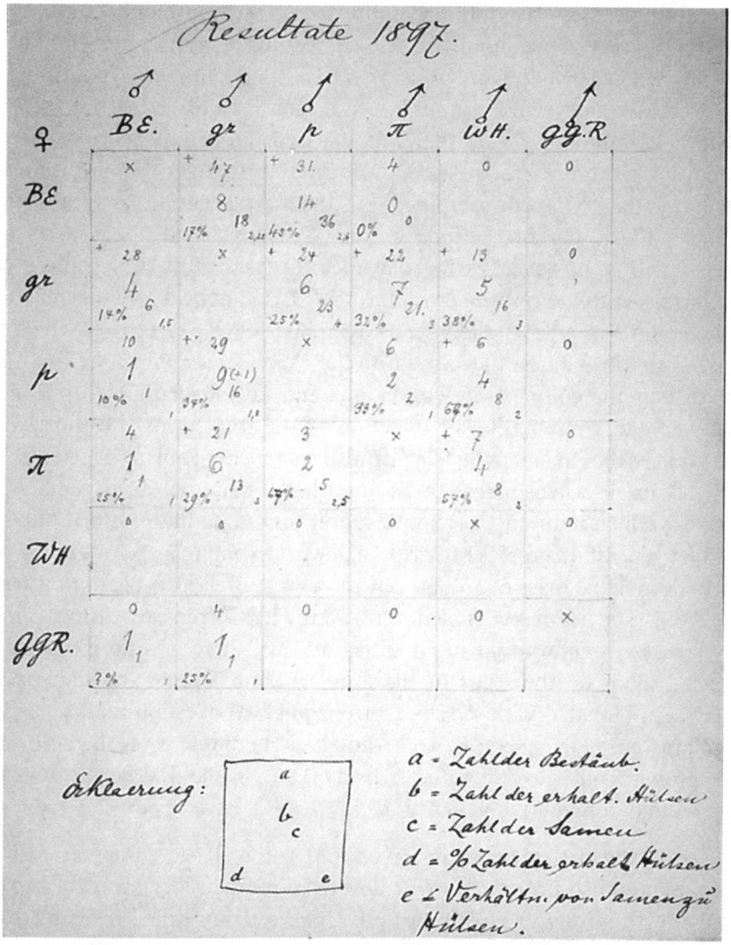

Resultate 1897.

♀ \ ♂	BE.	gr	p	π	WH.	gg.R
BE	x	+47, 8, 18, 17%, 2,25	+31, 14, 36, 45%, 2,6	4, 0, 0, 0%	0	0
gr	+28, 4, 6, 14%, 1,5	x	+24, 6, 23, 25%, 4	+22, 7, 21, 32%, 3	+13, 5, 16, 38%, 3	0
p	10, 1, 1, 10%, 1	+29, 9(+1), 16, 34%, 1,8	x	6, 2, 2, 33%, 1	+6, 4, 8, 67%, 2	0
π	4, 1, 1, 25%, 1	+21, 6, 13, 29%, 2	3, 2, 5, 67%, 2,5	x	+7, 4, 8, 57%, 2	0
WH	0	0	0	0	x	0
ggR.	0, 1, 1, ?%, 1	4, 1, 1, 25%, 1	0	0	0	x

Erklaerung: [a, b, c, d, e]

a = Zahl der Bestäub.
b = Zahl der erhalt. Hülsen
c = Zahl der Samen
d = % Zahl der erhalt. Hülsen
e = Verhältn. von Samen zu Hülsen.

Figure 5. "Resultate 1897." Archive for the History of the Max Planck Society, III. Abt., Rep. 17, Nr. 115, folder "Pisum Resultate 1897."

The three varieties P_{BE} (Bohnenerbse), P_{gr}, and P_p were crossed once again, the three new ones P_{π}, P_{WH}, and P_{ggR} were varieties which came to replace P_{pm} (Pride of the Market), P_w (Alliance weisse Zwergerbse), and P_{go} (späte Golderbse) from the previous year. The information is similar to that given in Figure 3, although it has become a bit more complex. The letters of the explanatory scheme refer to: a (top center) = number of pollinations; b (middle) = number of pods; c (to the right below b) = number of seeds contained in the pods; d (lower left) = % of pods obtained as compared to the number of pollinations; e (lower right) = the average number of seeds in one pod. Taken together, this multitude of numbers represents a kind of efficiency control for the artificial pollinations; it has nothing to do with a representation of the transmission of characters.

From the overview we gather that in 1897 Correns did a new series of extended first generation crosses. He obtained more seeds, but the numbers are again far from reflecting statistical significance as a result of conscious planning. For comparison, whereas in the previous year the total number of seeds from the reciprocal crosses between P_{gr} and P_p had been 14 (10 + 4), in 1897 it was altogether 39 (23 and 16). Again the 23 P_{gr+p} seeds looked all yellow, whereas the P_{p+gr} seeds looked more orange with an inclination to green.[42] With that, we end up with altogether 53 seeds obtained for the first generation. In the table of "Experiment I" in his first *Pisum* paper, Correns listed 51 yellow seeds, a difference which is, however, of no relevance for the results. For some unknown reason, Correns counted only 14 from the 1897 P_{p+gr} hybrid seeds instead of 16.[43] But he did count as "yellow" the 4 seeds from his 1896 P_{p+gr} crosses which he had rated "greenish" in the first place because the character on which he finally based his analysis was the color of the cotyledons and no longer the coloration of the seed as a whole (a combination of the color of the cotyledons and the seed coat).

Figure 6 lists the "results 1897" with respect to the problem of "xenia."[44]

From this list it is reasonable to assume that it was the failed xenia from the previous year which induced Correns to repeat and in addition to do a number of different first generation crosses rather than, a wish to obtain larger numbers of seeds in order to do reliable mass counting. But there were obviously again no clear xenia to be observed. The list makes evident, however, why Correns' experiments included so many crosses: It shows that in all cases but one, there was no influence of the pollen giving plant on the seeds of the mother plant. Behind the single potential exception – gr pollinated with WH –, Correns put two question marks.

At the bottom of the sheet, Correns noted as a comment on the coloration of the seeds grown from self-pollinated hybrids P_{gr+p} in the first generation, that is, from the second consecutive generation:

> gr + p in the first generation gave seeds whose basic color varied between green and reddish-orange and which were all more or less (scarcely and only in places up to strongly) purply dotted. Consequently [they] resembled neither the xenia nor the father nor the mother, but rather almost ggR (in which however the pods looked completely different).[45]

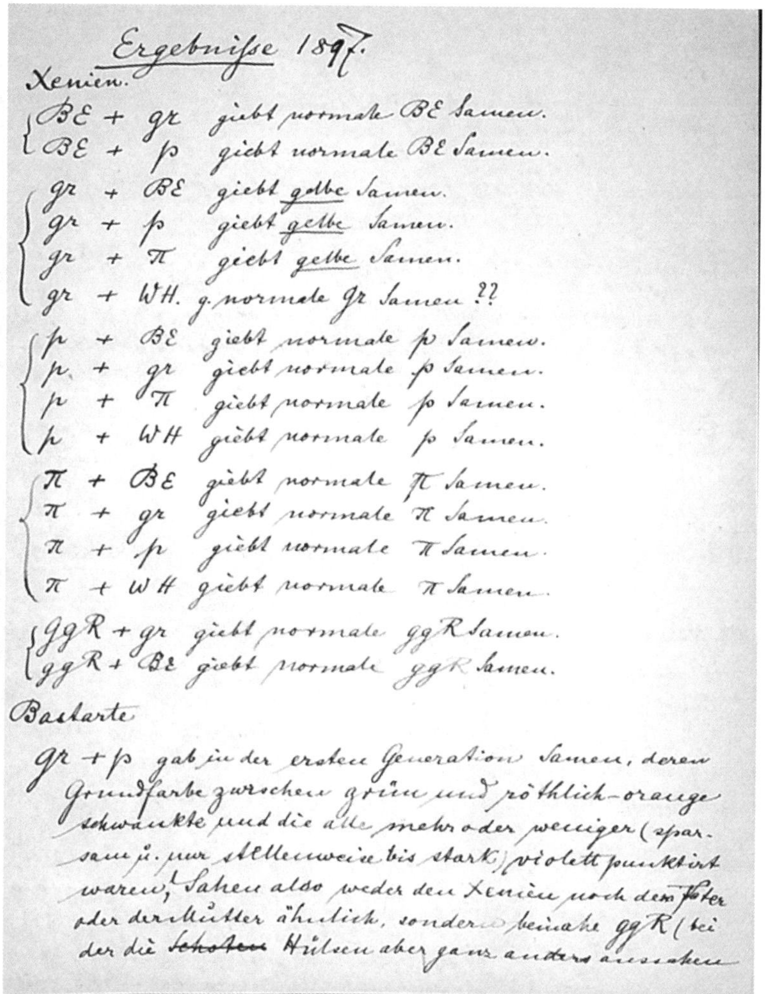

Ergebnisse 1897.
Xenien.
BE + gr giebt normale BE Samen.
BE + p giebt normale BE Samen.
gr + BE giebt gelbe Samen.
gr + p giebt gelbe Samen.
gr + π giebt gelbe Samen.
gr + WH. g. normale gr Samen ??
p + BE giebt normale p Samen.
p + gr giebt normale p Samen.
p + π giebt normale p Samen.
p + WH giebt normale p Samen.
π + BE giebt normale π Samen.
π + gr giebt normale π Samen.
π + p giebt normale π Samen.
π + WH giebt normale π Samen.
ggR + gr giebt normale ggR Samen.
ggR + BE giebt normale ggR Samen.
Bastarte
gr + p gab in der ersten Generation Samen, deren Grundfarbe zwischen grün und röthlich-orange schwankte und die alle mehr oder weniger (sparsam u. nur stellenweise bis stark) violett punktirt waren, sahen also weder den Xenien noch dem Vater oder der Mutter ähnlich, sondern beinahe ggR (bei der die ~~Schoten~~ Hülsen aber ganz anders aussehen

Figure 6. "Ergebnisse 1897." Archive for the History of the Max Planck Society, III. Abt., Rep. 17, Nr. 115, folder "Pisum Resultate 1897."

This last note brings me to a closer inspection of the protocols on which the results from the first self-pollinated hybrid generation are recorded. On April 20, 1897, Correns sowed seven yellow seeds from the 10 which he had obtained from the previous year's P_{gr} pollinated with P_p. As a control, he planted twelve seeds derived from the same individual plants which had resulted from self-pollination.[46] All plants and apparently all but one of the controls grew. The hybrids gave altogether 147 seeds contained in altogether 38 pods.[47] Correns meticulously noted the form and the color of every seed individually so that every seed could be traced back to the individual plant and even to the particular pod from which it had resulted. They were "orange yellowish green," "orange-greyish," "almost purely

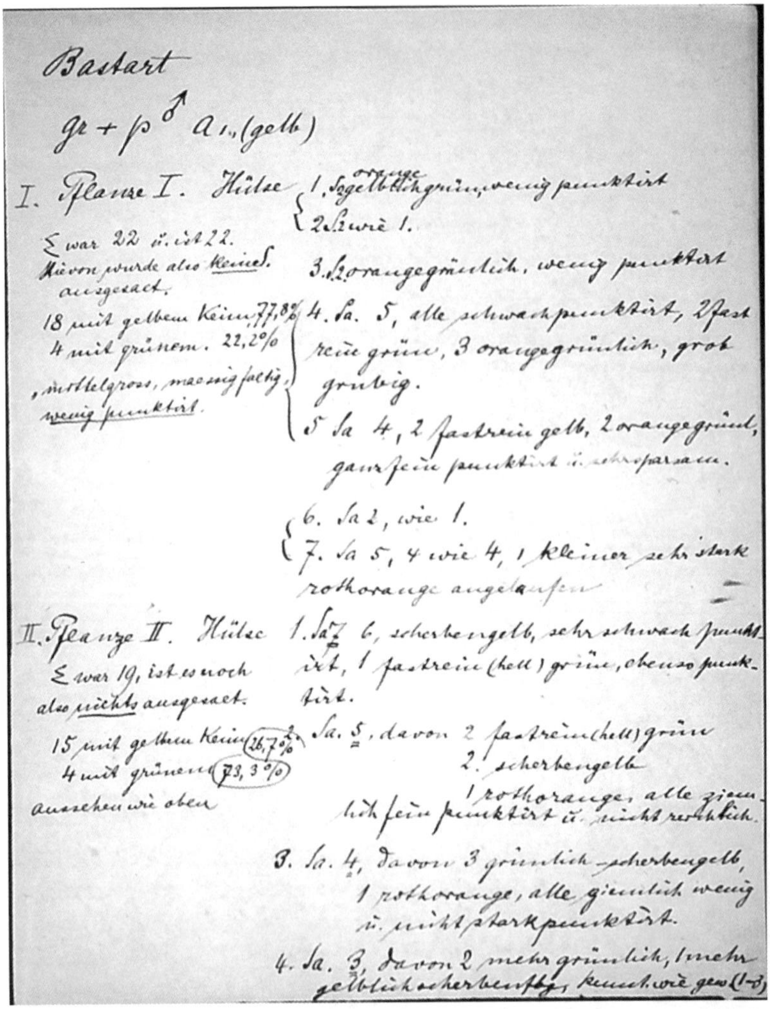
Bastart

gr + p ♂ A 1, (gelb)

I. Pflanze I. Hülse 1. Sa gelblichgrün [orange], wenig punktirt
2. S. wie 1.

Σ war 22 u. ist 22.
Hievon wurde also Keines ausgesaet.

3. S. orangegrünlich, wenig punktirt

18 mit gelbem Keim 77,8%
4 mit grünem. 22,2%
, mittelgross, maessig faltig,
wenig punktirt.

4. Sa. 5, alle schwach punktirt, 2 fast reine grüne, 3 orangegrünlich, grob grubig.

5 Sa 4, 2 fastrein gelb, 2 orangegrünl, ganz fein punktirt u. sehr sparsam.

6. Sa 2, wie 1.
7. Sa 5, 4 wie 4, 1 kleiner sehr stark rothorange angelaufen

II. Pflanze II. Hülse 1. Sa 7 6, scherbengelb, sehr schwach punktirt, 1 fastrein (hell) grün, ebenso punktirt.

Σ war 19, ist es noch
also nichts ausgesaet.

15 mit gelbem Keim 26,7%
4 mit grünem 73,3%
aussehen wie oben

2. Sa. 5, davon 2 fastrein (hell) grün
2. scherbengelb
1 rothorange, alle ziemlich fein punktirt u. nicht reichlich.

3. Sa. 4, davon 3 grünlich-scherbengelb, 1 rothorange, alle ziemlich wenig u. nicht stark punktirt.

4. Sa. 3, davon 2 mehr grünlich, 1 mehr gelblich scherbenfarbig, Punkt. wie gew. (1-3)

Figure 7. "Bastart gr + p ♂ A 1, (gelb)." Archive for the History of the Max Planck Society, III. Abt., Rep. 17, Nr. 115, folder "Pisum Resultate 1897."

green," "almost purely yellow" and came in another number of nuances. In a word, as noted above, their "basic color varied between green and reddish-orange." There is no clear sorting out into either green or yellow here, but there are many nuances which, as already mentioned, "resembled neither the xenia nor the father nor the mother."

Whereas in the protocols of the first generation, Correns registered the self-pollinated controls on the left part of the sheet and the cross-fertilized hybrids on the right side, the protocols of the second generation contain the detailed description of the seeds on the right side thus leaving an empty space to the left (Figure 7).

On this margin, we read for instance with respect to the results from the first of the seven plants, in a smaller and compressed handwriting: "$\sum$ was 22 and is 22. Consequently, *no* seed has been sowed from them. 18 with yellow germ, 77,8% 4 with green, 22,2%, medium large, moderately wrinkled, *only slightly dotted*." For plant 4, we find the annotation:

> $\sum$ 18, only 15 of them are still there. Therefore 3 are missing. From the rest (15) 3 have green germs = 20%, 12 yellow = 80%. Otherwise like I. Sowing yellow 2 green 1, therefore 14 yellow 77,8%, 4 green 22,2%.[48]

Obviously, Correns had turned back to these protocols at a later date, at the earliest some time after the sowing season of the next year, 1898, the date being derived from the fact that he must have already used the seeds which were now missing and whose number he sought to restore in order to arrive at correct percentages. And we see that now he is talking about the color of the germ ("Keim") and no longer of that of the seed ("Samen"). The calculated ratios for each of the seven plants were based on small numbers, and they varied considerably. But a trend becomes visible toward a segregation ratio of 3:1 in the first hybrid generation with respect to the dominant yellow color of the cotyledons inherited from P_p. This is, obvious from its spatial separation, clearly an ex post interpretation of the 1897 results. We will have to turn now to the protocols of the following year 1898 in order to see whether we can find hints as to what might have induced Correns to give up his xenia search and start to re-interpret his data according to segregation behavior.

THE PROTOCOLS OF THE YEAR 1898

In 1898, Correns grew a second generation of the hybrid $P_{gr♀+p♂}$. He took nine seeds from plant III (A1) from the hybrid generation of the previous year and planted them in three pots. Six of them were yellowish-orange, three of them had a strong green component. In addition, he planted three seeds from plant IV (A1) (two yellow and one green), three seeds from plant II (B1) (again two yellow and one green), and three seeds from plant I (B1) (all three yellow) in another three pots.[49] When Correns later reconstructed the number of yellow and green germs which he had obtained from the harvest of the preceding year 1897, there was a precise match of numbers. There is however one interesting discrepancy. From plant III of A1 Correns had sowed three green seeds – the protocol of 1898 says "seeds conspicuously *green*."[50] But when he reconstructed the numbers of yellow and green germs on the margin of the protocol of 1897, he notes as "green" only two germs. The reason is that only two individuals of the second generation hybrid plants in 1898 gave germs which were all green, whereas one of them gave a mixture of yellow and green germs. Correns thus concluded in retrospect that this greenish looking seed had contained a yellow germ and now counted it as yellow. This is yet another hint that the additional marginalia could not have been made before the fall of 1898, or later.

Another set of second generation hybrids was raised from the reciprocal cross, that is, $P_{p♀+gr♂}$. Correns stated that the germs from the seeds of these plants were "at times green, at times yellow," most of the individuals "represent the normal hybrid." Two of the individual plants however were "pure p" and therefore left out from further consideration.[51]

It is worth mentioning at this point that Correns drafted all these protocols in such a way that he later was able to trace back each individual seed and the plants raised therefrom to the individual plants and their seeds of the foregoing generation. He kept thus a virtually complete record of *individual* seeds on paper as well as in the form of a material repository. This observation is again in line with the assumption that the starting point of the series of *Pisum* experiments was not to corroborate a statistical regularity – as we might expect if Correns had derived such an idea from his first reading of Mendel's paper in 1896 and subsequently tested it – but that he was looking for expected or else for unknown characteristics of seeds that were emerging in this broadly conceived series of crossings. The protocols were thus not mere and inert records of data. Their very design and structure related to and translated the object of investigation as well as it focussed the attention on certain aspects of it. On the other hand, they were drafted such that there was enough redundancy and excess of possible information in both directions so as to allow for a reorientation of the experimenter's gaze at a later stage.

Just as the protocols of 1897, those of 1898 carry additional notes which must have been added later. Whereas the 1897 protocols had the additional notes on the margins, the 1898 protocols start with a short summarizing description of the seed which had been sowed and of the form and color of the seeds from the resulting harvest on top of the page, followed by additional notes, this time at the bottom of the pages. There Correns noted the total number of seeds obtained from each individual plant, the number of green and yellow embryos in them, and the percentages of each of them. And again, he complemented the actual number of seeds left over in his boxes with those which he had used as sowing material for the next generation. For instance, we read on the protocol from pot III, regarding individual plant number II (Figure 8):

> Pods with 3, 4, 4 seeds, green to light reddish-orange. From the 30 preserved there are 10 with green germs, 20 with yellow germs, from the seeds sowed were 1 with a green germ, 2 with yellow germs, therefore altogether 33 seeds, of them 11 = 33,3% with a *green* germ, 22 = 66,7% with a *yellow* germ.[52]

Another of these remarks on the protocol from pot VI, regarding individual plant number II reads: "∑ now: 7 green 20 yellow. Sowing: 1 green 2 yellow. ∑∑ 30, from them 8 green = 26,7%, 22 yellow 73,3%."[53] From these remarks it appears that we have to conclude that these additional remarks – and therefore most probably also those of the previous year – were only added after the next generation of plants had been sowed in 1899.

Let us now turn to another set of experiments carried out in 1898. In two more

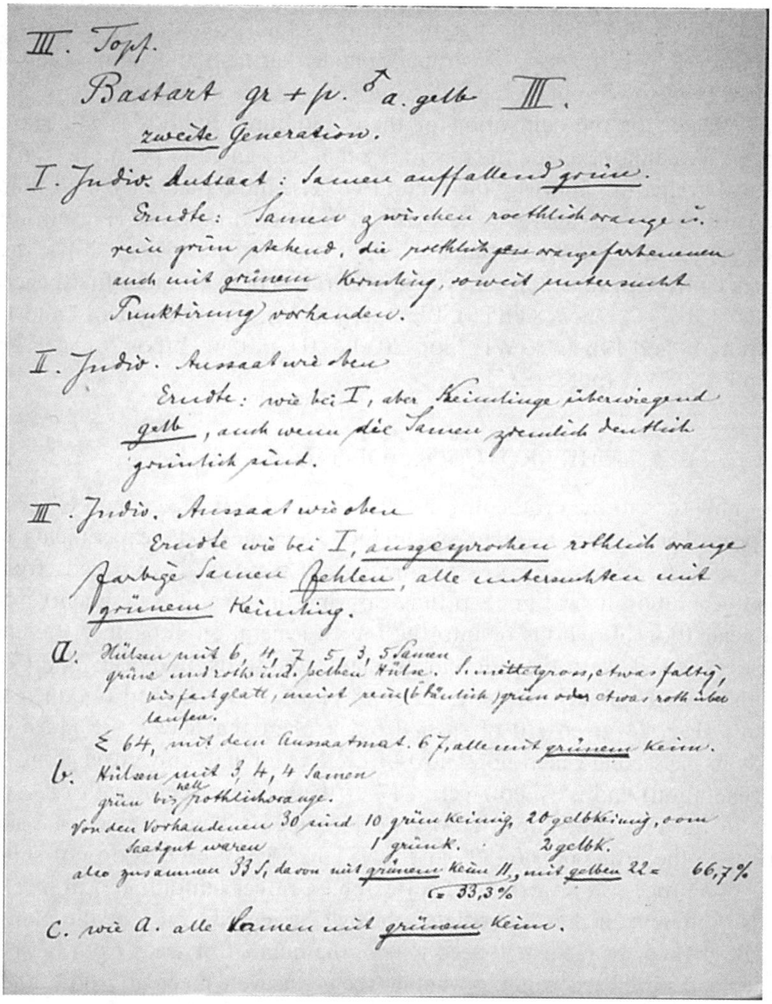

III. Topf.
Bastart gr + p. ♂ a. gelb III.
zweite Generation.
I. Individ. Aussaat: Samen auffallend grün.
Erndte: Samen zwischen röthlich orange u. rein grün stehend, die röthlich orange gefärbten auch mit grünem Keimling, soweit untersucht Punktirung vorhanden.
II. Individ. Aussaat wie oben
Erndte: wie bei I, aber Keimlinge überwiegend gelb, auch wenn die Samen ziemlich deutlich grünlich sind.
III. Individ. Aussaat wie oben
Erndte wie bei I, ausgesprochen röthlich orange farbige Samen fehlen, alle untersuchten mit grünem Keimling.
a. Hülsen mit 6, 4, 7, 5, 3, 5 Samen
grüne und roth und gelben Hülse. Samen mittelgross, etwas faltig, bis fast glatt, meist rein (bläulich) grün oder etwas roth überlaufen.
Σ 64, mit dem Aussaatmat. 67, alle mit grünem Keim.
b. Hülsen mit 3, 4, 4 Samen
grün bis hell röthlich orange.
Von den Vorhandenen 30 sind 10 grün keimig, 20 gelb keimig, vom Saatgut waren 1 grün K. 2 gelb K.
also zusammen 33 S, davon mit grünem Keim 11, mit gelben 22 = 66,7%
(= 33,3%
c. wie a. alle Samen mit grünem Keim.

Figure 8. "III. Topf, Bastart gr + p ♂ A1 gelb III." Archive for the History of the Max Planck Society, III. Abt., Rep. 17, Nr. 115, folder "Pisum Resultate 1898."

pots, Correns raised six additional individuals of the first generation of hybrids, apparently from seeds which he had obtained from the 1897 hybridization of P_{gr} with pollen from P_p. He gave again a short description of the resulting seeds. On one plant, he noted that he had made a backcross with the pollen of P_p: "1 ovary pollinated with p (wool red) (the other pollinations remained without result)." This pollination gave 5 seeds with exclusively yellow embryos.[54] The protocols do not tell us from which individual plant of 1897 the seeds had been taken. The plants, as Correns states, yielded "germs yellow or green," but he did not mention the numbers of each type.[55] Here again it is only on the lower half of the page that we find notes which indicate that the numbers and percentages there given for green and yellow germs were put to paper no earlier than after some seeds of this crop had been sowed out in the following year, that is in 1899.

The reason for the cultivation of these additional hybrids is not stated by Correns. We cannot exclude the possibility that he wanted to get more seeds for a statistical evaluation, but why then didn't he raise more plants? At that point, he could easily have done a large scale screening for the 3:1 ratio and concentrate on a few examples of his hybrids. But this is not what the protocols tell us. Instead, Correns cultivated, together with the appropriate controls, more first generation hybrids of BE + gr (pots 8 and 9), BE + p (pot 10), gr + BE (pots 14 and 15), gr + π (pots 18 and 19), gr + WH (pots 20 and 21), p + WH (pot 26), π + gr (pot 28), and π + WH (pot 29).

THE PROTOCOLS OF THE YEAR 1899

Let us now turn to the protocols for 1899, the final year of Correns' experiments with peas. They report a rather systematic extension of the experiments of the preceding year. A first set of experiments was performed with seeds from the second generation hybrid gr + p thus carrying this line of experiments into the third generation of hybrids or into the fourth generation altogether. In sum, 54 seeds were sowed, 49 of them developed plants and seeds. It appears that Correns took three seeds of each of the total of 18 plants which he had obtained in the previous year. 24 green (10 of them from a plant that gave only green germs already in the second generation and 14 from a plant with only green germs in the first generation) and 25 yellow germs (7 of them from a plant with only yellow germs in the first generation) grew and developed a huge number of seeds. In contrast to the protocols of 1897 and 1898, the 1899 protocols do not show the marks of a process of reworking. They exhibit a rather uniform appearance. They are also different in that immediately below the identification of the plant, e.g. "1.a. B, gr + p, II. Gen. I. 1. seed *green*" the number of seeds from the pods is counted and sorted according to whether the germs were green or yellow, and only then, instead of the other way round as earlier, further informations about the form, size and additional characteristics of the seeds are given (Figure 9).[56]

A second cluster of similar protocols (pot numbers 46 to 48) describes a second generation of gr + p hybrids derived from the first generation hybrids of 1898, a third cluster (pot numbers 55 to 58) reports the results from a second generation of reciprocal p + gr hybrids. These plants yielded altogether an impressive number of seeds which were all registered according to the yellow or green color of their cotyledons. The last hint about the xenia question stems from two hybrids raised in 1899 and termed "BE + gr ("xenia")." A questionable case in all probability, since the notion appears now in inverted commas and in the end receives no further comment, except that considering the results, Correns annotated: "Obviously gr + BE!"[57] The paper finally resulting from the *Pisum* experiments simply states that "my experiments on the formation of xenia [gave] here only negative results."[58]

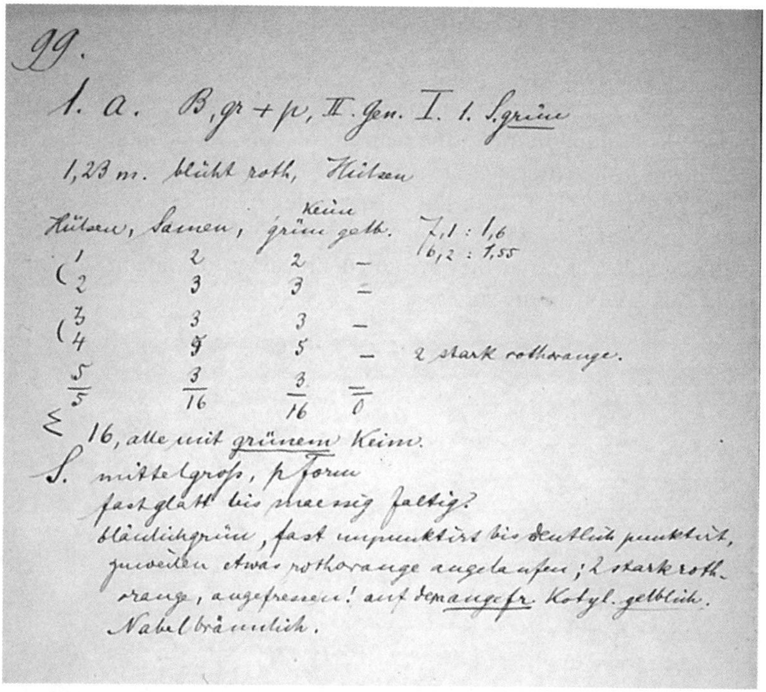
99.
1. a. B, gr + p, II. Gen. I. 1. S. grün
1,23 m. blüht roth, Hülsen

Hülsen,	Samen,	Keim grün	gelb.	
1	2	2	—	
2	3	3	—	
3	3	3	—	
4	5	5	—	2 stark rothorange.
5	3	3	—	
Σ 5	16	16	0	

7,1 : 1,6
6,2 : 1,55

16, alle mit grünem Keim.
S. mittelgroß, p Form
fast glatt bis schwach faltig?
bläulichgrün, fast unpunktirt bis deutlich punktirt,
zuweilen etwas rothorange angelaufen; 2 stark roth-
orange, angefressen! auf dem angefr. Kotyl. gelblich.
Nabel bräunlich.

Figure 9. "99. 1. a. B, gr + p, II. Gen. I. 1. S. *grün*." Archive for the History of the Max Planck Society, III. Abt., Rep. 17, Nr. 115, folder "Erbsen 99."

As usual, no date is entered on the individual protocol sheets, but in all events they can be assumed to have been written in the Fall of 1899. It is, in addition, quite reasonable to assume that the retrospective additions to the protocols from the years 1897 and 1898 date from the same time. At that time, Correns was fully aware of the segregation pattern with its pure lines on the one hand and its 3:1 ratio for the hybrids on the other hand. There are no direct hints from the survey of this whole set of protocols as to whether Correns was already on the track to a solution after he had contemplated the results of his 1898 harvest or even earlier. In principle he could have guessed from looking at his second generation hybrids that all the germs resulting from his green peas remained green and that some resulting from his yellow ones also remained yellow but others split again into yellows and greens. On the one hand, as already stated, the notes added to the 1898 protocols suggest that the numbers and percentages were inserted only after the 1899 crop had been sowed. On the other hand, the huge and rather systematically conducted effort of raising hybrids of the second and the third generation for the gr + p and p + gr combinations as well as for a number of others suggest that Correns must have had some good reasons for engaging in such an extended endeavor in the Spring and Summer of 1899, the time he was completing his book on mosses.

THE IMPACT OF ZEA MAYS

From the notes accompanying the set of *Pisum* protocols it does not appear possible to close this information gap. There is, however, an interesting and decisive note to be found among the maize protocols. It is contained in a folder labeled "Theoretical Matter etc." (Figure 10).[59]

The note is dated 2. I. 97 with the 7 overwritten by an 8. Correns may not yet have been accustomed to write the new year 98 on the second of January. The accompanying notes, as far as they are dated, are all written late in 1897 and early in 1898. In this note Correns states:

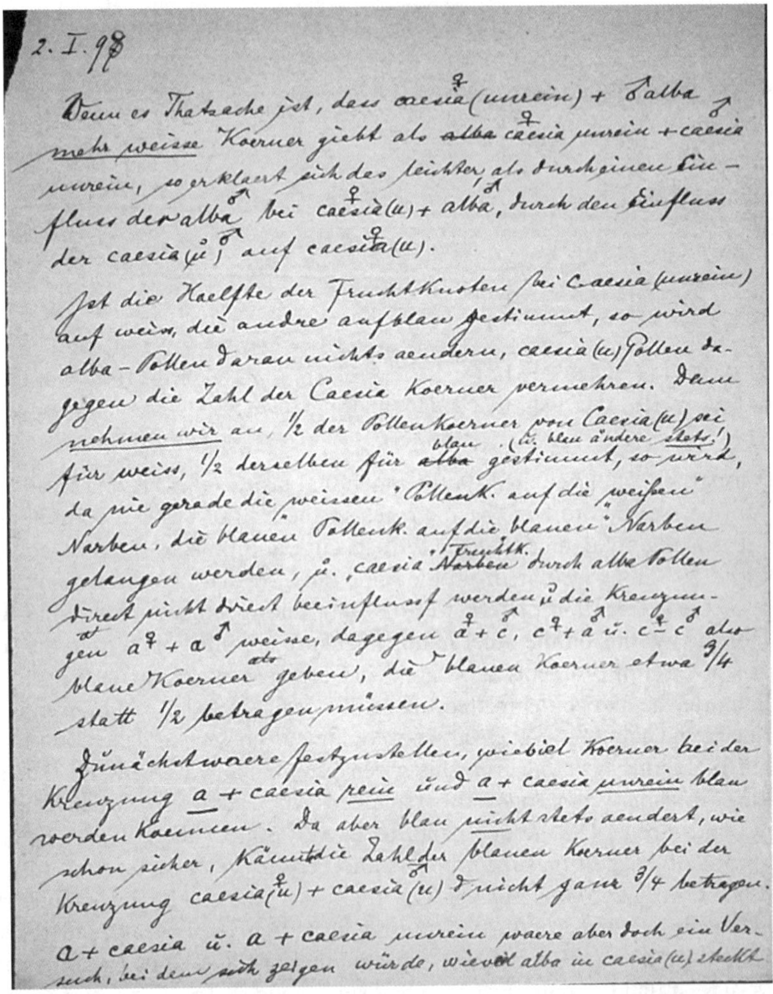

2. I. 98

Denn es Thatsache ist, dass caesia ♀ (unrein) + ♂ alba mehr weisse Koerner giebt als ~~alba~~ caesia ♀ unrein + caesia ♂ unrein, so erklaert sich das leichter, als durch einen Einfluss der alba ♂ bei caesia ♀ (u) + alba ♂, durch den Einfluss der caesia (u) ♀ auf caesia ♂ (u).

Ist die Haelfte der Fruchtknoten bei caesia (unrein) auf weiss, die andre auf blau bestimmt, so wird alba-Pollen daran nichts aendern, caesia (u) Pollen dagegen die Zahl der Caesia Koerner vermehren. Denn nehmen wir an, ½ der Pollenkoerner von Caesia (u) sei für weiss, ½ derselben für ~~alba~~ blau bestimmt, (u. blau ändere stets!) so wird, da sie gerade die weissen Pollenk. auf die weissen Narben, die blauen Pollenk. auf die blauen Narben gelangen werden, u. caesia ~~Narben~~ Fruchtk. durch alba Pollen direkt nicht direkt beeinflusst werden, u. die Kreuzungen a ♀ + a ♂ weisse, dagegen a ♀ + c ♂, c ♀ + a ♂ u. c ♀ + c ♂ also blaue Koerner geben, die blauen Koerner etwa 3/4 statt 1/2 betragen müssen.

Zunächst waere festzustellen, wieviel Koerner bei der Kreuzung a + caesia rein und a + caesia unrein blau werden koennen. Da aber blau nicht stets aendert, wie schon sicher, könnte die Zahl der blauen Koerner bei der Kreuzung caesia ♀ (u) + caesia ♂ (u) nicht ganz 3/4 betragen.

a + caesia u. a + caesia unrein waere aber doch ein Versuch, bei dem sich zeigen würde, wieviel alba in caesia (u) steckt.

Figure 10. Note on maize, 2. I. 98. Archive for the History of the Max Planck Society, III. Abt., Rep. 17, Nr. 85, folder "Theoretisches etc."

> If it is a fact that caesia ♀ (impure) + alba ♂ gives *more white* kernels than caesia ♀ impure + caesia ♂ impure, this is more easily explained through an influence of caesia ♂ (impure) on caesia ♀ (impure) than through an influence of alba ♂ on caesia ♀ (impure) + alba ♂.[60]

Obviously he is inclined to exclude xenia effects of *alba* pollen in this case. Note also that Correns gives no definite ratio of white and blue kernels. "Caesia impure" is the first generation hybrid between the corn variety *caesia* and the variety *alba* yielding (impurely) blue grains. What Correns compares here is the outcome of a second generation self-fertilization of an *alba* + *caesia* hybrid to a backcross of the hybrid with an alba parent. He goes on with a hypothetical alternative explanation:

> If one half of the ovaries in caesia (impure) is disposed for white and the other half for blue, then the pollen of alba will change nothing, but caesia (impure) pollen will increase the number of caesia grains. For *if we assume* that one half of the pollen grains of caesia (impure) be disposed for white, one half of them for blue (and that blue alters *always*!), then, because the "white" pollen grains never just will come to the 'white' stigmas, the blue pollen grains to the 'blue' stigmas, and because 'caesia' ovaries are not directly influenced by alba pollen and the crosses a ♀ + a ♂ give white, whereas a ♀ + c ♂, c ♀ + a ♂ and c ♀ + c ♂ therefore give blue grains, the blue grains must amount to approximately 3/4 instead of 1/2. First it would have to be tested how many grains can become blue when *a* + caesia *pure* and *a* + caesia *impure* are crossed.[61]

But then Correns immediately goes on to qualify the supposition "that blue alters *always*" and suspects that therefore "the number of blue grains ... could be slightly less than 3/4)."[62] Here we have, for the first time, in a rather clearcut albeit conditioned form, the supposition of a 3:1 segregation in the first hybrid generation and concomitantly the assumption of an underlying disjunction of the factors responsible for the blue and white color of the grains, respectively. It is quite clear that Correns in this case excludes the intervention of xenia – "'caesia' ovaries are not directly influenced by alba pollen," he says – and that therefore there is something different to be contemplated in these corn crosses. He comes back to that point in a note on February 24, 1898:

> With respect to a dissociation of the characters *in the pollen* experiments with the hybrid vulgata + dulcis would have to be conducted. (Because the grains are either smooth *or* wrinkled, whereas with alba + caesia intermediate formations occur),[63]

that is, seeds with a more or less intense color. Thus the conclusion is at hand that Correns was already ruminating about the later solution to the problem when he was preparing himself for the crosses of the season 1898.

The first suggestion obviously came, as the words in this note indicate, from *caesia* and *alba* corn crosses in which one of the crossing partners revealed itself to be "impure." And indeed, if we go back to the maize protocols we are able to trace a few of the elements that came to play a role in the explanatory proposition just quoted. As early as 1894, Correns had noted that the corn variety *alba* pollinated with *caesia* gave "*all* grains, except one, +/- 'caesia.'"[64] Blue thus "alters always," as Correns stated in the note quoted above. In 1896, he again found the "influence of *caesia* rather unquestionable," although he had come to be suspicious about the "purity of the *alba*,"[65] that is, the constitution of the mother plant. A self pollination experiment carried out with another caesia plant led him to the conclusion that "the caesia was *not pure*."[66] The reciprocal cross, where he fertilized a *caesia* plant with pollen from *alba* showed Correns that, if he excluded one of his experiments where obviously one of the *caesia* crossing partners again was not pure, "*no influence of the pollen on the ovary is recognizable*."[67] Thus he knew that *alba* pollen does not give rise to xenia, and that blue suppresses white. In a series of further fertilizations of *alba* plants with *caesia* pollen, however, the expected alteration toward blue grains was not complete. What he found was a mixture of blue and white grains. Yet by that time Correns was so convinced of his previous results that he concluded that the *caesia* which he used "was not pure, but c + a!"[68] The reciprocal cross between a *caesia* mother plant and an *alba* pollen plant also delivered "grains approximately 1/2 *alba*, 1/2 *caesia*."[69] As it were, these were inadvertent backcrosses because one of the crossing partners happened to be a hybrid itself. Thus we come to the surprising result that it were unintended backcrosses of corn varieties, together with a fairly established dominant character in a context where xenia could be ruled out, that condensed into a turning point at which Correns' experiments began to take a different orientation.

MERGING LINES

At that point, the *Pisum* crosses with their notorious absence of xenia took over from maize where phenomena of intermediacy of characters and xenia themselves complicated matters considerably. In an undated note that can be assumed to have been written almost two years later, in the Winter of 1899/1900, Correns finally comes back, now with a clear picture in mind of the regularities having taken shape from the 1898 and the 1899 crosses with *Pisum*, to the crosses between the maize varieties *alba* and *caesia*. As if he had forgotten the starting point just described, the relation between the two plant groups is now inverted, and Correns even emphasizes the following question with a second mark: "Is the behavior of the hybrids of the Pisum races also applicable to the hybrids between races of maize??"[70] In a similar vein, in another note from about that same time, he talks about the possibility of an "application of Mendel's theory to the hybrids between races of corn."[71]

The relevant notes belonging to the *Pisum* series that appear to have been written during the same time, in the Winter of 1899/1900, are unfortunately all undated. On one of these sheets, we find Correns make statistical assumptions on the generative process underlying his results:

Combination of these cells with one disposition each to *pairs* (sexual act). Comparison with a sac of 2000 balls (1000 yellow, 1000 green). Probable: 500 unequal pairs, 500 equal ones, from them 250 times yellow and yellow, 250 green and green. Therefore 250 ge + ge, 500 gr + ge, 250 gr + gr [Correns mistakenly writes ge + ge]. ... *Repeated* hybridization 250 gr, 750 ge. 500 gr + ge and 250 ge + ge only distinguishable through the *experiment* (progeny). Veritable genealogical tree (see Figure 11). Does this also hold for intermingling characters?[72]

Correns then goes on to make the respective calculations for backcrosses with the parent plants. These considerations obviously formed the basis for the

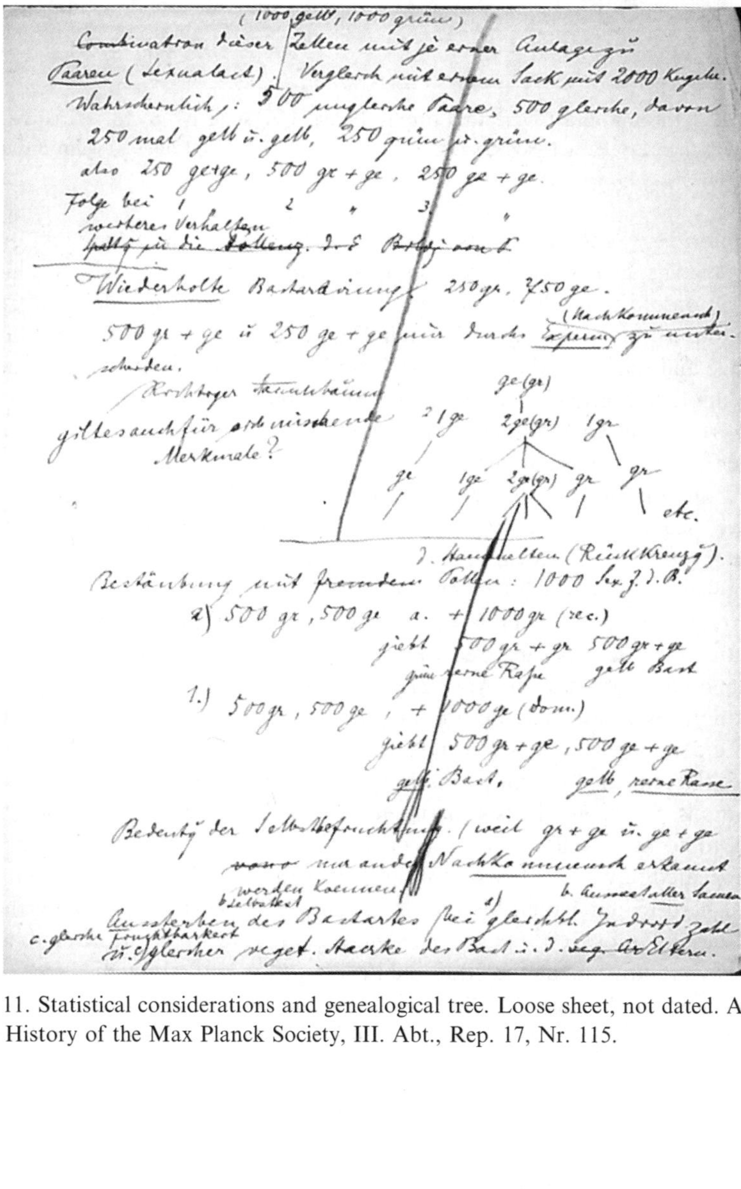

Figure 11. Statistical considerations and genealogical tree. Loose sheet, not dated. Archive for the History of the Max Planck Society, III. Abt., Rep. 17, Nr. 115.

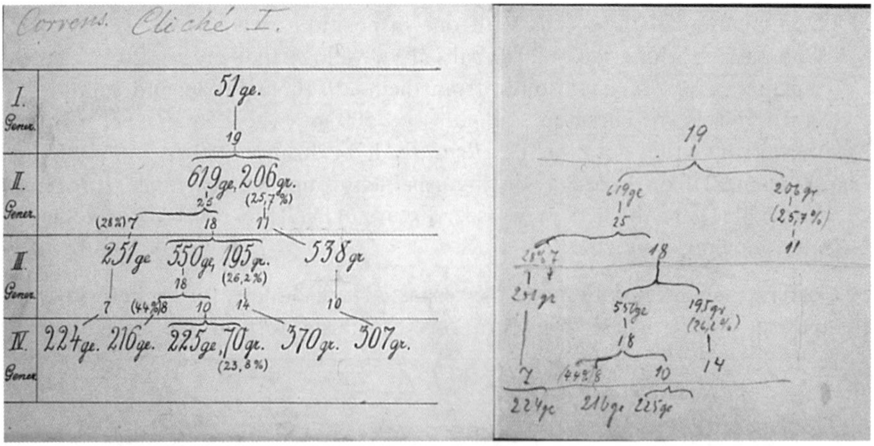

Figure 12. Drawing and Cliché to Correns 1900a, "Versuch I," p. 162; Archive for the History of the Max Planck Society, III. Abt., Rep. 17, Nr. 115, folder "Pisum-Kreuzungen 1896–1900."

description he was to give in his *Pisum* paper which he sent to the press on April 24, 1900. The genealogical tree on this sketch forms the basis for his later represention of the results (see Figure 4).[73] We can see its form develop from the sketch to the draft of the cliché shown in Figure 12, to the cliché itself on the left of this page and finally to the rectilinear form it assumes in print ("Experiment I").

On another drafted sketch (Figure 13) Correns tries a representation of the results from three generations of P_{gr+p} and P_{p+gr} hybrids which was never taken up later and was never published.[74]

And on still another of these sketches (Figure 14) we find a drawing of the putative "sequential order" of "dispositions" in hybrids differing in more than one character pair.[75] This drawing became the basis for the explanatory scheme of the chromosome theory which Correns proposed in 1902.[76]

Correns was thus well prepared to put together his *Pisum* manuscript within two days when he was surprised by the reprint of Hugo de Vries' "Sur la loi de disjonction des hybrides" on April 21, 1900.[77] All the relevant pieces, including the calculations, were in place. Before this event, he does not appear to have been in such a great hurry to publish his results. As can be seen from the record of notes, between December 1899 and March 1900 Correns was extensively revising the voluminous protocols of his experiments with maize varieties. This he did with respect to xenia as well as with respect to the calculation of progeny ratios where they appeared obvious.[78] It was not until he received de Vries' reprint that he felt pressed hard to publish all his results as quickly as possible, especially those on maize besides those on *Pisum*. This may also have been the reason why he obviously did not pursue the additional experiments he had planned to carry out with peas in the Spring of 1900.[79] He now had to position himself not only in relation to his predecessor Mendel, but also in relation to his contemporary and

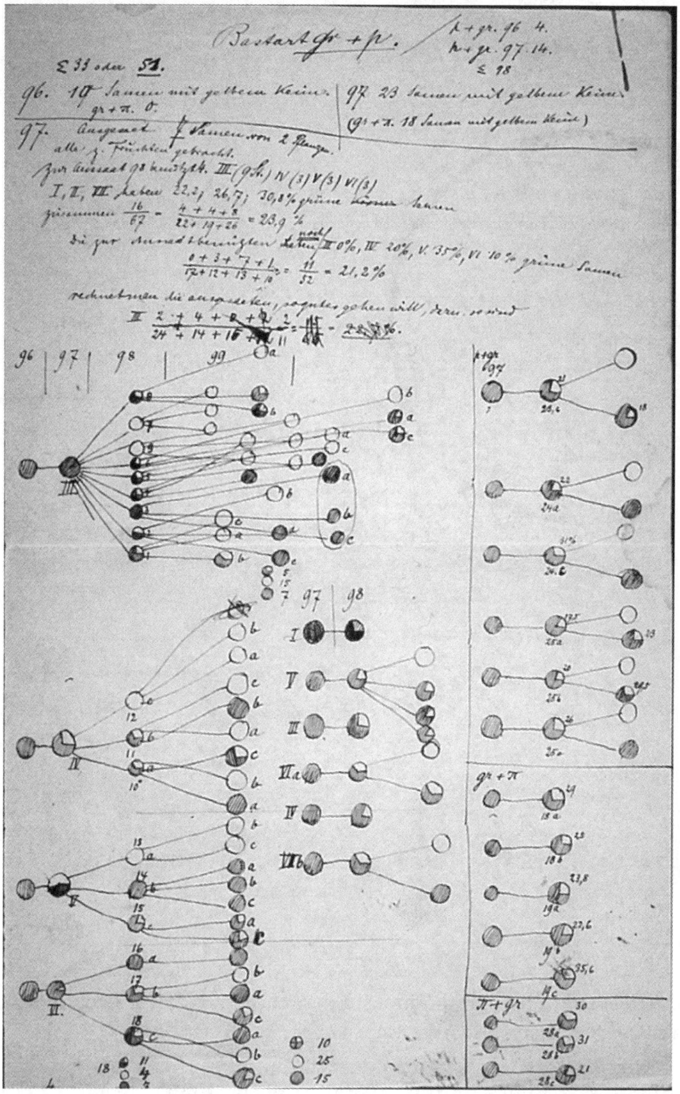

Figure 13. Representation of segregation ratios. Loose sheet, not dated. Archive for the History of the Max Planck Society, III. Abt., Rep. 17, Nr. 115.

competitor de Vries. He did that in a subtle, amphibolic way in the introduction to his *Pisum* paper. The passage reads as follows:

> When I had found the lawful behavior and its explanation ... I fared as de Vries obviously does now: I held all this to be something new. But then I had to convince myself that the abbot Gregor Mendel in Brünn in the 1860s ... had come to the same result as de Vries and myself .... [80]

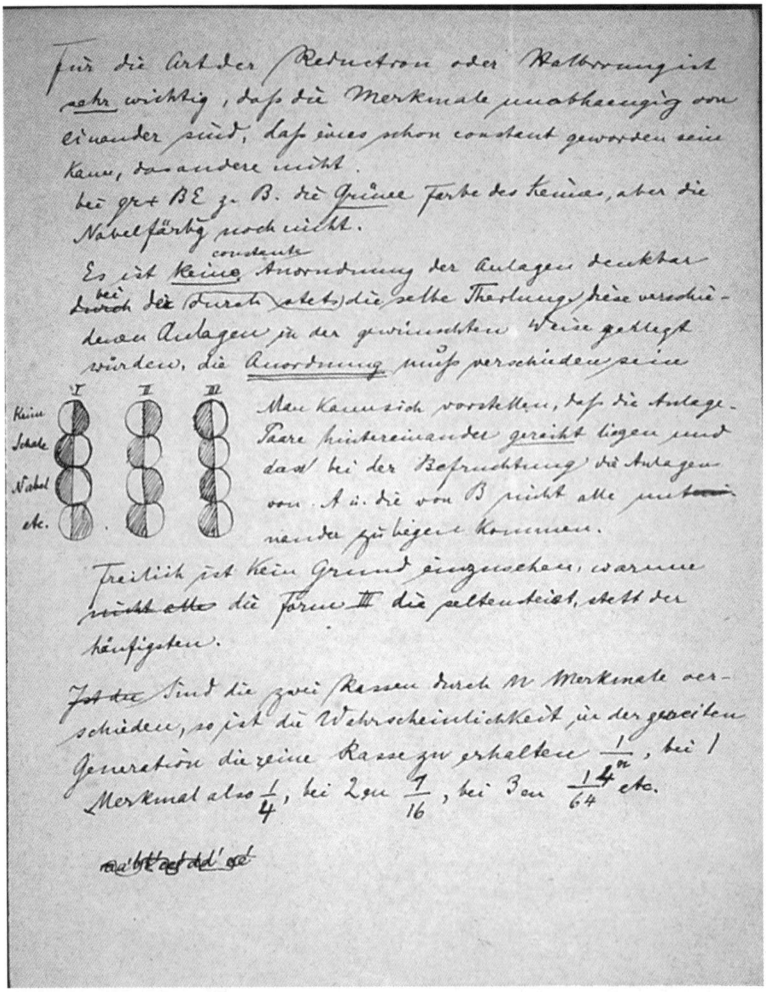

Für die Art der Reduction oder Halbirung ist sehr wichtig, daß die Merkmale unabhaengig von einander sind, daß eines schon constant geworden sein kann, das andere nicht.
bei gr + BE z. B. die Grüne Farbe des Keimes, aber die Nabelfärbg noch nicht.
Es ist keine constante Anordnung der Anlagen denkbar bei der (durch stets) die selbe Theilung diese verschiedenen Anlagen in der gewünschten Weise getheilt würden. Die Anordnung muß verschieden sein.

I II III
Keim
Schale
Nabel
etc.

Man kann sich vorstellen, daß die Anlage-Paare hintereinander gereiht liegen und dass bei der Befruchtung die Anlagen von A u. die von B nicht alle neben einander zu liegen kommen.
Freilich ist kein Grund einzusehen, warum die Form III die seltenste ist, statt der häufigsten.
Sind die zwei Rassen durch n Merkmale verschieden, so ist die Wahrscheinlichkeit, in der zweiten Generation die reine Rasse zu erhalten $\frac{1}{4^n}$, bei 1 Merkmal also $\frac{1}{4}$, bei 2en $\frac{1}{16}$, bei 3en $\frac{1}{64}$ etc.

Figure 14. Sequential order of dispositions. Loose sheet, not dated. Archive for the History of the Max Planck Society, III. Abt., Rep. 17, Nr. 115.

This formulation is interesting in several respects. First, Correns dissociates the finding that the hybrids behave "lawful[ly]" from the "explanation" of that finding. Second, he implies that he had found both before he "had to convince himself" that Mendel had arrived at these results already several decades ago. But Correns clearly avoids any explicit statement as to how and when he had hit upon Mendel. Third, he claims implicitly that he had arrived at the solution before de Vries by stating that "I fared as de Vries obviously does now." And finally, with that same formulation in which Correns claims priority over de Vries he implicitly reproaches de Vries for "obviously" believing his results to be new although they went back to Mendel. The ambiguity of the expression "obviously" even keeps open the possibility that de Vries did know about Mendel but did not tell. And Correns reinforced the contrast by taking himself back and publishing his own paper under the title "G. Mendel's Regel über das Verhalten der Nachkommenschaft der Rassenbastarde."

During that same year, Correns rushed to publish two more papers on hybridization.[81] In contrast, after a preliminary publication on xenia in maize a year earlier (1899),[82] his huge maize monograph on "Hybrids between races of maize with special attention to xenia" came finally only in 1901.[83] On the field of xenia, too, de Vries had been a little quicker than Correns to publish analogous results both as a preliminary account in 1899 and as a more extended report in 1900,[84] after which Correns decided grudgingly to "postpone" his own publication for another year.[85]

CONCLUSION

My reconstruction leads to the conclusion that, despite Correns' early note on his reading of Mendel's paper in April 1896, the segregation behavior did not take on a clearcut meaning for Correns immediatly. It must, however, have imposed itself upon him in its prospective significance at some point at the beginning of 1898, as a result of the outcome of inadvertent backcrosses of corn varieties, and then gradually have hardened into a matter of fact by the fall of 1899. Concomitantly, in the case of *Pisum* the xenia problem lost importance for him and was completely dropped at the end because the results proved to be negative. From the notes, it cannot be excluded that the puzzle finally fell in place for Correns only late in 1899. But that the solution hit him like "a sudden flash" must definitely be rejected. Likewise, the assertion that Correns read Mendel's paper only toward the end of 1899 does not hold. The assertion can at best mean that he reread it late in 1899 and that this time, he saw it with decidedly different eyes.

My interest in reconstructing Correns' path of investigation has not been to demystify yet another of the so-called rediscoverers of Mendel's laws. My general interest in this case study, as well as in others,[86] has been and is to shed light on the question of how entering into and establishing a particular experimental system can develop an agency of its own and can lead researchers into directions which they did not anticipate. Correns, it seems to me, is a particularly interesting example just because he *could* have known more in advance. He even had an important source of information before his eyes when he started to set up his experimental system. However, it did obviously not make sense to him from the perspective of his own starting point, which was not to find the rules of hybridization, but to elucidate the processes which lead to the formation of xenia.

It can be seen as a fortunate byproduct of this starting point which in the course of the experiments, however, became central, that Correns was concentrating on characters that he expected to become visible on the seeds. Paradoxically speaking, we could say that the xenia both prevented him from an early recognition of the transmission ratios *and* enabled him to do just that after all. For the seeds acted, in his system, insofar as he collected and kept them for sowing in the following generations, as a kind of naturally digitalized material protocol: they had either green or yellow cotyledons. To this 'protocol,' namely his boxes filled with peas, Correns could come back at any time, even after a year or two, and he could contemplate and reconsider the results even after all the other characters – to which at the beginning he might have paid no attention – had for long perished together with the plants which carried them. Not until Correns had worked himself deeply into the breeding system of *Zea mays* and *Pisum* for about four years did this aspect of the system become relevant, at a time when he realized that his results were starting to point in a different direction. It appears to me that such material characteristics of experimental systems and the recurrent moves they allow represent an important feature of experimental exploration. Insofar as researchers rely on them in their painstaking endeavor to attribute meaning to experimental data, historians of science should pay attention to them in their reconstructions.

NOTES

[1] I wish to thank Christoph Hoffmann, Frederic Holmes, and Friedrich Steinle for valuable comments on an earlier draft of this paper. A German version of the paper appeared in *History and Philosophy of the Life Sciences* 22 (2000): 187–218.

[2] H. F. Roberts, *Plant Hybridization before Mendel* (Princeton: Princeton University Press, 1929); Robert C. Olby, *Origins of Mendelism* (New York: Schocken Books, 1966). Vítezslaw Orel, *Gregor Mendel: The First Geneticist* (Oxford: Oxford University Press, 1996).

[3] Ilse Jahn, "Zur Geschichte der Wiederentdeckung der Mendelschen Gesetze," *Wissenschaftliche Zeitschrift der Friedrich-Schiller Universität Jena, mathematisch-naturwissenschaftliche Reihe* 7 (1957–1958): 215–227; Raphael Falk, "The struggle of genetics for independence," *Journal of the History of Biology* 28 (1995): 219–246, esp. pp. 221–225.

[4] Conway Zirkle, "Some oddities in the delayed discovery of Mendelism," *Journal of Heredity* 55 (1964): 65–72; Onno G. Meijer, "Hugo de Vries no Mendelian?," *Annals of Science* 42 (1985): 189–232.

[5] Archive for the History of the Max Planck Society (hereafter AHMPS), III. Abt., Rep. 17, Nr. 115.

[6] Emmy Stein, "Dem Gedächtnis von Carl Erich Correns nach einem halben Jahrhundert der Vererbungswissenschaft," *Die Naturwissenschaften* 37 (1950): 457–463, p. 457, footnote 1.

[7] Hugo de Vries, "Sur la loi de disjonction des hybrides," *Comptes rendus de l'Académie des Sciences Paris* 130 (1900a): 845–847.

[8] Erich von Tschermak-Seysenegg, "Über künstliche Kreuzung bei Pisum sativum," *Berichte der Deutschen Botanischen Gesellschaft* 18 (1900): 232–239.

[9] Carl Correns, "G. Mendel's Regel über das Verhalten der Nachkommenschaft der Rassenbastarde," *Berichte der Deutschen Botanischen Gesellschaft* 18 (1900a): 158–168.

[10] Robert Olby, *The Path to the Double Helix* (New York: Dover, 1994), p. 390.

[11] Hans-Jörg Rheinberger, "When did Carl Correns read Gregor Mendel's paper? A research note," *Isis* 86 (1995): 612–616.

[12] "Der Bastard bildet Sexualkerne, die in allen möglichen Combinationen die Anlagen für die einzelnen Merkmale der Eltern vereinigen, nur die *desselben* Merkmalspaares nicht. Jede Combination kommt annähernd gleich oft vor." (Correns 1900a), p. 166.

[13] Carl Correns, "Über Dickenwachstum durch Intussusception bei einigen Algenmembranen," Dissertation, Universität München, *Flora* 72 (1889): 298–347.
[14] Carl Correns, *Über die Abhängigkeit der Reizerscheinungen höherer Pflanzen von der Gegenwart freien Sauerstoffs*, Habilitationsschrift (Tübingen 1892).
[15] Carl Correns, *Bastarde zwischen Maisrassen. Mit besonderer Berücksichtigung der Xenien* (Stuttgart: Verlag von Erwin Nägele, 1901), Preface.
[16] Carl Correns, "Zur Kenntnis der inneren Struktur einiger Algenmembranen," in A. Zimmermann, ed., *Beiträge zur Morphologie und Physiologie der Pflanzenzelle I* (Tübingen: Verlag der Laupp'schen Buchhandlung, 1893): 260–305.
[17] Carl Correns, "Zur Physiologie der Ranken," *Botanische Zeitung* 54 (1896): 1–20.
[18] Carl Correns, *Untersuchungen über die Vermehrung der Laubmoose durch Brutorgane und Stecklinge* (Jena: G. Fischer, 1899a).
[19] Carl Correns, "Über die Membran und die Bewegung der Oscillarien (Vorläufige Mittheilung)," *Berichte der Deutschen Botanischen Gesellschaft* 15 (1897): 139–148.
[20] Carl Correns, "Floristische Bemerkungen über das Ursernthal," *Berichte der Schweizerischen Botanischen Gesellschaft* 5 (1895): 86–93.
[21] "Während der zehn Jahre, die ich in Tübingen Privatdozent war, gingen meine Arbeiten und Veröffentlichungen grossenteils auf den schon bereits betretenen Pfaden. ... Daneben fing ich, sobald ich durch meine Habilitation die Möglichkeit hatte, einen botanischen Garten benutzen zu können, an, nebenher allerlei Versuche zu machen, die wir jetzt 'genetisch' nennen würden, und die mir schon vorher durch den Kopf gegangen waren." "Selbstbiographie Correns." AHMPS, III. Abt., Rep. 17, Nr. 1, p. 3.
[22] "Selbstbiographie Correns." AHMPS, III. Abt., Rep. 17, Nr. 1, p. 4.
[23] Sergej G. Navashin, "Resultate einer Revision der Befruchtungsvorgänge bei Lilium Martagon und Fritillaria tenella," *Bulletin de l'Académie Impériale des Sciences de St. Petersburg* 9 (1898): 377–382; Sergej G. Navashin, "Neue Beobachtungen über Befruchtung bei Fritillaria tenella und Lilium Martagon," *Botanisches Centralblatt* 77 (1899): 62; Léon Guignard, "Sur les anthérozoïdes et la double copulation sexuelle chez les végétaux angiospermes," *Comptes rendus de l'Académie des Sciences*, Paris 128 (1899): 864–871; Léon Guignard, "Les découvertes récentes sur la fécondation chez les végétaux angiospermes," *Cinquantenaire de la Société de Biologie* (1899): 189–198.
[24] "Experimente Gaertners mit *Pisum*." AHMPS, III. Abt., Rep. 17, Nr. 115, folder "Pisum-Kreuzungen 1896–1900."
[25] Carl Friedrich v. Gärtner, *Versuche und Beobachtungen über die Bastarderzeugung im Pflanzenreich. Mit Hinweisung auf die ähnlichen Erscheinungen im Thierreiche* (Stuttgart: K. F. Hering, 1849).
[26] "Also – wie auch G. selbst hervorhebt, eine ganz auffallende Beeinflussung des befruchteten Samens (nicht der befruchteten Frucht!) Sehr auffaellig, dass gewohnlich *alle* Samen einer Hülse, event. alle Hülsen einer Kreuzung (c. 5) diese Beeinflussung zeigten! Beruht die Faerbg. des Samens auf der der Cotyledonen? dann waere das Resultat erklaerlich, als *Mittelbildg.* aufzufassen!" AHMPS, III. Abt., Rep. 17, Nr. 115, folder "Pisum-Kreuzungen 1896–1900."
[27] "Rimpau gelangen mehrere Kreuzungen leicht, alle zeigten große Neigung zu Rückschlaegen, selbst nach 6 Jahren mehrere nicht konstant." AHMPS, III. Abt., Rep. 17, Nr. 115, folder "Pisum-Kreuzungen 1896–1900." The paper referred to is: Wilhelm Rimpau, "Die Kreuzung als Mittel zur Erzeugung neuer Varietäten von landwirthschaftlichen Culturpflanzen," in *Tageblatt der 57. Versammlung Deutscher Naturforschung und Ärzte* (Magdeburg: Faber, 1884), pp. 179–186.
[28] AHMPS, III. Abt., Rep. 17, Nr. 115, folder "Pisum-Kreuzungen 1896–1900." ♀ and ♂ denote female and male germ cells, respectively. The note refers to Gregor Mendel, "Versuche über Pflanzenhybriden," *Verhandlungen des Naturforschenden Vereins in Brünn* 4 (1866): 3–47.
[29] "Es war in einer schlaflosen Novembernacht gegen Morgen, als mir auf einmal die Erklärung für die Beobachtungen an Pisum und Zea aufgingen. Erst als ich dann die Veröffentlichung vorbereitete, sah ich die Literatur systematisch durch und fand nun, mit Hilfe von Fockes Pflanzenmischlingen, dass Mendel das alles schon vor 35 Jahren gefunden und publiziert hatte. ...

Hätte ich die Erklärung früher gefunden, so hätte ich sie, trotz der Arbeit an dem Moosbuche in einem vorläufigen Bericht veröffentlicht. Denn die Bedeutung der Ergebnisse war mir gleich ziemlich weitgehend klar." "Selbstbiographie Correns." AHMPS, III. Abt., Rep. 17, Nr. 1, pp. 4–5.

[30] (Roberts 1929), p. 335.

[31] (Meijer 1985), p. 194.

[32] Wilhelm Olbers Focke, *Die Pflanzen-Mischlinge. Ein Beitrag zur Biologie der Gewächse* (Berlin: Borntraeger, 1881).

[33] It is now located in the library of the biological Max Planck Institutes in Tübingen. An inspection of the reprint gave no useful further hints with respect to relevant annotations attributable to potential readers, or dates of reading. I thank Heinz Schwarz for making photographs of the pages carrying a few light pencil marks available to me.

[34] "Erbsenaussat." AHMPS, III. Abt., Rep. 17, Nr. 115, folder "Pisum-Kreuzungen, Resultate 96."

[35] "Bestäub. u. Ertrag 96." AHMPS, III. Abt., Rep. 17, Nr. 115, folder "Pisum-Kreuzungen, Resultate 96."

[36] The corresponding paper was eventually published four years later. Carl Correns, "Ueber den Einfluss, welchen die Zahl der zur Bestäubung verwendeten Pollenkörner auf die Nachkommenschaft hat," *Berichte der Deutschen Botanischen Gesellschaft* 18 (1900b): 422–435.

[37] AHMPS, III. Abt., Rep. 17, Nr. 80.

[38] "1. VIII. 96 Pisum pm + gr: Alle Xen missglückt." AHMPS, III. Abt., Rep. 17, Nr. 115, folder "Pisum-Kreuzungen, Resultate 96."

[39] Drawing and Cliché to Correns 1900a, "Versuch I," p. 162; AHMPS, III. Abt., Rep. 17, Nr. 115, folder "Pisum-Kreuzungen 1896–1900."

[40] "31. VII. Pisum p + gr." AHMPS, III. Abt., Rep. 17, Nr. 115, folder "Pisum-Kreuzungen, Resultate 96."

[41] "Resultate 1897." AHMPS, III. Abt., Rep. 17, Nr. 115, folder "Pisum Resultate 1897."

[42] Protocols "gr + p" and "p + gr."AHMPS, III. Abt., Rep. 17, Nr. 115, folder "Pisum Resultate 1897."

[43] AHMPS, III. Abt., Rep. 17, Nr. 115, folder "Pisum-Kreuzungen 1896–1900."

[44] "Ergebnisse 1897." AHMPS, III. Abt., Rep. 17, Nr. 115, folder "Pisum Resultate 1897."

[45] "gr + p gab in der ersten Generation Samen, deren Grundfarbe zwischen grün und röthlich-orange schwankte und die alle mehr oder weniger (sparsam u. nur stellenweise bis stark) violett punktirt waren. Sahen also weder den Xenien noch dem Vater oder der Mutter ähnlich, sondern beinahe ggR (bei der die Hülsen aber ganz anders aussahen." "Ergebnisse 1897." AHMPS, III. Abt., Rep. 17, Nr. 115, folder "Pisum Resultate 1897."

[46] "Pisum Bastarte." AHMPS, III. Abt., Rep. 17, Nr. 115, folder "Pisum Resultate 1897."

[47] "Bastart gr + p ♂ A 1, (gelb)" and "gr + p B.1. gelb." AHMPS, III. Abt., Rep. 17, Nr. 115, folder "Pisum Resultate 1897."

[48] "$\sum$ war 22 u. ist 22. Hiervon wurde also *kein* S. ausgesaet. 18 mit gelbem Keim 77,8%, 4 mit grünem, 22,2%, mittelgross, maessig faltig, *wenig punktirt*." "$\sum$ 18, davon sind nur noch 15 vorhanden. Es fehlen also 3. Vom Rest (15) haben 3 grüne Keime = 20%, 12 gelbe = 80%. Sonst wie I. Aussaat gelb 2 grün 1, also 14 gelb, 77,8%, 4 grün, 22,2%." "Bastart gr + p A 1, (gelb)." AHMPS, III. Abt., Rep. 17, Nr. 115, folder "Pisum Resultate 1897."

[49] "I-III Topf, Bastart gr + p ♂ A1 gelb III; IV Topf, Bastart gr + p ♂ A1 IV; V Topf Bastart gr + p ♂ B1 (II); VI Topf Bastart gr + p ♂ B1 (I)." AHMPS, III. Abt., Rep. 17, Nr. 115, folder "Pisum Resultate 1898."

[50] "III Topf, Bastart gr + p A1 gelb III." AHMPS, III. Abt., Rep. 17, Nr. 115, folder "Pisum Resultate 1898."

[51] "Keimling bald grün, bald gelb" – "sind der gewöhnliche Bastart" – "ist reines p." "Topf XXIII-XXV, p + gr." AHMPS, III. Abt., Rep. 17, Nr. 115, folder "Pisum Resultate 1898."

[52] "Hülsen mit 3, 4, 4 Samen, grün bis hell rothlichorange. Von den vorhandenen 30 sind 10 grünkeimig, 20 gelbkeimig, vom Saatgut waren 1 grünk., 2 gelbk. also zusammen 33 S, davon mit

grünem Keim 11 = 33,3%, mit *gelben* 22 = 66,7%." "III Topf, Bastart gr + p A1 gelb III." AHMPS, III. Abt., Rep. 17, Nr. 115, folder "Pisum Resultate 1898."

[53] "∑ jetzt: 7 grün 20 gelb. Aussaat: 1 grün 2 gelb. ∑∑ 30, davon 8 grün = 26,7%, 22 gelb 73,3%." "VI Topf, Bastart gr + p B1 (I)." AHMPS, III. Abt., Rep. 17, Nr. 115, folder "Pisum Resultate 1898."

[54] "1 Fruchtknoten mit p bestäubt (Wolle roth) (die übrigen Bestäubungen verliefen resultatlos)." "Topf XVI gr + p." AHMPS, III. Abt., Rep. 17, Nr. 115, folder "Pisum Resultate 1898."

[55] "Keimlinge gelb oder grün." "Topf XVI gr + p." AHMPS, III. Abt., Rep. 17, Nr. 115, folder "Pisum Resultate 1898."

[56] "99. 1. a. B, gr + p, II. Gen. I. 1. S. *grün*." AHMPS, III. Abt., Rep. 17, Nr. 115, folder "Erbsen 99."

[57] "T. 21. a, BE + gr ("Xenien") Pfl. XI. 97." AHMPS, III. Abt., Rep. 17, Nr. 115, folder "Erbsen 99."

[58] "Meine Versuche über die Bildung von Xenien [ergaben] hier nur negative Resultate." Correns 1900a, p. 159.

[59] AHMPS, III. Abt., Rep. 17, Nr. 85, folder "Theoretisches etc."

[60] "Wenn es Thatsache ist, dass caesia ♀ (unrein) + alba ♂ *mehr weisse* Koerner giebt als caesia ♀ unrein + caesia ♂ unrein, so erklaert sich das leichter, als durch einen Einfluss der alba ♂ bei caesia ♀ (u) + alba ♂, durch den Einfluss der caesia ♂ (u) auf caesia ♀ (u).

[61] Ist die Haelfte der Fruchknoten bei caesia (unrein) auf weiss, die andre auf blau gestimmt, so wird alba-Pollen daran nichts aendern, caesia (u) Pollen dagegen die Zahl der caesia Koerner vermehren. Dann *nehmen wir* an, 1/2 der Pollenkoerner von caesia (u) sei für weiss, 1/2 derselben für blau gestimmt (u. blau ändere *stets*!), so wird, da nie gerade die 'weissen' Pollenk. auf die 'weissen' Narben, die blauen Pollenk. auf die 'blauen' Narben gelangen werden, u. 'caesia' Fruchtk. durch alba Pollen direct nicht direct beeinflusst werden u die Kreuzungen a ♀ + a ♂ weisse, dagegen a ♀ + c ♂, c ♀ + a ♂ u. c ♀ + c ♂ also blaue Koerner geben, die blauen Koerner etwa 3/4 statt 1/2 betragen müssen. Zunächst waere festzustellen, wieviel Koerner bei der Kreuzung a + caesia *rein* und *a* + caesia *unrein* blau werden koennen.

[62] Da aber blau *nicht* stets aendert, wie schon sicher, könnte die Zahl der blauen Koerner ... nicht ganz 3/4 betragen." "2. I. 98." AHMPS, III. Abt., Rep. 17, Nr. 85, folder "Theoretisches etc."

[63] "Mit Bezug auf eine Spaltung der Eigenschaften *im Pollen* waeren Versuche mit dem Bastart vulgata + dulcis anzustellen. (Weil die Koerner entweder glatt *oder* runzlig sind, waehrend bei alba + caesia Mittelbildungen vorkommen.)" "24. II. 98." AHMPS, III. Abt., Rep. 17, Nr. 85, folder "Theoretisches etc."

[64] "Kurzer, aber voller Kolben, *alle* Koerner bis auf eines +/– 'caesia,' z. Th. gefleckt." "Resultate 1894." AHMPS, III. Abt., Rep. 17, Nr. 84.

[65] "Trotz der fraglichen Reinheit der *alba* der Einfluss der *caesia* wohl unzweifelhaft." "Vers: Sommer 96, ♀ Pflanze alba, Pollen: caesia." AHMPS, III. Abt., Rep. 17, Nr. 84, folder "Erndte 96."

[66] "Die caesia war *nicht rein*." "Vers: Sommer 96, ♀ Pflanze caesia 'λ', Pollen: caesia 'λ'." AHMPS, III. Abt., Rep. 17, Nr. 84, folder "Erndte 96."

[67] "Wenn 3. ausgeschlossen wird, ist *kein Einfluss des Pollens auf den Fruchtknoten erkennbar*." "Vers: Sommer 96, ♀ Pflanze *caesia*, Pollen: alba." AHMPS, III. Abt., Rep. 17, Nr. 84, folder "Erndte 96."

[68] "Die caesia war nicht rein, sondern c + a!." "1897 a + c. alba ♀ + caesia ♂." AHMPS, III. Abt., Rep. 17, Nr. 84, folder "Resultate 1897."

[69] "Koerner etwa 1/2 *alba*, 1/2 *caesia*." "1897 c + a. caesia ♀ + alba ♂." AHMPS, III. Abt., Rep. 17, Nr. 84, folder "Resultate 1897."

[70] "Ist das Verhalten der Erbsenrassenbast. auch auf die Rassenbastarte beim Mais anwendbar??" Undated loose sheet. AHMPS, III. Abt., Rep. 17, Nr. 85, folder "Notizen."

[71] "Zur Anwendung der Mendelschen Theorie auf die Rassenbastarte des Maises." Undated loose sheet. AHMPS, III. Abt., Rep. 17, Nr. 85, folder "Notizen."

[72] "Combination dieser Zellen mit je einer Anlage zu *Paaren* (Sexualact). Vergleich mit einem Sack mit 2000 Kugeln (1000 gelb, 1000 grün). Wahrscheinlich: 500 ungleiche Paare, 500 gleiche, davon 250 mal gelb u. gelb, 250 grün u. grün. Also 250 ge + ge, 500 gr + ge, 250 gr + gr [Correns mistakenly writes ge + ge]. . . . *Wiederholte* Bastardirung 250 gr, 750 ge. 500 gr + ge u 250 ge + ge nur durch *Experm* (Nachkommensch) zu unterscheiden. Richtiger Stammbaum. Gilt es auch für sich mischende Merkmale?" Loose sheet, not dated. AHMPS, III. Abt., Rep. 17, Nr. 115.

[73] (Correns 1900a), pp. 163–164.

[74] Loose sheet, not dated. AHMPS, III. Abt., Rep. 17, Nr. 115.

[75] Loose sheet, not dated. AHMPS, III. Abt., Rep. 17, Nr. 115.

[76] Carl Correns, "Ueber den Modus und den Zeitpunkt der Spaltung der Anlagen bei den Bastarden vom Erbsen-Typus," *Botanische Zeitung* 60 (1902): 65–82. See also Hans-Jörg Rheinberger, "Mendelian inheritance in Germany between 1900–1910. The case of Carl Correns (1864–1933)," *Comptes rendus de l'Académie des Sciences, Série III, Sciences de la Vie* 323 (2000): 1089–1096.

[77] (Roberts 1929), p. 337; (de Vries 1900a).

[78] AHMPS, III. Abt., Rep. 17, Nr. 84.

[79] Loose sheet, not dated. AHMPS, III. Abt., Rep. 17, Nr. 115.

[80] "Als ich das gesetzmässige Verhalten und die Erklärung dafür . . . gefunden hatte, ist es mir gegangen, wie es de Vries offenbar jetzt geht: ich habe das alles für etwas Neues gehalten. Dann habe ich mich aber überzeugen müssen, dass der Abt Gregor Mendel in Brünn in den sechziger Jahren . . . zu demselben Resultat gekommen ist . . . ." (Correns 1900a), p. 158.

[81] Carl Correns, "Gregor Mendel's 'Versuche über Pflanzen-Hybriden' und die Bestätigung ihrer Ergebnisse durch die neuesten Untersuchungen," *Botanische Zeitung* 58 (1900c): 229–235; Carl Correns, "Ueber Levkojenbastarde," *Botanisches Centralblatt* 84 (1900d): 97–113.

[82] Carl Correns, "Untersuchungen über die Xenien bei Zea Mays," *Berichte der Deutschen Botanischen Gesellschaft* 17 (1899b): 410–417.

[83] (Correns 1901).

[84] Hugo de Vries, "Sur la fécondation hybride de l'albumen," *Comptes rendus hebdomadaires des séances de l'Académie des Sciences* 129 (1899): 973–975; Hugo de Vries, "Sur la fécondation hybride de l'endosperme chez le Maïs," *Revue générale de Botanique* 12 (1900b): 129–137.

[85] (Correns 1901), Introduction.

[86] Hans-Jörg Rheinberger, *Toward a History of Epistemic Things. Synthesizing Proteins in the Test Tube* (Stanford: Stanford University Press, 1997); Hans-Jörg Rheinberger, "Ephestia: The experimental design of Alfred Kühn's physiological developmental genetics," *Journal of the History of Biology* 33 (2000): 535–576.

JÜRGEN RENN* AND TILMAN SAUER**

ERRORS AND INSIGHTS: RECONSTRUCTING THE GENESIS OF GENERAL RELATIVITY FROM EINSTEIN'S ZURICH NOTEBOOK

In this paper we formulate some historiographical reflections on a cooperative research project, undertaken jointly with Michel Janssen, John Norton, and John Stachel, that is still under way.[1] Having worked on a notebook by Einstein for several years now, we take the occasion to ask ourselves the question: what is it that we can and did learn from the analysis of this notebook that we could not have learned by the analysis of other sources only? Introducing four epistemological leitmotives that have guided us in our reconstruction of Einstein's path towards General Relativity, we shall try to characterize an epistemological approach to the history of science for which research notebooks provide an essential historical source.

TELLING THE GENESIS OF GENERAL RELATIVITY FROM PUBLISHED SOURCES

Consider an account of the history of General Relativity that is based on an analysis of the published sources only. Such accounts may be found especially in the older literature.[2] To state the obvious, historians relied primarily on an analysis of the publications for the trivial reason that those sources are most readily available. This way of proceeding, however, was often, explicitly or not, also justified on grounds of the assumption that published sources are taken to represent the rational core of scientific achievements. In fact, this approach is interested in public, categorical knowledge which is not tainted by single scientists' idiosyncrasies and the contingent features of an individual pathway of discovery – as documented in research notebooks – and it therefore turns to those historical documents in which this knowledge is laid out and justified publicly. Drawing on published sources only, the received account of the history of General Relativity, in broad terms, then is as follows.

The theory of General Relativity as we understand it today was completed in a memoir from November 25, 1915, with the title "The Field Equations of Gravitation" (Einstein 1915a), see Figure 1.

* Max Planck Institute for the History of Science, Berlin
** Einstein Papers Project, California Institute of Technology

Frederic L. Holmes, Jürgen Renn and Hans-Jörg Rheinberger (eds.), Reworking the Bench: Research Notebooks in the History of Science, 253–268 © *2003 Kluwer Academic Publishers. Printed in Great Britain*

844 Sitzung der physikalisch-mathematischen Klasse vom 25. November 1915

Die Feldgleichungen der Gravitation.

Von A. Einstein.

In zwei vor kurzem erschienenen Mitteilungen[1] habe ich gezeigt, wie man zu Feldgleichungen der Gravitation gelangen kann, die dem Postulat allgemeiner Relativität entsprechen, d. h. die in ihrer allgemeinen Fassung beliebigen Substitutionen der Raumzeitvariabeln gegenüber kovariant sind.

Der Entwicklungsgang war dabei folgender. Zunächst fand ich Gleichungen, welche die Newtonsche Theorie als Näherung enthalten und beliebigen Substitutionen von der Determinante 1 gegenüber kovariant waren. Hierauf fand ich, daß diesen Gleichungen allgemein kovariante entsprechen, falls der Skalar des Energietensors der »Materie« verschwindet. Das Koordinatensystem war dann nach der einfachen Regel zu spezialisieren, daß $\sqrt{-g}$ zu 1 gemacht wird, wodurch die Gleichungen der Theorie eine eminente Vereinfachung erfahren. Dabei mußte aber, wie erwähnt, die Hypothese eingeführt werden, daß der Skalar des Energietensors der Materie verschwinde.

Neuerdings finde ich nun, daß man ohne Hypothese über den Energietensor der Materie auskommen kann, wenn man den Energietensor der Materie in etwas anderer Weise in die Feldgleichungen einsetzt, als dies in meinen beiden früheren Mitteilungen geschehen ist. Die Feldgleichungen für das Vakuum, auf welche ich die Erklärung der Perihelbewegung des Merkur gegründet habe, bleiben von dieser Modifikation unberührt. Ich gebe hier nochmals die ganze Betrachtung, damit der Leser nicht genötigt ist, die früheren Mitteilungen unausgesetzt heranzuziehen.

Aus der bekannten Riemannschen Kovariante vierten Ranges leitet man folgende Kovariante zweiten Ranges ab:

$$G_{im} = R_{im} + S_{im} \tag{1}$$

$$R_{im} = -\sum_{l} \frac{\partial \begin{Bmatrix} im \\ l \end{Bmatrix}}{\partial x_l} + \sum_{l\rho} \begin{Bmatrix} il \\ \rho \end{Bmatrix} \begin{Bmatrix} m\rho \\ l \end{Bmatrix} \tag{1a}$$

$$S_{im} = \sum_{l} \frac{\partial \begin{Bmatrix} il \\ l \end{Bmatrix}}{\partial x_m} - \sum_{l\rho} \begin{Bmatrix} im \\ \rho \end{Bmatrix} \begin{Bmatrix} \rho l \\ l \end{Bmatrix} \tag{1b}$$

[1] Sitzungsber. XLIV, S. 778 und XLVI, S. 799, 1915.

Figure 1. The theory of General Relativity was completed with the publication of generally covariant field equations of gravitation, nowadays called the "Einstein-equations," in a communication to the Prussian Academy dated 25 November 1915.

The first full account of the theory of General Relativity was then published in a comprehensive paper on "The Foundation of General Relativity" by Einstein shortly after, in early 1916 (Einstein 1916). It expounds a new theory of the gravitational field that is "generally covariant." It presents a theory formulated by means of equations which do not change their form if an arbitrary change of coordinates is performed. Classical, Newtonian theory in contrast, was formulated as being covariant with respect to so-called Galilean coordinate transformations only, i.e. those which correspond to a change of frames of reference which are moving uniformly and rectilinearly with respect to each other. The equations of 1915, now, do *not* change their form even if *arbitrary* coordinate transformations were performed, e.g. also those corresponding to accelerated or rotating frames of reference where in Newtonian mechanics inertial and centrifugal forces arise. The theory of 1915 hence was "generally relativistic."

As the title of the November 1915 paper indicates, generally covariant equations for the gravitational field are a crucial ingredient of the new theory. Going back from this paper to earlier publications, the prehistory of General Relativity seems to be characterized by dead ends, flaws, and incomprehensible sideways, and may appear as a comedy of errors. Even to Einstein himself, the prehistory of the final November memoir in hindsight appeared a "jungle,"[3] About his published papers on the topic, he wrote in January 1916:

> The series of my works on gravitation is a chain of aberrations which, however, nevertheless eventually led closer to the aim.[4]

Gravitational field equations that differed substantially from the correct, final ones had indeed been published by Einstein only two weeks before publication of the final memoir in a "Nachtrag" (Einstein 1915c). And this "Nachtrag" was, in fact, only written for modification of still different field equations, published only seven days before, in a memoir "On the General Theory of Relativity" ("Zur allgemeinen Relativitätstheorie") (Einstein 1915b).

This crooked path appears all the more puzzling, as these earlier papers already contain almost every ingredient of the final theory of General Relativity. Those ingredients already involve serious breaks with the classical, Newtonian theory of gravitation. In particular, in Newtonian theory the gravitational potential is represented by a *scalar* potential, i.e. a *single* function. In Einstein's papers of 1915, on the other hand, the gravitational potential is represented by the metric tensor, a very different and much more complicated object, comprising, in fact, ten different functions. In addition, the introduction of the metric tensor as the mathematical representation of the gravitational potential established a conceptual connection between gravitation, space, and time, implying that the motion of heavy bodies in a gravitational field is no longer conceived as being deviated by a force from a straight inertial path but has to be conceived as inertial motion along geodesics in a curved spacetime.

Already in these papers, the metric tensor comes along with the mathematical representation of later General Relativity, differential geometry and tensor calculus, at Einstein's time provided by the mathematical framework of the absolute differential calculus that had been developed only a few years earlier by the Italian mathematicians Ricci and Levi-Civita. In this mathematical representation the Riemann tensor figures prominently as a central generally covariant object formed from the metric tensor and its derivatives.

Going back in the reconstruction of the genesis of General Relativity on the basis of the published documents, the earliest paper where all these crucial ingredients are assembled is a publication with the title "Entwurf einer verallgemeinerten Relativitätstheorie und einer Theorie der Gravitation" ("Outline of a Generalized Theory of Relativity and a Theory of Gravitation") (Einstein and Grossmann 1913), published jointly two and a half years earlier in spring 1913 by Einstein and his mathematician friend Marcel Grossmann, see Figure 2. While this paper, in fact, introduces the metric tensor and the associated mathematics of the tensor calculus, it does not contain the correct, generally covariant field equations.

A particularly puzzling feature of this account is the fact that the theory of the "Entwurf" paper presents *wrong* and, from a modern point of view, *senseless* field equations, even though everything that is needed for finding the correct field equation was already there. In fact, Einstein admitted that his failure to find generally covariant field equations was a "*dark spot*" in his theory, a dark spot which *delayed* the final solution for more than two years. In the "Entwurf," he wrote:

> But it has to be emphasized that it turns out to be impossible to find, under this presupposition [that the equations should be of second-order in correspondence to the Newton-Poisson law], a differential expression $\Gamma_{\mu\nu}$ which is a generalization of $\Delta\varphi$, and which turns out to be a tensor with regard to arbitrary transformations.[5]

To make things worse, in that paper, Einstein explicitly stated that he actually had considered what we now immediately recognize to be the *correct* solution – but had discarded it. In a footnote, we are referred to Marcel Grossmann's part of the paper which deals with the mathematical aspects of the theory more explicitly. There, Grossmann writes:

> One succeeds indeed, at first, to name a covariant differential tensor of second rank and second order which could enter these equations, namely, $G_{im} = \sum_{kl} \gamma_{kl}\,(ik, lm) = \sum_{k} \{ik, km\}$. However, it turns out that this tensor does not, in the special case of the infinitely weak static gravitational field, reduce to the expression $\Delta\varphi$.[6]

From a modern perspective this claim is clearly wrong. The expression G_{im}, readily recognized as the Ricci tensor, would be the expression for the generally relativistic field equation for the matter free case, and it *does* reduce to the Laplacian operator Δ of Newtonian gravitation theory for the case of weak and static gravitational fields.

ENTWURF EINER
VERALLGEMEINERTEN RELATIVITÄTSTHEORIE
UND EINER
THEORIE DER GRAVITATION

I. PHYSIKALISCHER TEIL
VON
ALBERT EINSTEIN
IN ZÜRICH

II. MATHEMATISCHER TEIL
VON
MARCEL GROSSMANN
IN ZÜRICH

LEIPZIG UND BERLIN
DRUCK UND VERLAG VON B. G. TEUBNER
1913

Figure 2. The "Outline of a Generalized Theory of Relativity and a Theory of Gravitation," published jointly with Marcel Grossmann in spring 1913, contained all the crucial ingredients of the final theory, except the field equations. Instead, the "Outline" advanced field equations that, from a modern point of view, are senseless.

The received approach to General Relativity, proceeding along the lines of an analysis of the published sources that are taken to represent the rational core of scientific achievements, is here confronted with a puzzle. This puzzle was traditionally explained by taking recourse to the assumption that the historical path as laid out in the published documents displays imperfections of the individual actors. The explanation, as put forth, e.g., in the standard scientific biography of Einstein by Abraham Pais (Pais 1982), proceeds as follows. We know today perfectly well how to reduce the Ricci tensor to the Laplacian operator. The procedure involves the stipulation of some *coordinate conditions*, additional relations to be imposed which restrict the choice of admissible coordinates. But, according to Pais, since Einstein and Grossmann were not yet really familiar with dealing with generally covariant equations, they had not yet realized this as their freedom to stipulate those additional coordinate conditions.

This explanation, however, has two shortcomings. First, it obviously represents a speculation about Einstein's and Grossmann's knowledge and understanding at the time. Second and perhaps more disturbing, the explanation has to charge both Einstein and Grossmann with a rather *trivial* error. The stipulation of coordinate condition is, in fact, a rather natural feature of the calculation under consideration. Generally covariant equations by definition hold in *all* coordinate systems whereas the equations of Newtonian gravitation theory do *not*. Hence we *must* stipulate those coordinate conditions and we are also perfectly *free* to do so.

There are, of course, additional sources that may be consulted to check whether the speculation about Einstein's and Grossmann's knowledge and understanding is historically accurate: correspondence, for instance, and research notebooks. Fortunately, it so happens that we do have a research notebook by Einstein dating precisely from the period immediately predating the "Entwurf." It also deals explicitly and extensively with just the problem of finding general relativistic gravitational field equations. This is the "Zurich Notebook," so-called since its entries date from Einstein's Zurich period, more precisely from about summer 1912 to early 1913.[7]

The relevance of the Zurich notebook for a historical reconstruction of Einstein's path to General Relativity was first recognized by John Stachel. In 1984, John Norton based his, now classic account of "How Einstein Found His Field Equations" (Norton 1984) on information drawn from that notebook and from correspondence, in addition to using all of Einstein's pertinent publications of that period.

The surprising consequence of taking into account Einstein's Zurich Notebook was that some entries in the notebook clearly and unmistakably showed that Einstein knew perfectly well how to stipulate and use conditions that, from the perspective of the final theory of General Relativity, appear to be coordinate conditions. In fact, he used several different such conditions and also the one that is needed to reduce the Ricci tensor to the Laplacian operator in a calculation performed exactly to that end, see Figure 3.

However, this result entails a new difficulty: an apparent conflict between the

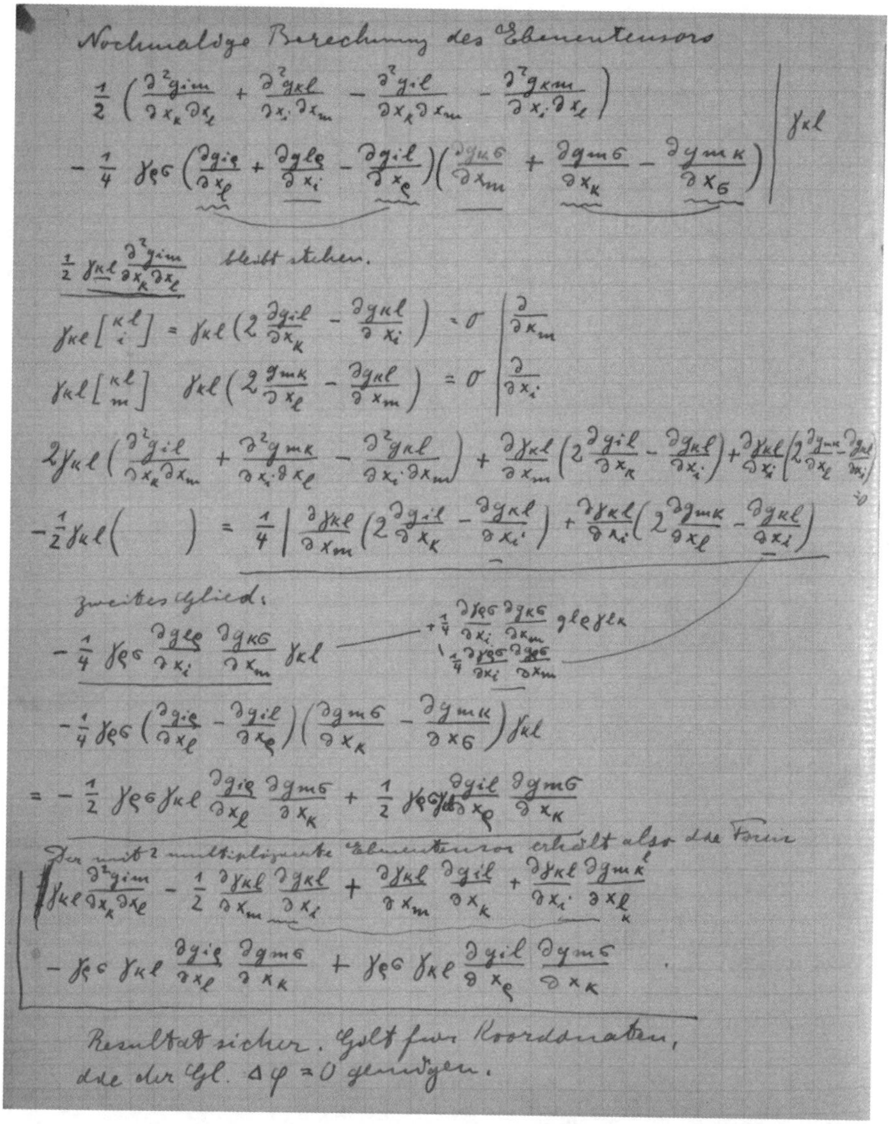

Figure 3. This page (19L) of the Zurich Notebook proves that Einstein knew how to apply the condition $\gamma_{kl}\left[\begin{matrix} kl \\ i \end{matrix}\right] = 0$ to reduce the Ricci tensor to a d'Alembertian-like operator that would reduce to the Lapace operator for weak, static fields. It refutes the speculation that Einstein did not know at that time, i.e. around 1912–13, how the Newtonian limit could be recovered from the Ricci tensor by stipulating this condition.

knowledge documented by Einstein's publications and that revealed by his notebook. Taking the evidence in the notebook seriously, one can no longer read the "Entwurf" literally. Einstein and Grossmann in the "Entwurf" no longer wrote what they must have meant.

Consequently, the notebook, if taken seriously, cannot be read as merely supplying a footnote to the received approach, essentially based on recounting Einstein's publications. It delivers more than just additional evidence corroborating some speculative reconstruction that is necessary to fill the gaps left by such a narrative. It has to be understood in its own right.

AN EPISTEMOLOGICAL APPROACH TO THE HISTORY OF GENERAL RELATIVITY

The treatment of unpublished sources as being merely of secondary importance may be justified on epistemological grounds, for instance in terms of the distinction between a logically structured context of justification and a psychologically structured context of discovery. In our view, however, such epistemological dividing lines tend to bury the question of the relation between the structure of scientific knowledge and the nature of the thinking processes involved in an individual contribution to it. If one accepts such an epistemological dividing line, it seems to us, one unavoidably misses the chance to exploit the reconstruction of an individual investigative pathway for insights into the structures of the scientific knowledge of an historical period, and conversely one renounces the opportunity to understand an individual cognitive achievement in terms of a historically informed theory of thinking, rather than mystifying it as an ultimately inaccessible act of "creativity."

A *historical* epistemology, on the other hand, does not accept such *a priori* given dividing lines because it considers the epistemological question of the structures of scientific knowledge as an open one, to be historically investigated itself. A reconstruction of Einstein's research notebook from the point of view of a historical epistemology leads in fact to insights which we believe would remain out of sight if the manuscript were perceived only as a source complementary to published work. In the following, we will characterize four epistemological *Leitmotive* that have guided us in our analysis of the Zurich notebook:

- Zoom out!
- Zoom in!
- Watch the representation take over! and:
- Take errors as clues!

Zoom out!

"Zoom out!" – this first epistemological *Leitmotiv* requires taking into account the shared knowledge resources of a thinking process, the "public knowledge," as one may call it. In fact, if one interprets the entries in Einstein's research notebook as traces of a thinking process shaped by the scientific knowledge available to him, the question arises as to how to characterize the structures and the function of this knowledge. Answering this question requires us to "zoom out" and turn away from the details of the notebook to the reconstruction of the physical knowledge rooted in classical physics and involved in the creation of General Relativity. We describe this knowledge in terms of heuristic requirements which had to be checked for each of the candidate gravitational field equations Einstein was considering.[8] One of these requirements is the so-called "equivalence principle," which turned the simple insight that all bodies fall with the same speed into a powerful heuristic device guiding Einstein's search for a new theory of gravitation. Since all bodies fall with the same speed, one would not notice, for instance, the motion of a falling apple as long as oneself is also falling along with it, say in a falling elevator. The falling and therefore accelerated elevator thus may temporarily serve as an inertial frame of reference in which the apple is at rest and in which no gravitational field is acting, at least until the elevator crashes. Such a mental construct suggested to Einstein that his new theory of gravitation should incorporate a generalized principle of relativity in which accelerated frames of reference are admitted for describing physical phenomena on the same level as the inertial frames of classical and special relativistic physics. Another heuristic principle was his "correspondence principle" according to which the new theory should describe, under certain limiting conditions, the kind of gravitational effects familiar from Newtonian physics. Finally, he reinterpreted the fact that a principle of energy conservation can be formulated for all known physical theories as a further heuristic requirement for the new theory. But since this theory was, of course, not known to him at the beginning of his search, all his heuristic requirements corresponded to attempts to transform the knowledge structure of classical physics into qualitative building blocks of this new theory, thus exploiting the potential of the shared knowledge resources available at the time.[9]

Zoom in!

"Zoom in!" – this second epistemological *Leitmotiv* requires a reconstruction of the "hands on"- working knowledge involved in a research process; it turns the attention to the "private knowledge," so to say. In the research process documented by the Zurich notebook, the knowledge of classical physics, guiding Einstein's search for a gravitational field equation, materialized in the form of mathematical elaborations of candidate field equations. Obviously, these elaborations also depended on the mathematical knowledge available to him. But the precise dynamics of Einstein's exploration of his candidate field equations was governed by the way in which these generic knowledge resources were brought together in the concrete calculations of his notebook. In order to reconstruct this dynamics, it was therefore not sufficient to reconstruct Einstein's overall resources but also necessary to redo every single calculation in his notebook, even the apparently most irrelevant ones.[10] Only by such a "zooming-in" on Einstein's actual working experience it was possible to reconstruct the cognitive dynamics of his search for the gravitational field equation.

It thus turned out that, in the course of Einstein's work, two distinct and complementary strategies emerged for the construction of suitable candidate gravitational field equations. These two strategies take his heuristic requirements in a complementary manner either as points of departure or as touchstones. His "physical" strategy started from candidates whose specialization to the Newtonian limit was obvious. These candidates did therefore not present a problem of physical interpretation – but they had to be checked for Einstein's other heuristic criteria, the conservation of energy and the generalized relativity principle. On one page of the notebook, for instance,[11] Einstein started an investigation of the compatibility of such a "physical" candidate with his heuristic condition of conservation of energy. He observed that this condition amounts to the introduction of additional requirements which he could not satisfactorily interpret for the time being and therefore did not pursue this line of thought any further. Instead, on the next page he started to follow a different approach and switched from the "physical" to what we have called the "mathematical" strategy.

This complementary strategy started from Einstein's generalized relativity principle translated into a clearly defined mathematical requirement, general covariance. The mathematical strategy itself consisted in searching first for suitable candidates satisfying this requirement and then testing them against Einstein's correspondence principle and the principle of conservation of energy. While the "mathematical" candidates were well-behaved mathematical objects from the beginning, it was their physical interpretation which still had to be explored in the course of the research process. Along the mathematical strategy, it was the generally covariant Riemann tensor which could be taken as a natural starting point. Einstein became aware of this tensor through the help of his mathematician friend Marcel Grossmann whose name is written next to it on the subsequent page of the notebook, see Figure 4.

In order to obtain a field equation of the appropriate form, Einstein had to extract a two-index object from the Riemann tensor. Indeed, we see that he immediately contracted the Riemann tensor once in order to obtain such a two index object, the covariant Ricci tensor. A short consideration, however, made him realize that this candidate does not comply with his correspondence principle. In addition to the term that does satisfy this principle, the Ricci tensor contains three other terms which obviate a smooth transition to the Newtonian limit. In the notebook, Einstein set these terms equal to zero, commenting them by the words "should vanish" ("Sollte verschwinden"). Consequently, the upshot of this first investigation was that the Ricci tensor violated Einstein's heuristic principle of

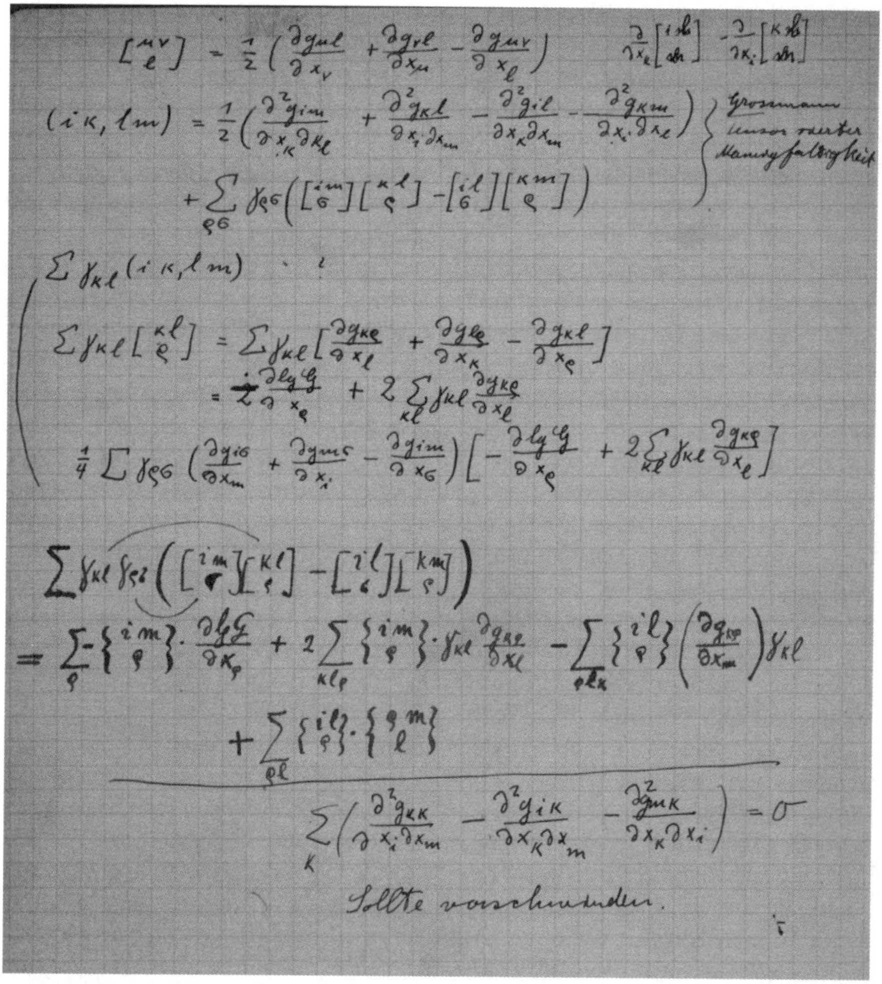

Figure 4. The generally covariant Riemann tensor *(ik, lm)*, introduced on this page (14L) of the Zurich Notebook was the natural starting point for considerations along Einstein's "mathematical strategy." Contraction of the Riemann tensor gave the Ricci tensor but Einstein discarded it here because it contained three second-derivative terms that "should vanish" ("Sollte verschwinden") if his "correspondence principle" was to be satisfied.

correspondence. As it turned out, the reconstruction of this internal dynamics of Einstein's heuristics made it possible not only to understand, step-by-step, his investigative pathway in the notebook but also the puzzles concerning the emergence of General Relativity raised by his published papers.

Watch the Representation Take Over!

"Watch the representation take over!" – this third epistemological *Leitmotiv* required taking into account that the medium, conceived as an external representation of thinking, itself plays an active role in shaping the thinking process and in generating unanticipated novelties. In fact, if the calculations in Einstein's notebook are considered to be a just a passive record of his thinking, which itself is supposed to remain unaffected by its external representation, then the development of his thinking and, in particular, the acquisition of fundamentally new insights in the course of his calculations must remain a mystery.

We have seen that the consideration of the Ricci tensor revealed a conflict between the Generalized Relativity Principle and the Correspondence Principle. This conflict did not, however, force Einstein to give up the Riemann tensor as his starting point, as it was precisely the unforeseeable novelties generated by his tentative calculations that provided still more paths, the exploration of which made good sense from the perspective of Einstein's heuristics. The expression of the conflict between Correspondence and Relativity Principles by the "disturbing terms" suggested, in particular, that these terms could be brought to vanish by imposing an additional condition on the admissible coordinates. Coordinates satisfying this condition are just the "harmonic" coordinates mentioned above. For these coordinates the Ricci tensor reduces to the appropriate form, cp. Figure 3. It remained to be checked, however, whether this additional condition was compatible with Einstein's other heuristic requirements.

There was, indeed, a problem which arose from the consideration of energy-momentum conservation. Together with the field equation this condition implies a further condition on the coordinates. In other words, attempting to represent his *two* heuristic requirements of correspondence and conservation both on the level of his concrete calculations, Einstein had to deal with *two* restrictive conditions on the coordinates. These two conditions now turned out to be mutually incompatible. This was an insight that could not have been foreseen on the level of the heuristic requirements taken by themselves. The dilemma of two incompatible restrictions then triggered a further exploration of the formalism, this time with the effect of generating a breath-taking novelty. On a subsequent page (cp. Figure 5), Einstein considered a slight modification of his candidate field equation by which he hoped to avoid the dilemma between his heuristic principles. This modification brought him to consider what appears to be in essence the correct field equation of his definitive theory of General Relativity – three years before he published this final version. This striking fact may well have remained unknown had it not been for a "zooming in" analysis of Einstein's calculations in the notebook.

Take Errors as Clues to Thinking Processes!

"Take errors as clues to thinking processes!" – this fourth epistemological *Leitmotiv* suggests to interpret what appear to be errors of reasoning in an

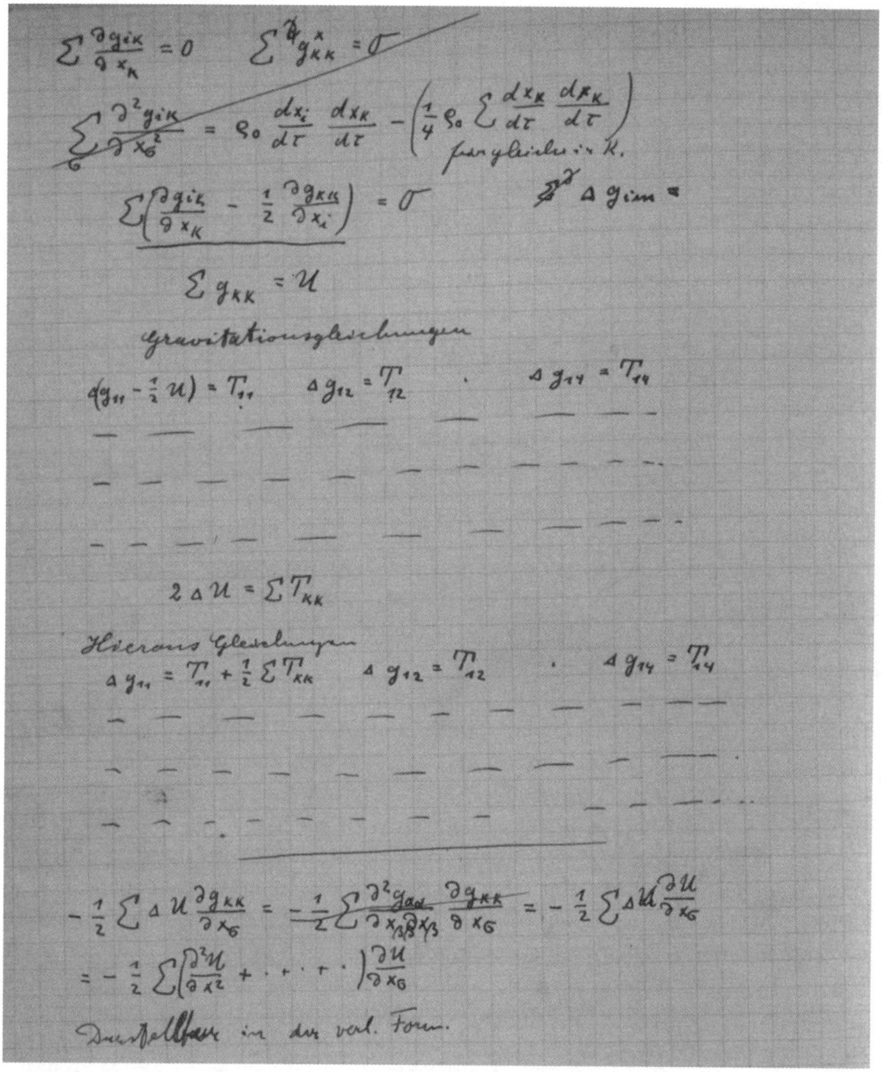

Figure 5. In order to reconciliate the two conflicting conditions that he had encountered before, Einstein here ad hoc introduced a trace term $\sum g_{kk} = U$, and thus effectively invented the correct gravitational field equations ("Gravitationsgleichungen") in linear approximation, three years before their publication. However, since this modification was introduced only on the level of the mathematical representation, Einstein at this point did not succeed in giving these equations satisfactory physical meaning.

historical document, if possible, not as deficiencies with respect to the modern understanding of the subject but as hints at a different conceptual organization of the past scientific knowledge. Of course, apparent "errors" from a modern perspective are not the same thing as real errors by contemporary standards. This *Leitmotiv* has therefore to be distinguished from a simple "principle of charity," which would only require that a good historical reconstruction is characterized by ascribing as few blatant errors as possible to the historical actor and thus renounces "stupidity" as an *explanans* of a historical interpretation. But while under the principle of charity "error" remains a rather abstract notion, the fourth *Leitmotiv* requires this notion to be related to the historical change of conceptual systems so that what appears to be an oddity in a more advanced system need not be an error in the earlier one.

As we have seen, in the winter of 1912/13, three years before the publication of his theory of General Relativity in late 1915, Einstein found, driven by his explorative calculations, what appears to be the essential ingredient of his definitive field equation, the later so-called Einstein tensor in linear approximation. But only shortly afterwards, as is also documented by the Zurich notebook, he discarded this candidate just as he had done with the Ricci tensor before. From the perspective of the received approach to the history of science, both rejections can only be explained by identifying the one or several "errors" which initially detracted Einstein's reasoning from the right path and later had to be eliminated. "Errors" are here inevitably defined with respect to the modern understanding of General Relativity. Such an explanation is confronted, however, with the paradox that the structure of General Relativity, which is only the outcome of Einstein's research process, is nevertheless supposed to determine – *ex negativo* – already the process of its discovery, as if this structure would act like a Platonic idea.

If Einstein's supposed "errors" are instead taken as expressions of a way of thinking conceptually different from the modern one, then we are called upon to reconstruct this way in its own internal coherence, differing from that of modern General Relativity. At the same time one thus has a chance of understanding the process by which Einstein's initial thinking about the problem of gravitation eventually led to the establishment of General Relativity without somehow supposing that the result of this process was already implicit in its beginning. On the basis of our reconstruction of Einstein's research we have found that the interplay between his heuristic requirements, on the one hand, and the exploration of the deductive consequences of the mathematical formalism in which these requirements materialized, on the other hand, is the crucial intellectual process that prepared the ground for the discovery of the field equation of General Relativity. The unforeseen deductive consequences are the reason why the delicate balance between Einstein's heuristic components turned out to favor different candidates in the course of his research process. Most strikingly, the same heuristic requirements, still rooted in the conceptual structures of classical physics, which induced Einstein to discard the essentially correct field equation in 1912/13 eventually brought him back to it three years later, while excluding all other candidates he had considered in the meantime. The conceptual novelties of General Relativity in its modern understanding were thus not the presupposition but only the result of the research begun by Einstein. These novelties were eventually obtained by readjusting the concepts of classical physics to the outcome of Einstein's work in a process of reflection that still was to take years after the initial formulation of his theory in 1915.

NOTES

[1] Preliminary results of the project were reported in (Castagnetti et al. 1994), (Renn and Sauer 1996), and (Renn and Sauer 1999) as well as in the annotation of the edited version of the Zurich Notebook in (Klein et al. 1995). A comprehensive report on our investigations, including a full facsimile edition of Einstein's Zurich Notebook, is in preparation (Renn, forthcoming).

[2] The literature on the history of General Relativity is abundant. Some pertinent references will be found in (Renn and Sauer 1999). For a general survey of the history of (Special and General) Relativity, see (Stachel 1995).

[3] Einstein to Lorentz, 1 January 1916, in (Schulmann et al. 1998), Doc. 177.

[4] "Die Serie meiner Gravitationsarbeiten ist eine Kette von Irrwegen, die aber doch allmählich dem Ziele näher führten." Einstein to Lorentz, 17 January 1916, in (Schulmann et al. 1998), Doc.183.

[5] "Es muβ aber hervorgehoben werden, daβ es sich als unmöglich erweist, unter dieser Voraussetzung [daβ die Gleichungen dem Newton-Poissonschen Gesetz entsprechend zweiter Ordnung sein sollen] einen Differentialausdruck $\Gamma_{\mu\nu}$ zu finden, der eine Verallgemeinerung von $\Delta\varphi$ ist, und sich beliebigen Transformationen gegenüber als Tensor erweist." (Einstein and Grossmann 1913), p. 11.

[6] "Es gelingt in der Tat zunächst, einen kovarianten Differentialtensor zweiten Ranges und zweiter Ordnung anzugeben, der in jene Gleichungen eintreten koennte, nämlich

$$G_{im} = \sum_{kl} \gamma_{kl}\,(ik, lm) = \sum_{k} \{ik, km\}$$

Allein es zeigt sich, dass sich dieser Tensor im Spezialfall des unendlich schwachen statischen Schwerefeldes nicht auf den Ausdruck $\Delta\varphi$ reduziert." (Einstein and Grossmann 1913), p. 36. In the quote, (ik, lm) denotes the fully covariant Riemann tensor, and γ_{kl} the contravariant metric. Hence G_{im} is the Ricci tensor. $\Delta\varphi$ denotes the Laplacian operator acting on the scalar Newtonian potential φ.

[7] The notebook is deposited at the Einstein Archives at Hebrew University, Jerusalem, Call No. 3-006. It comprises 83 pages with calculations, of which 58 pages deal with the problem of gravitation. These latter pages have been published in transcription as Doc. 10 of (Klein et al. 1995).

[8] For preliminary accounts, see (Renn and Sauer 1996, 1999).

[9] Renn and Sauer.

[10] A detailed commentary of the calculations along with a facsimile and a transcription of the notebook will be published as part of (Renn, forthcoming).

[11] P. 13R, i.e. [p. 26] of Doc. 10 in (Klein et al. 1995).

REFERENCES

Castagnetti et al. (1994): Castagnetti, Giuseppe, Peter Damerow, Werner Heinrich, Jürgen Renn, and Tilman Sauer, *Wissenschaft zwischen Grundlagenkrise und Politik: Einstein in Berlin. Arbeitsbericht der Arbeitsstelle Albert Einstein 1991–1993* (Berlin: Max-Planck-Institut für Bildungsforschung, 1994).

Einstein (1915a): Einstein, Albert, "Die Feldgleichungen der Gravitation," *Sitzungsberichte der Preussischen Akademie der Wissenschaften* (1915): 844–847.

Einstein (1915b): Einstein, Albert, "Zur allgemeinen Relativitätstheorie," *Sitzungsberichte der Preussischen Akademie der Wissenschaften* (1915): 778–786.

Einstein (1915c): Einstein, Albert, "Zur allgemeinen Relativitätstheorie (Nachtrag)," *Sitzungsberichte der Preussischen Akademie der Wissenschaften* (1915): 799–801.

Einstein (1916): Einstein, Albert, *Die Grundlage der allgemeinen Relativitätstheorie* (Leipzig: Barth, 1916).

Einstein and Grossmann (1913): Einstein, Albert and Marcel Grossmann, *Entwurf einer verallgemeinerten Relativitätstheorie und einer Theorie der Gravitation* (Leipzig: Teubner, 1913).

Klein et al. (1995): Klein, Martin et al., eds., *The Collected Papers of Albert Einstein*, vol. 4 (Princeton, N.J.: Princeton University Press, 1995).

Norton (1984): Norton, John D., "How Einstein Found His Field Equations, 1912–1915," *Historical Studies in the Physical Sciences* 14 (1984): 253–316.

Pais (1982): Pais, Abraham, *'Subtle is the Lord ...' The Science and the Life of Albert Einstein* (Oxford: Oxford University Press, 1982).

Renn, ed. (forthcoming): *The Genesis of General Relativity*, 4 vols. (Dordrecht: Kluwer).

Renn and Sauer (1996): Renn, Jürgen and Tilman Sauer, "Einstein's Züricher Notizbuch," *Physikalische Blätter* 52 (1996): 865–872.

Renn and Sauer (1999): Renn, Jürgen and Tilman Sauer, "Heuristics and Mathematical Representation in Einstein's Search for a Gravitational Field Equation," in *The Expanding Worlds of General Relativity*, ed. Hubert Goenner et al. (Boston, Basel, Berlin: Birkhäuser, 1999): 87–125.

Schulmann et al. (1998): Schulmann, Robert et al., eds., *The Collected Papers of Albert Einstein*, vol. 8: *The Berlin Years: Correspondence, 1914–1918* (Princeton, N.J.: Princeton University Press, 1998).

Stachel (1995): Stachel, John, "History of Relativity," in *Twentieth Century Physics*, vol. 1, ed. Laurie M. Brown et al. (Bristol, Philadelphia, New York: Institute of Physics Publishing and American Institute of Physics Press, 1995): 249–356.

GERD GRAßHOFF* AND MICHAEL MAY**

HANS KREBS' AND KURT HENSELEIT'S LABORATORY NOTEBOOKS AND THEIR DISCOVERY OF THE UREA CYCLE – RECONSTRUCTED WITH COMPUTER MODELS[1]

DISCOVERY OF THE UREA CYCLE

The Case

In 1932 Hans Krebs and Kurt Henseleit explained the urea synthesis in animal liver by the urea cycle – the first cyclic metabolic pathway discovered in biochemistry. This discovery was a milestone in the history of the discipline. For his subsequent studies of a similar process, the tricarboxylic acid cycle, Hans Krebs was later awarded the Nobel Prize. Already in the mid-19th century new analytical techniques showed that the rate of urea synthesis in living animals increased when they were fed an additional supply of glycine and leucine. Schultzen and Nencki assumed in 1869 that amino acids are intermediates in the reaction chain from proteins to urea.[2] The introduction of the perfusion method marked an essential refinement of the experimental procedures. Reagents are guided through an organ outside the living organism, where the chemical composition of the leaving liquid is determined. In this way one found that not only glycine and leucine but almost all known proteins and amino acids increase the urea production in the liver. Until the twenties one attempted to optimize the utilization of the perfusion method; yet one did not succeed in decrypting the details of the chemical reactions leading to the formation of urea. At this time Hans Krebs was working as an assistant in the laboratory of Otto Warburg in Berlin. During these years he conducted basic research and obtained a practical and biochemical knowledge that would play an important role during his discovery of the urea cycle. Krebs owed to Otto Warburg especially the adaptation of the tissue slice method and the use of manometric devices for sensitive measurements of small amounts of substances. In summer 1931, after Krebs had moved to Freiburg, he started his research project on urea synthesis with his doctoral student Kurt Henseleit. It lasted nearly a full year and took him almost 200 experiments until the urea cycle could be established.

---

* University of Bern
** GMD – Research Center for Information Technology, St. Augustin

Frederic L. Holmes, Jürgen Renn and Hans-Jörg Rheinberger (eds.), Reworking the Bench: Research Notebooks in the History of Science, 269–294

Case Studies

The formation of scientific theories as a paradigmatic case of creative problem solving has been the subject of the interdisciplinary research project "Epistemic Systems" that was initiated with the support of the Deutsche Forschungsgemeinschaft at the University of Hamburg in 1989. One of the first case studies has been Krebs' discovery of the urea cycle. It is an exceptional case, because fortunately all laboratory notebooks of both Hans Krebs and his assistant Kurt Henseleit could be located and served as an excellent historical basis for their daily work in the laboratory.[3] We started the project as a reaction to a publication of Herbert Simon and Deepak Kulkarni,[4] who had published their version of a scientific discovery program in which they claimed a historically adequate simulation of Krebs' discovery of the urea cycle based on the historical studies of Frederic Holmes.[5] It soon turned out that we disagreed with both the historical and the methodological claims of Simon and Kulkarni.[6] This initiated a long-term research project into the case history, the methodology of scientific discovery, causal reasoning and the development of its computer models.[7] The case of Krebs' discovery turned out to be particularly fruitful not only because it can be viewed as a paradigmatic case of well documented experimental research. Parallel to our study Frederic Holmes published his seminal work on Hans Krebs, in which among many other aspects of Krebs' life he describes his experimental work supplemented by Krebs' own recollections, after Holmes had discussed the notebooks with Krebs in a long series of interviews.[8]

A new framework called "epistemic systems" has been developed for the methodological analysis of the formation of scientific theories, which views the succession of scientific episodes as a series of both cognitive and physical actions. These actions could be the construction of a particular scientific hypothesis, the design of an experiment, the purchase of needed instruments or the draft of a publication. The conditions required for such actions to be conducted, e.g. the material research equipment, belong also to an epistemic system. The rationale for such actions stands at the centre of a historical reconstruction of epistemic systems.

LABORATORY NOTEBOOKS – WHAT THEY SHOW AND WHAT THEY CONCEAL

Notebook Entries

Figures 1 and 2 show the first and last (out of three) pages of Henseleit's entry on his experiment with ornithine. These pages are typical of the style in which both researchers recorded their experimental work. It begins with a dated, underlined title. In most cases it mentions the name of the substances to be tested. It is followed by a characterization of the organism from which cell tissues had been taken for the experiments. A short phrase is added to describe the procedures

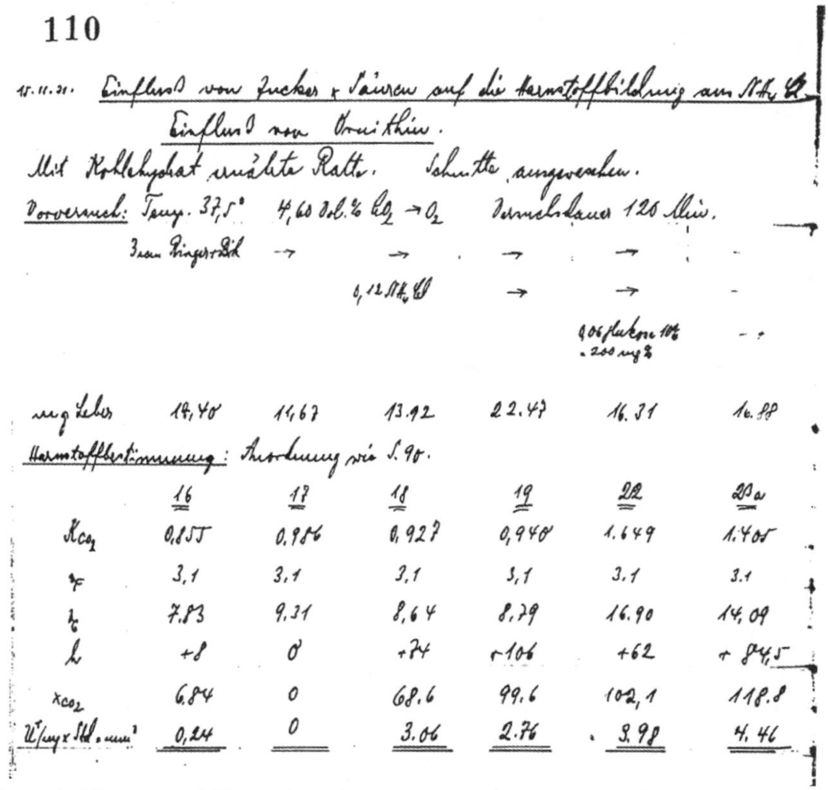

Figure 1. First page of Henseleit's laboratory entry on the ornithine experiments.

applied to the cell material before it was planted into the experimental tube (e.g. "Schnitte ausgewaschen").

The next sections on the page record the specific experimental properties such as temperature, the chemical composition of the solution and the time during which the tissues were exposed to substances in the solution.

The lower two thirds of the page display a table structure. The table is divided into two sections: on top there are rows beginning with the amount and name of substances added to the solution (e.g. first row "Ringer Lösung", second row one column to the right "NH_4Cl"). Little arrows to the right of the name of the substances indicate that the same amount of substance was added to the next experimental setup. Hence each column defines the specific attributes of one experimental arrangement within the comparable series of tests. The lower section of the table starts with the record of the weight of the liver tissue relative to which the amount of urea is measured. Since Krebs and Henseleit used manometric methods for their measuring devices, they recorded differences of pressure as measuring data. Together with auxiliary data they calculated the specific amount of urea formed in the reaction tube. This value comes at the bottom of the column

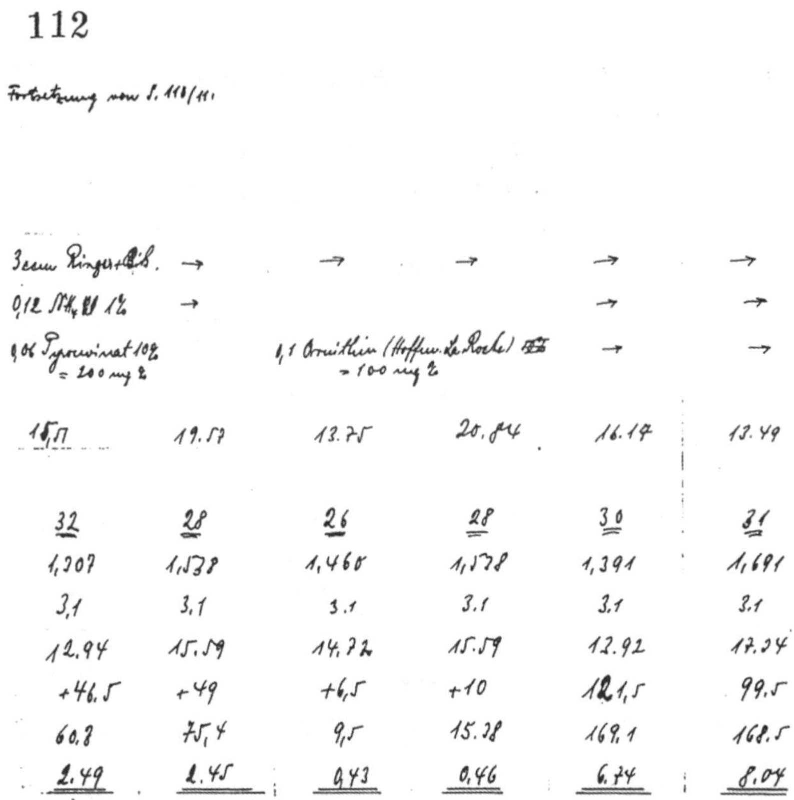

112

Fortsetzung von S. 110/111

3 ccm Ringer + Bic. →
0,12 NH4Cl 1% →
0,06 Pyruvinat 10% = 200 mg %
1,1 Ornithin (Hoffm.-La Roche) = 100 mg %

15,57	19.57	13.75	20.84	16.14	13.49
32	28	26	28	30	31
1,207	1,538	1,460	1,578	1,391	1,691
3,1	3,1	3.1	3.1	3.1	3.1
12.94	15.59	14.72	15.59	12.92	17.24
+46.5	+49	+6,5	+10	121,5	99.5
60.8	75,4	9,5	15.28	169.1	168.5
2.49	1.45	0.43	0.46	6.74	8.04

Figure 2. Third and last page of Henseleit's laboratory entry about the ornithine experiments.

and is typically underlined by either Henseleit, who recorded most of the experiments, or Krebs.

In many cases the experimental record in the notebook is concluded with a very short summary titled "Ergebnis". But the historian who had hoped to find here a detailed document of an ongoing thought-process of the researcher will be disappointed. In most cases a result is noted, e.g. that substance A or B leads to an increase in the formation of urea, or else there is a brief conclusion, for instance that next time one should wash the tissues for a longer time. In the case of the ornithine experiment only the last two columns exhibit an experimental setting, in which ornithine *together* with ammonia produces much more manometric pressure – and hence urea – than all other settings, even with ornithine alone. This is in fact a very puzzling result because according to the standard hypothesis, with which Krebs was operating at the time, even ornithine alone should have produced urea. This experiment showed the contrary (see the fourth and third last columns in Figure 2); ornithine alone did not produce urea. Further implications of the experiment are not recorded under the title "Ergebnis". But Krebs and Henseleit did draw conclusions: the notebook carefully records all experimental conditions and those of the measuring process, which led to the measured data and from which they derived the putative findings in terms of the specific formation of urea.

The experiment is set up as a causal test for factors which are either already known to play a causal influence in the formation of urea or which are tested for that role. Typically the notebook entry records such conditions. Krebs set up the experiments as a causal difference test: if two experimental situations are equal in causally relevant aspects and one additional factor exhibits an effect which is missing in the comparable situation without the presence of that factor, then Krebs rightly concludes the causal relevance of that factor. Krebs and Henseleit operated throughout their work with this methodological machinery. The assumption that the comparable situations should be equal to each other in causally relevant aspects is crucial. The experimenter's skill is revealed in his ability to realize such a condition in an experiment. Should small differences occur, one must be able to control them and account for them. The only way to test the validity of that condition (which we call *condition of homogeneity*) is to repeat an experiment under conditions which one can control, and check whether the observed effects remain the same. If this is not the case, the condition of homogeneity is not satisfied and the experiment does not allow causal conclusions. Krebs had to learn it the hard way, when he conducted experiments with thymine. He measured some increase of pressure with his manometer and jumped to the conclusion that thymine is causally relevant for the formation of urea. It took him a month's work to discover his error: he had not controlled the validity of the condition of homogeneity with sufficient care. He rectifies his error by repeating the experiment right from the beginning. The notebook entries for the ornithine experiment therefore contain two columns with an identical arrangement. The result is that there are small fluctuations in the outcome but not in the dimension of the "ornithine effect" (i.e. the production of urea from ornithine for the given conditions). For this range of error Krebs could thus treat his experimental procedures as causally equal.

What is Concealed

The notebook pages shown here are typical for the work of Krebs and Henseleit. Only occasionally does one find entries for bibliographical findings. Even the lines headed "Ergebnis" describe only the most obvious causal conclusions. Yet, information of crucial importance for the reconstruction of the discovery process is missing in the notebooks:

- In the laboratory notebooks both researchers never record their working hypothesis of the reaction pathway that they entertain during a certain phase of their work.
- They never outline their research strategy or the strategic goals by which they hope to discover the reaction path.

- They do not differentiate between experiments which test their working hypothesis and experiments which check alternative pathways as suggested in the literature.
- They record neither a change of belief in a certain hypothesis nor any methodological considerations.

The historian's dilemma is that there are no alternative documents from which such events could be directly established. No correspondence relevant to those issues is available. The first scientific publications of Krebs and Henseleit on their discovery even describe an idealized experimental setup that in all its aspects was never performed as a single experiment. Krebs and Henseleit gathered on the basis of their experimental experience the *ideal* conditions for an experiment – but this series of experiments was not described in the original publication. Even the fortunate circumstance that Frederic Holmes was able to go over all pages of the notebooks with Krebs while eliciting his comments could not fill that gap. The interviews supply much additional information but it is difficult to qualify which recalled episode is reliable. In particular, research goals – intentions to act – are not memorized reliably and are subject to a retrospective reconstruction. They could be distorted in a systematic fashion, when later knowledge is used to interpret one's own past intentions. Given this dilemma one needs to look for a different historiographical approach. Here we want to present a rather non-orthodox technique of computer modelling of the discovery process.

ELEMENTS OF AN EPISTEMIC WORLD

Epistemic Systems

Scientific discovery is seen as a complex kind of problem solving activity that is directed by scientific goals, heuristics, and methodological rules.

How do we decide whether a model of discovery correctly represents a particular historical discovery process? As with other scientific theories, this is done by testing models empirically. Models of historical processes are tested by comparing the sequence of the model's states with documents which the researcher produced during his work. In the case of the discovery of the urea cycle these were the laboratory notebooks of Hans Krebs and his assistant Kurt Henseleit. A reconstruction of a scientific achievement has to be *historically adequate*, i.e. it must not contradict any of the historical data. Even more, all elements of an epistemic system must be relevant in producing the observed sequence of events. The test of relevance can only be done by counterfactual historical scenarios. Only then one can show that a given element brings about a specific situation. Computer models are especially well suited for the study of counterfactual history.

The set of cognitive components relevant for the discovery process forms an *epistemic system*. An epistemic system is a dynamic model of a scientific discovery. It contains: *epistemic goals* that direct the agent to the solution of his problem;

propositional attitudes such as beliefs, assumptions, or considerations without judgement regarding their truth value; *heuristics* which are rules of action, according to which either propositional attitudes (e.g., belief in a causal hypothesis) are generated or physical actions (e.g., experiments) are performed; *epistemic actions* that realize the proposed steps in the discovery process. And it includes the material requirements to perform those scientific activities, e.g. instrumentation and manpower.

Causal Models and Causal Reasoning

We assume that Krebs was guided by some general principles of causal reasoning. Crucial for this kind of reasoning is a structure called a "causal graph", which is a complex network of cause-effect relationships. Knowledge of causal relationships, we assume, is essential for scientific activities such as explaining, predicting and controlling natural processes. In the following discussion, we isolate the causal aspects. In practice, causal reasoning is of course intertwined with other considerations. However, focusing on causal reasoning is justified since it was a kind of driving force of Krebs' investigations, and it allows us to investigate those aspects in a purified manner.

We interpret a biochemical pathway as an instantiation of a causal graph. As an example, take the graph from Figure 3. It shows the complicated paths of degradation of amino acids, according to Neubauer (1928), a standard textbook available to Hans Krebs. The types of causal relationships in this graph are manifold:

Directly causally relevant factors:
e.g. alanine→pyruvic acid
The presence of alanine (Alanin) causes the presence of pyruvic acid (Brenztraubensäure). This is the basic type of a causal relation.

Causal chains:
e.g. alanine→pyruvic acid→lactic acid
The path from alanine to pyruvic acid and to lactic acid (Milchsäure) is a causal chain. The relevance of alanine for the formation of lactic acid is mediated by pyruvic acid. The figure contains many such pathways.

Multiple effects:
e.g. acetaldehyde→acetic acid, acetaldehyde→acetone bodies
The presence of acetaldehyde has more than one effect: it causes not only the presence of acetic acid, but also that of an acetone body. Different effects of a common cause tend to occur together. This makes them diagnostically relevant for each other.

Multiple causes:
e.g. alanine→pyruvic acid, oxaloacetic acid→pyruvic acid

Pyruvic acid can be formed on different pathways, e.g. by alanine as well as by oxaloacetic acid.

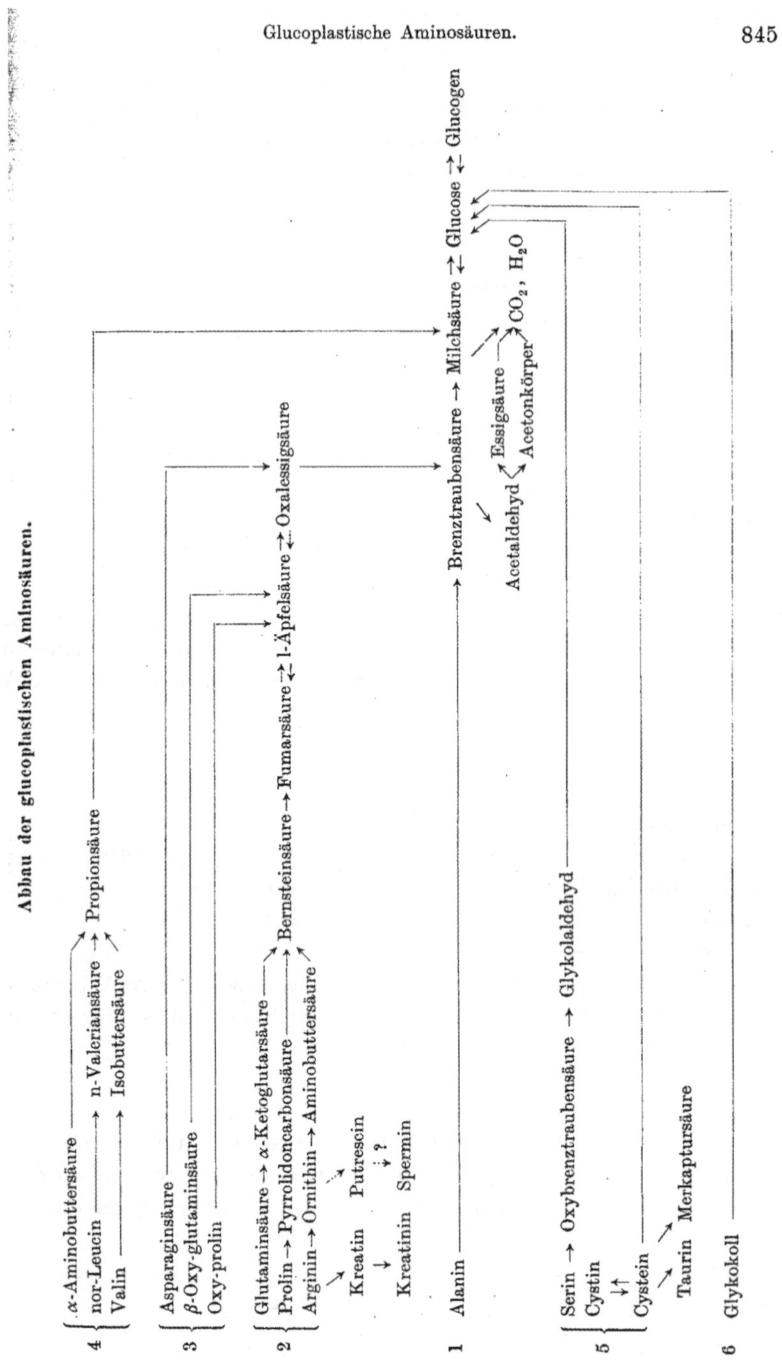

Figure 3. A certain type of degradation path of amino acids in Neubauer (1928), p. 845.

An additional type of relationship are complex causes, where several substances have to be present for an effect; e.g. both hydrogen and oxygen have to be present for the formation of water. All relationships in Figure 3 turn out to be complex, if the graph is specified in greater detail. Some of the reactions are such that two substances have to be combined to form a third. And all reactions depend on additional factors such as temperature, pressure and concentration.

A second additional type is an inhibiting factor that inhibits an effect that would be present if the inhibiting factor were absent. An example is a toxic substance that inhibits cellular respiration.

A third additional type, crucial for our case, is a cyclic process, where some specific event of type A is relevant for the occurrence of another event of the same type. Krebs' discovery was that degradation of amino acids in mammals is a cyclic process with ornithine acting as a catalyst. A reversible reaction, e.g. between glucose and glycogen in Figure 3, is also a rudimentary type of a cyclic structure.

Figure 4 summarizes the possible causal relationships representable in a causal graph.

A formal analysis of causal relationship has been one of the major philosophical challenges. Although causal reasoning governs most of our daily life and scientific reasoning, it is all but clear whether one could formulate a set of rules which capture our intuitive understanding of causality. David Hume's analysis is mostly negative and Stuart Mill's concept has been refuted early on. Even the most advanced theories of causal regularity by John Mackie were abandoned by its author.[9] It has been a substantial part of our research project to formulate an adequate causal theory and a theory of causal reasoning that is applicable to well documented cases of scientific reasoning. The graph structures which are discussed here can be given a logical interpretation.[10]

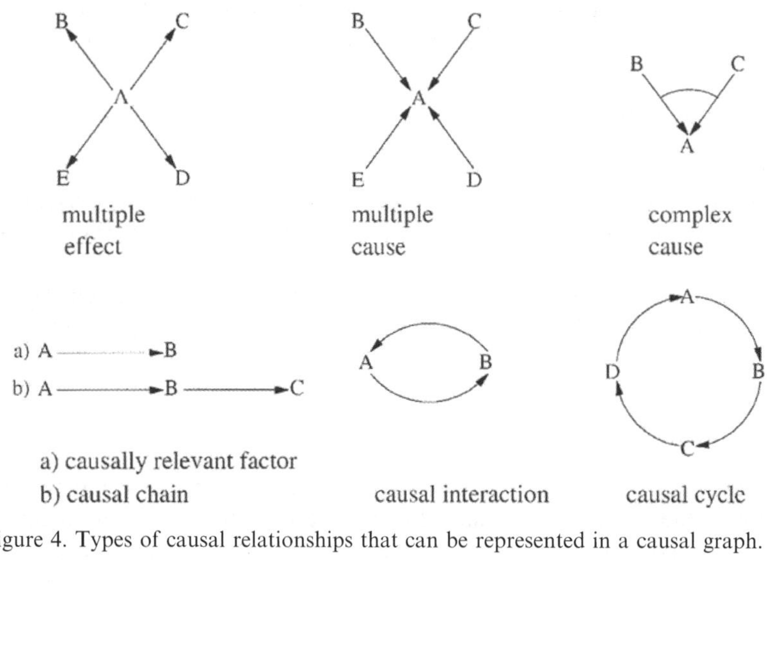

Figure 4. Types of causal relationships that can be represented in a causal graph.

The proposed account is based on two main ideas: uniformity and relevance. A complex cause is represented by a conjunction ("and"); alternative causes by a disjunction of conjunctions ("or"); a causal factor by a part of conjunction, and an inhibiting causal factor by a negation ("not"). The idea of relevance is implemented via the notion of minimality. Each complex cause has to be minimal, i.e. no part of it is sufficient. This ensures that every factor that is part of the complex cause plays an indispensable part in bringing about the effect. A complex cause is also called a minimal sufficient condition. Moreover, the disjunction of the minimal sufficient conditions is also required to be minimal: no part of it is necessary. This prevents redundant complex causes. We call a minimal necessary disjunction of minimal sufficient conjunctions a minimal theory. The presence of a complex cause is assumed to be sufficient for bringing about the effect, and the effect does not occur without one of its causes. This formal representation is very suitable for a computational representation and treatment of causal reasoning.[11]

Our knowledge of a complex set of regularities is often highly incomplete: What we know are small fragments of only some complex causes of a causal network. Justifying the causal relevance of those partial structures we have found is arguably more important than identifying fully sufficient causes.

Principle of Action

One of the key issues of cognitive models is the interaction of various types of cognitive elements. What is the connection between goals and actions? How are they represented and what is the decision process? Which preference order regulates the choice between alternative actions?

The central coordination of the elements of an epistemic system is described by a principle of action.[12] This principle reflects our understanding of actions, goals, preferences, abilities and knowledge. The following principle holds for any person *A*, action *H*, and goal *G*:

Principle of Action: If

1. *A* has the goal *G*,
2. *A* believes that doing *H* under the given circumstances is a way to reach *G*,
3. There is no other action besides *H*, which in *A*'s opinion has a higher preference for reaching *G*,
4. *A* has no other goal which diverts him from *G* under the given circumstances,
5. *A* knows how to do *H*,
6. *A* is able to do *H*

then
A concretizes H.

To *concretize* an action means in case of elementary actions to begin with its completion without further planning. In the case of complex actions it means generating the goal to perform the action in the future. Which actions are elementary is a question of the granularity of the historical model. The finer the intended explanatory structure of the historical model, the more complex actions must be considered. In this model goals are wishes to perform an act. One needs knowledge of which subordinate actions might help to complete a complex action. Preference orders decide between alternative actions.

Epistemic Action Spaces

In scientific contexts actions concern predominantly the modification and construction of models. Here a methodological rule specifies a *complex action* to achieve a certain epistemic goal, e.g., to expand a theoretical model to explain a newly discovered phenomenon. A complex action can be further specified by a sequence of simpler actions – which may turn out to be complex themselves. This yields a *hierarchy of actions*. The scientist's initial goal is to successfully perform the complex action of solving his research problem. For that he performs a set of simpler actions. Since there can be different ways of performing a complex action, an *action space* is unfolded as an and-or-graph.

The planning of actions by using a hybrid graph search strategy combines elements of heuristic search with some mechanisms to evaluate alternative plans and to alter the plan when necessary. This is achieved by incorporating global information about the problem space, by allowing the complete or partial re-evaluation of the research goals. In every planning step, a small portion of the action space is visited: a goal is expanded until every path ends in a node that is primitive or evaluable by applying a heuristic evaluation scheme. Because some nodes are neither primitive nor expandable, we get a *tree* of actions. This local action space is called the *horizon of action*. As long as the system stays in this local space it evaluates all alternatives in this space and chooses the most promising one. We therefore have a kind of *best-first-search* at a local level. But if the system sets up a new complex subgoal, a new local space is generated. The old action space is put on a stack; the system comes back to this space only after it has solved the subgoal. When choosing an action it only considers the local space.

Figure 5 shows the root node of complex actions, which organizes the modification of models. The tree links disjunctive actions: A modification of the model M is successfully performed, if either it is successfully expanded to a new model N, which is itself successfully modified; or the modification of M is performed successfully by a contraction followed by a modification of the new model, etc. This recursive definition of complex actions allows the generation of large hierarchies of goals, with nested research goals. For its modelling a language called *EpiLog* (*Epi*stemic Systems in Pro*log*) has been developed.

MODELLING SCIENTIFIC DISCOVERIES

One of the key aspects of the project is the implementation of epistemic systems as a computer model. For that purpose we have developed a graphical user interface

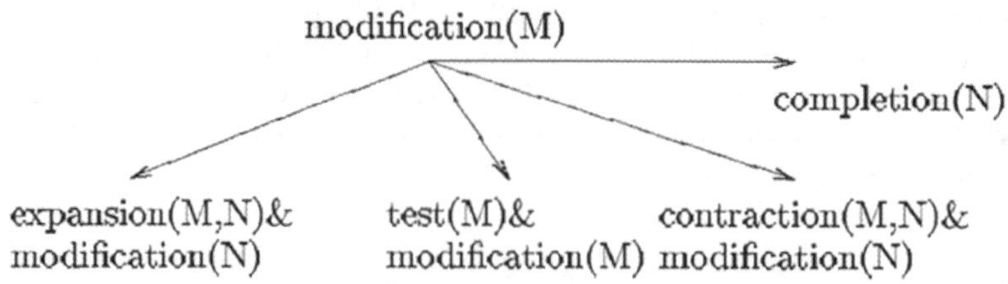

Figure 5. Hierarchy of actions.

that makes semantic structures and model behaviour transparent. This workbench allows rapid prototyping of the assumed epistemic state of a scientist, storing a specific state at a given stage of the discovery process and restarts from that level of knowledge. In this fashion also contra-historical sequences can be modelled and analyzed. The features implemented for the modelling of epistemic systems include:

- A representation of *action rules* or *heuristics*, which are relations between a complex actions and their subsequent actions. As an example, an action rule

  experiment ⟪– experimental design,
  experimentation,
  experiment evaluation.

  defines that in order to successfully perform an experiment, one has to perform an experimental design, perform the experiment and evaluate it.
- Preference rules establish in case of alternative actions, which is the best choice in relation to the initial goal and should hence be performed first. If there is an action rule for such action, the new goal is generated to perform the chosen action.
- The interaction of goals, action rules, preference rules and propositional attitudes related to the causal models of chemistry is governed by an implementation of the *Principle of Action*.
- Rules for the design and evaluation of experiments are derived from a general theory of causal regularities.[13]
- Chemical knowledge in the form of a dictionary of chemical components and a module to stoichiometrically balance reaction paths is implemented.
- The possible outcome of experiments is modelled in such a fashion that it should replicate the (noisy) data obtained in the historical experiments.

The case studies which have been investigated so far indicate that there is only a fairly restricted class of types of action at the "top level": Among them are complex actions like model expansion, model revision and causal test strategies. The variety of strategies scientists use is a specialization of these top level actions. *Hence, there is a general methodology underlying scientific research.* At this case study we substantiate our claims by proving the adequacy and fruitfulness of the general model.

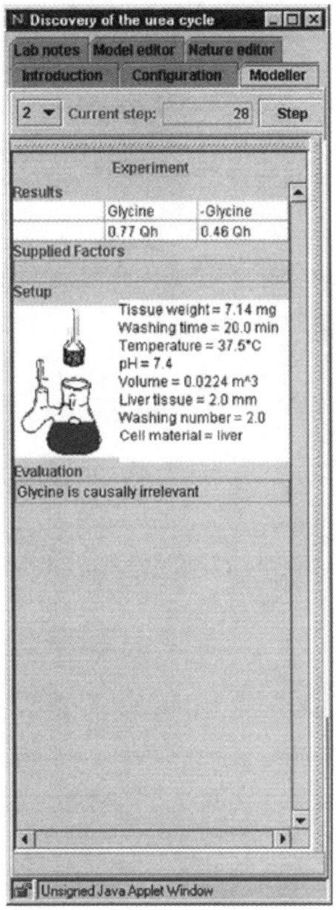

Figure 6. Modelling workbench. This window shows the information on the proposed experimental setup and the results of the simulation of the experimental outcome.

The information obtained by such a model is very rich. It produces a sequence of cognitive states concerning research hypotheses, goals, experimental plans and the methodological treatment of the empirical findings. Should the output of the model be confined to what one can read in a laboratory notebook, the screen would look like Figure 6. It just shows the features of the experimental setup and the simulated empirical finding. A much larger picture is shown in Figure 7. The big window shows the hierarchy of possible research actions and exhibits the state in which the model drafts research plans and performs their actions. Another window shows the current working hypothesis on which the experiments are planned. For the purpose of this paper we must restrict our discussion of the computer model to the examination of one interesting finding of the model: it performs experiments after the so-called "ornithine effect" as searches for the degradation path of ornithine, shown as graph in Figure 8.

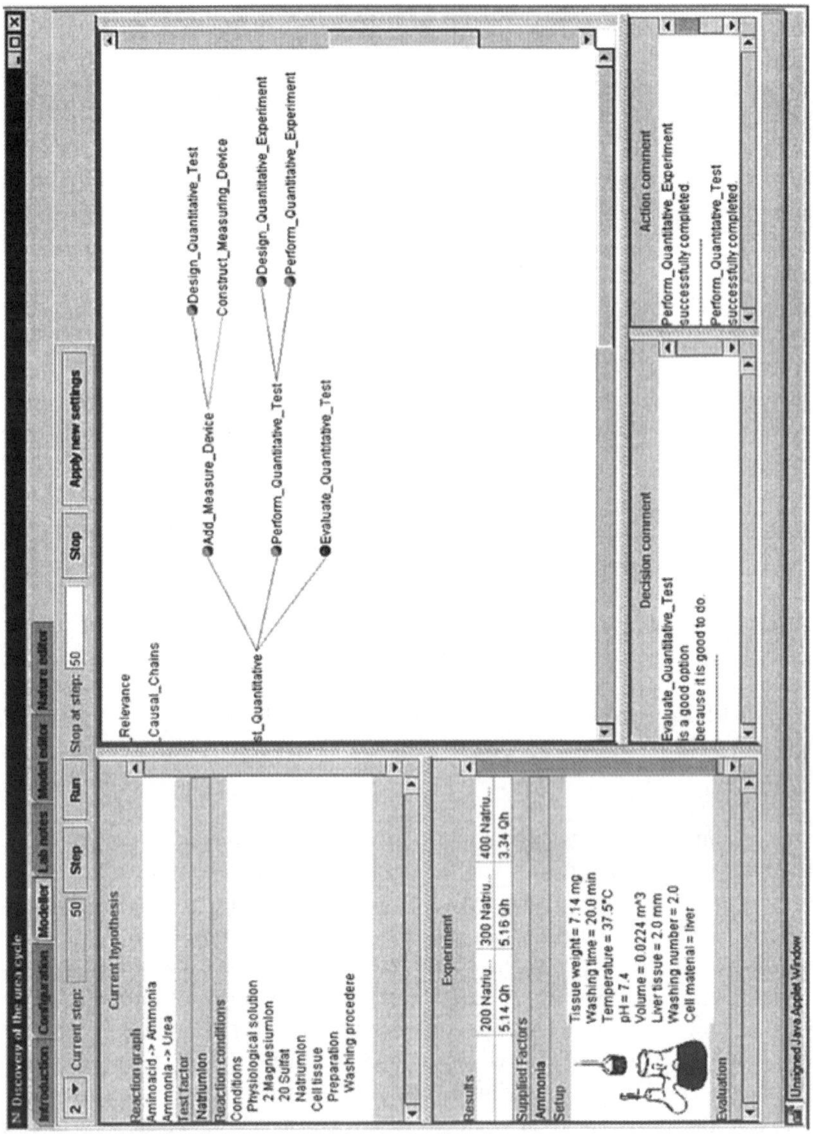

Figure 7. Full window of the discovery model.

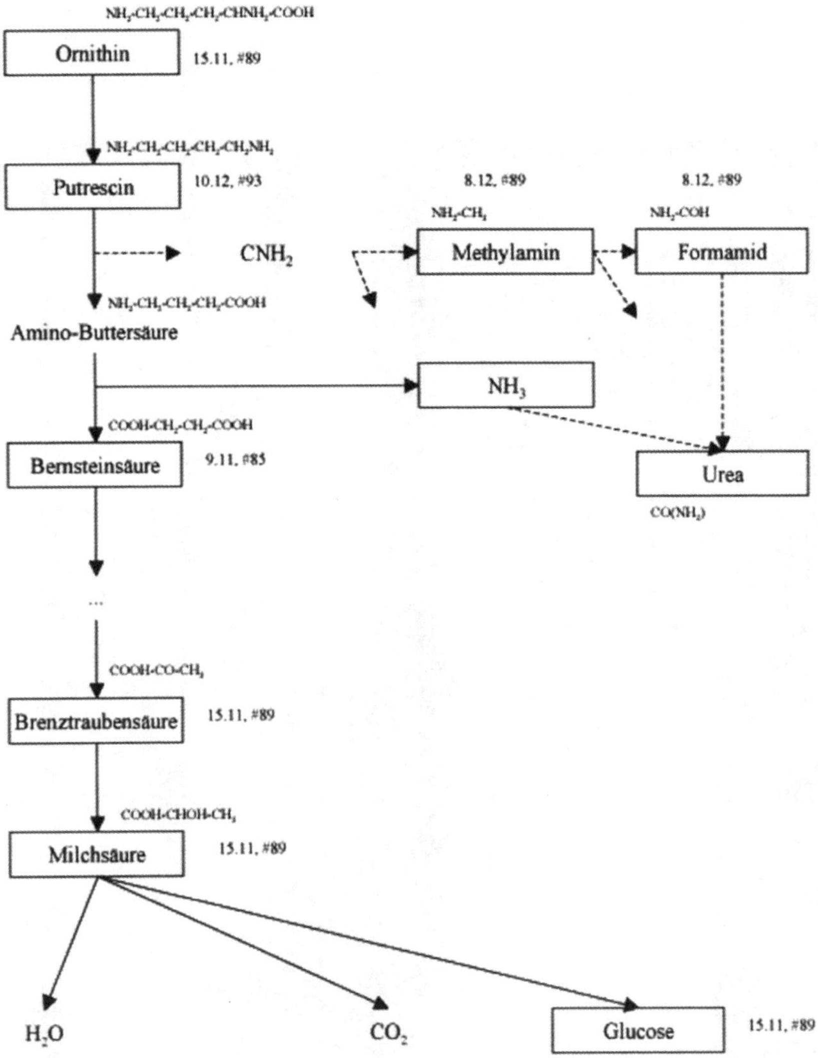

Figure 8. Postulated causal graph. Degradation path of ornithine that is compatible with the experimental data about the ornithine effect and the knowledge Krebs had at this point.

EXAMINING HISTORICAL HYPOTHESES AND SEARCHING FOR NEW EVIDENCE

In the previous sections we presented a model which helps to construct new historical hypotheses and allows exploring counterfactual historical sequences. For supporting the critical examination of hypotheses we developed a second tool

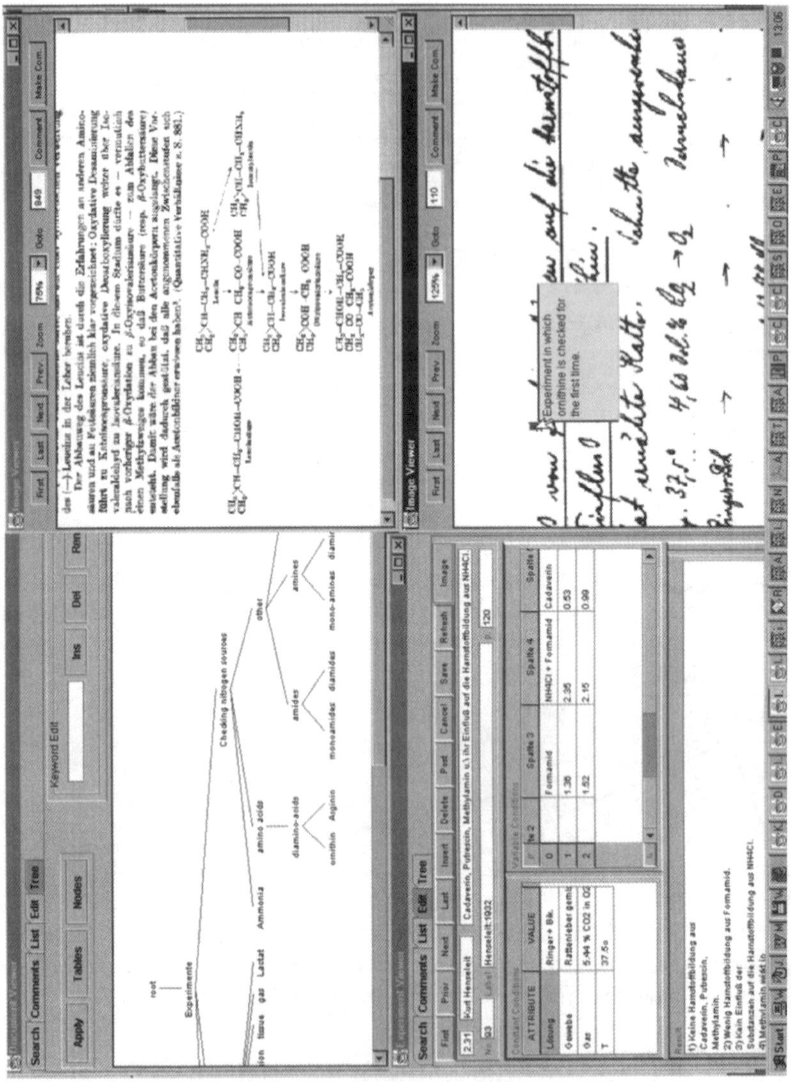

Figure 9. Electronic environment supporting access to data, transcription of documents, interpretation, and electronic publication.

(Figure 9) that allows to check hypotheses found by the model against the data by searching for new confirming or refuting evidence. The tools can be run independently.

The second tool assists the researcher in different stages of his research: accessing data collections, transcription of documents, interpretation, and electronic publication on the internet.

It comprises a database containing

- bibliographical data,
- scanned images of laboratory notebooks,
- articles,
- books,
- and diagrams.

It offers extensive search capabilities that allow to explore the historical data efficiently. Also provided are graphical tools for commenting historical documents – whole documents, individual pages, parts of pages – whereby the comments are also saved in a database. It is implemented in the Java programming language so that it is independent of a specific computing platform and can be operated over the internet using a web-browser.

Krebs' Reaction to the Discovery of the Ornithine Effect

According to the hypothesis generated by the computer model Krebs reacted to the discovery of the ornithine effect by forming a tentative causal graph (Figure 8), which he subsequently subjected to experimental tests. In doing so he was guided by the graph structures that he found in the textbooks (e.g. Neubauer) and by literature available to him. Note that we need not assume that the path depicted is the correct one (in fact, it is fundamentally wrong) or even that Krebs believed this path to be correct. The only assumption we need to make is that Krebs found it worthwhile to investigate the hypothesis by subjecting it to an experimental check (if only to refute it).

What evidence do we have for the general claim that Krebs conceived his problem in terms of causal graphs? What specific evidence do we have for the claim that it was this graph (or a similar one) which he had in mind?

Evidence comes from several sources: notebooks, textbooks, and articles. We will discuss the evidence step by step.

The substances shown in the box were tested by Krebs around November 1932 (Figure 9). A date is given to the right of the box and the experiment number is specified. The numbering is not part of the laboratory notebooks but is used in the electronic database. For the pathways depicted in solid lines direct evidence can be given (for details see discussion in text). The dashed lines represent more speculative pathways.

First Source of Evidence: Laboratory Notebooks

From the laboratory notebook entries, we know for certain that Krebs made experiments with most of the building blocks of this graph. To the right of each substance in Figure 8 we noted a date where Krebs checked this substance. The following table lists the dates on which he checked the substances, along with the corresponding page in the laboratory notebook. Krebs and Henseleit had already collected some empirical evidence about the behaviour of substances mentioned in the graph before discovering the ornithine effect.[14]

Experiment no. and page number in Henseleit's notebook	Date	Substances tested (all experiments were run by Kurt Henseleit, using German terms.)
89, p. 110	15.11.	Ornithin, Milchsäure, Brenztraubensäure, Glukose, Fruktose, Sorbose
90, p. 114	16.11.	Ornithin, Valeriansäure
91, p. 116	8.12.	Formamid, Methylamin
92, p. 119	8.12.	Formamid
93, p. 120	12.12.	Formamid, Methylamin, Putrescin, Cadaverin
94, p. 126	14.12.	Formamid, Ornithin, Asparagin
95, p. 130	16.12.	Formamid

Table 1: Substances tested after the discovery of the ornithine effect.

So Krebs knew the building blocks of the graph in Figure 8. But did he consider the pathways, the links, between these building blocks? This is the question to which we now have to turn our attention.

Second Source of Evidence: Textbooks

We already noted that Krebs used a textbook by Otto Neubauer, which summarized the knowledge about the synthesis of urea available at that time.[15] This textbook provides further important evidence.

Neubauer summarizes the knowledge about degradation pathways of amino acids. The part at the bottom of the graph (Figure 8) was thought to be a general degradation pathway of amino acids, not specific to ornithine. Krebs could have tested these substances to see whether the non-amino parts of amino acids have some indirect effect on the formation of urea.

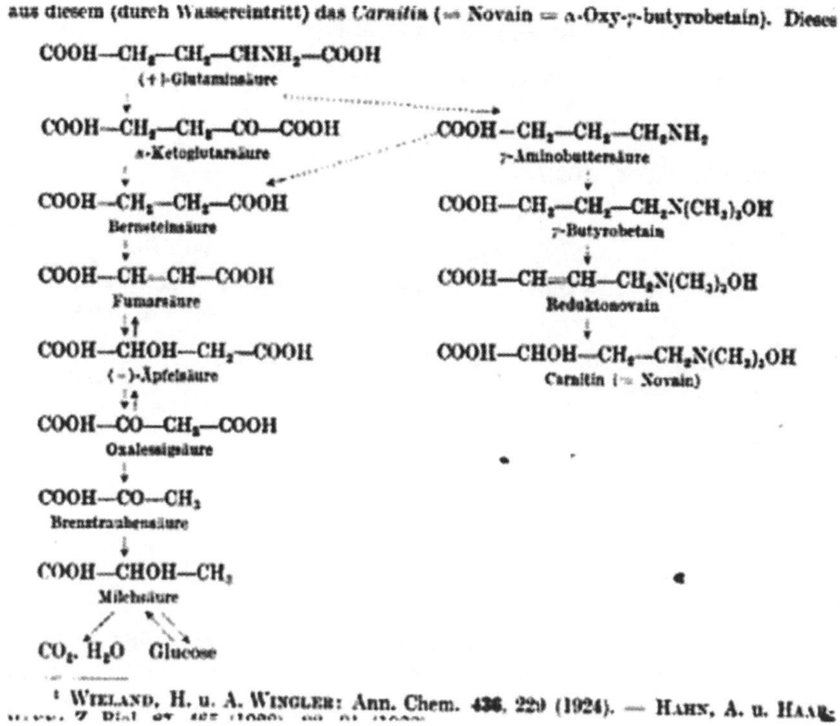

Figure 10. Degradation path of amino acids, Neubauer (1928), p. 836.

We see that the bottom part of this graph (Figure 10) directly corresponds to the bottom part in Figure 8.

Some pages later in the textbook[16] we find a second pathway, formulating possible ways of urea degradation from arginine to ornithine. Here we find the graph on the upper left part of Figure 11.[17] It was assumed to be a non-standard pathway, given the standard theory. But since the standard hypothesis about urea formation was hard to reconcile with the experimental data, it is plausible that Krebs had to look for non-standard explanations. The substance putrescine, which he tested, only appears on the non-standard path. This non-standard path is discussed on page 838 of Neubauer (Figure 12).

Third Source of Evidence: Articles

What evidence do we have for the third part of the graph containing formamide? This part is certainly the most interesting one.

Let us recall the situation: After several months of testing, Krebs eventually found an effect which was, as he recalled, totally unexpected. One would assume that after finding such a phenomenon, one would make it the centre of one's

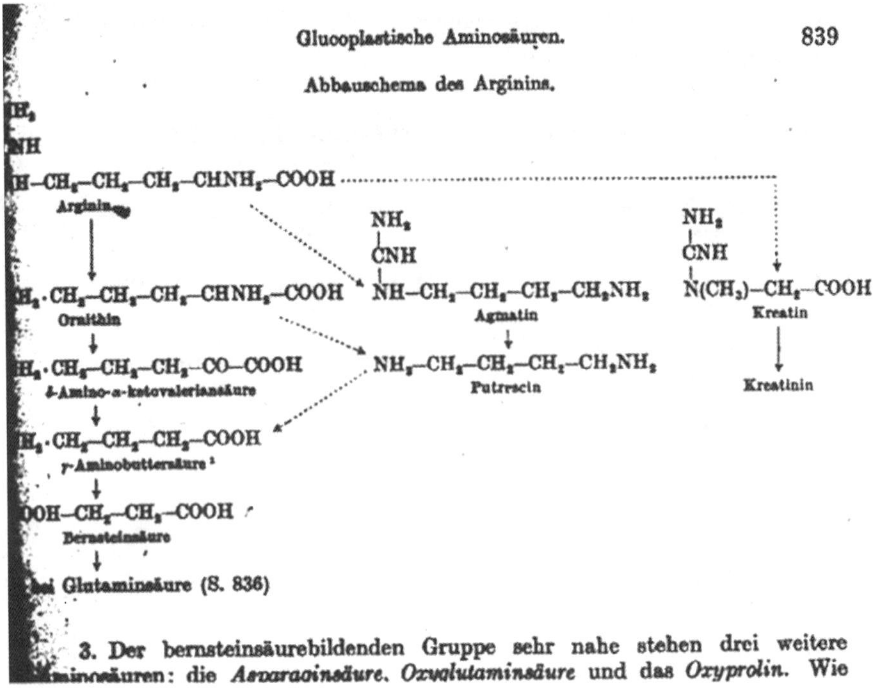

Figure 11. Description of degradation path from ornithine via putrescine in Neubauer (1928), p. 839.

attention, one would hurry to investigate whether it is an effect or an artifact and to determine its scope. Interestingly, Krebs indicates in his recollections that he was doing just that.[18] But as the notebooks demonstrate, in the next experiments, Krebs predominantly checked formamide (see Table 1). On the account given by Holmes (1991), checking of formamide was unrelated to the ornithine effect.[19] This leaves one with the puzzling fact that Krebs does not do what methodologically one would expect him to do: investigate the ornithine effect by checking a non-standard degradation path. On our account, testing of formamide finds a natural justification, since it was assumed to be a possible breakdown product of amino acids. But is there any evidence that Krebs knew about such a possible reaction with formamide/ammonia?

Such evidence can indeed be found in the literature that was available to Krebs. There is an article by J. T. Halsey, "Über die Vorstufen des Harnstoffs", *Hoppe-Seylers Zeitschrift für physiologische Chemie* 25 (1898), pp. 325–336, which discusses formamide as a possible precursor of urea. Did Krebs know this article?[20] We did not find a direct citation of Halsey's article or the article of Hofmeister in Krebs' work.[21] Therefore, we investigated whether the literature cited by Krebs contains references to this publication. This would increase the

838 O. Neubauer: Intermediärer Eiweißstoffwechsel.

sei es spontan oder nach Zufuhr von Arginin. Es ist klar, daß es hier aus dem Ornithinanteil des Arginins hervorgegangen ist. Man muß in diesen Fällen von Cystinurie eine allgemeinere Störung des Desaminierungsvorganges annehmen (s. S. 909). Für die Deutung des Vorkommens der beiden Diamine ergeben sich zwei Möglichkeiten. *Entweder* es geht bei den Diaminosäuren die Decarboxylierung schon normalerweise der NH_3-Abspaltung voraus, dann wären die Diamine physiologische Zwischenprodukte und es handelte sich um eine einfache *Hemmung* der nachher fälligen Desaminierung; *oder* aber es soll auch bei den Diaminosäuren die Ammoniakabspaltung zuerst stattfinden; weil sie aber wegen der Stoffwechselstörung verhindert ist, so erfolgt nun ein in diesem Stadium unphysiologischer Vorgang, die Decarboxylierung, der Abbau gerät auf einen falschen Weg („Parektropie infolge Hemmung").

Die erste Annahme würde sich in folgender Weise wiedergeben lassen:

Ornithin → Putrescin --Hemmung--→ γ-Aminobuttersäure → Bernsteinsäure → Milchsäure

die zweite:

Ornithin --Hemmung--→ Amino-α-ketovaleriansäure → γ-Aminobutters. → Bernsteins. → Milchsäure
⋯→ Putrescin

Eine sichere Entscheidung ist derzeit nicht möglich. Vielleicht werden auch unter physiologischen Bedingungen *beide* Wege begangen. Jedenfalls ist Putrescin (und ebenso Cadaverin) im normalen Organismus verbrennlich[1].

Nach den Durchblutungsversuchen von Felix und Tomita[2] ist Katzenleber imstande, den Ornithinanteil des Arginins weiter zu verändern; die Umwandlungsprodukte sind aber noch nicht festgestellt; in einzelnen Versuchen ließ sich Bernsteinsäure nachweisen (Felix u. Müller[2]). Fertiges Ornithin wird von überlebender Leber nicht angegriffen; ebensowenig Putrescin (Felix u. Röthler[2]).

Figure 12. Description of degradation path from ornithine via putrescine in Neubauer (1928), p. 838.

likelihood that Krebs knew the articles, since he could have followed those links. We found several such references:

1. Krebs repeatedly cites an important article by Kossel and Dakin.[22] In this article the authors refer to an article by Kossel, which was printed in the same volume as the article by Halsey.[23] So if Krebs followed this link, he could accidentally have come across the paper by Halsey. Since these articles are in the same journal, which was readily available to him, it would have been easy for Krebs to follow the link.
2. Krebs repeatedly cites an important article by Löffler.[24] The article contains a detailed overview on the work on urea synthesis. Krebs refers to page 230 of this article; on the corresponding page 231 – a page Krebs must have seen – Löffler mentions both the article by Halsey and the substance formamide. Krebs must have read this.

These findings establish that in one way or another – even by indirect reference – Krebs must have known the reaction path as described by the article of Halsey. What role could this piece of information have played in his reasoning? As we already mentioned, Halsey discusses formamide as a possible precursor of urea, and since Krebs measured the formation of urea from formamide it is likely that he saw it in such a function.

We come to the final step in our argumentation. This is the assumption that – contrary to the standard interpretation – Krebs considered a formation of urea where one amino group *does not go through the stage of ammonia.* Both Krebs[25] and following him Holmes[26] claim that Krebs at some point did entertain such a guess. It is convincing because it provides a reason for testing amino acids and ammonia in combination. *The graph shown in Figure 8 is just a specific instantiation of such a hypothesis.*

Krebs must have been led to such a hypothesis by the need to make sense of the fact that ammonia and ornithine in combination yield more urea than each of them on its own. This could not be reconciled with the standard assumption that all amino groups pass through the stage of ammonia. Under this assumption, as Krebs says, urea production from ammonia should be at least as fast as an amino acid, since it is shortened by some reaction steps.[27] Krebs also had evidence that the carbon skeleton of ornithine (or amino acids in general) had no indirect effects on the urea production, since he had already tested the known breakdown products of ornithine (as we saw in Table 1). The graph in Figure 8 explains Krebs' experimental findings and failures, yet it is not very far from the standard assumption. It is compatible with many of the phenomena known to Krebs. Additionally, the graph could be seen as a special degradation path for diamino-acids such as ornithine so that one could maintain the standard assumption for the other amino acids.

The correct interpretation, a cycle with ornithine functioning as a catalyst, was not available to Krebs until much later.

A further piece of evidence in support of this interpretation is again in Neubauer,[28] who mentions the possibility of a degradation of amino groups together with a carbon atom, although he considers it to be implausible. Yet Neubauer judged this path to be implausible in the light of the evidence available to him. Krebs, however, had just found new evidence, which seemingly was in conflict with the standard account. So if he thought his new evidence to be reliable, he was forced to reconsider the standard account and to look for alternatives. There is no need to assume that Krebs put much trust in this hypothesis. But at least he could try to eliminate this alternative possibility.

So let us sum up the argument. We assumed that Krebs used quite general strategies of causal reasoning about chemical reaction paths and experimental design and to this end he made intense use of causal graphs. To support this hypothesis we identified pieces of knowledge in the notebooks, textbooks and literature which were both known to Krebs and related to his case. Combining those general assumptions with the specific domain-dependent knowledge allows

C-Atomen keine Milchsäure, somit auch keine Acetonkörper liefern.

Bei Fettsäuren mit einer *verzweigten C-Kette* muß angenommen werden, daß die Seitenkette zunächst abgespalten wird (dabei scheint an dem C-Atom der Verzweigungsstelle eine OH-Gruppe aufzutreten); sodann unterliegen sie ebenfalls dem Gesetz der β-Oxydation[2].

[1] Damit soll die Möglichkeit, daß der Amino-N u. U. doch auch in anderer Form als in der von NH_3 abgespalten werden könnte, nicht durchaus in Abrede gestellt werden. FRARON [Physiologie. Rev. 6, 399 (1926)] vertritt die Ansicht, der N werde als *Cyansäure* abgespalten, wobei der um ein C-Atom verkürzte Rest der Aminosäure als Aldehyd zurückbleibe

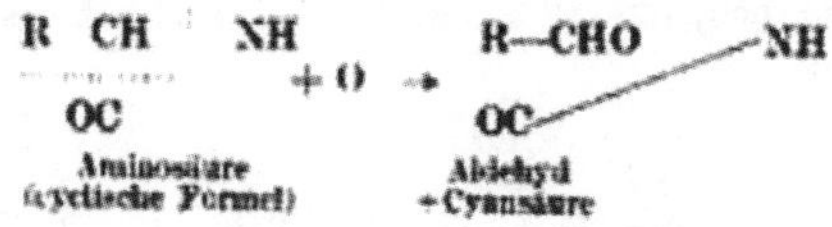

Die abgespaltene Cyansäure könnte sofort zur Harnstoffbildung Verwendung finden (s. S. 815 Harnstoffbildung); andererseits könnte sie bei Bedarf (Gegenwart von Säuren) leicht NH_3 bilden. Gegen die allgemeine Gültigkeit dieser Abbauweise sprechen die unten für die Ketonsäurebildung angegebenen Gründe; besonders das erwähnte Vorkommen N-freier Aminosäurenabkömmlinge mit intaktem C-Skelett. — Eine Abspaltung von *Glykokoll* aus manchen Aminosäuren ist unwahrscheinlich (s. oben S. 764). — Über die Frage einer intermediären Bildung von *Uraminosäuren* s. unten, Nebenwege, S. 964.

[2] BAER, J. u. L. BLUM: Arch. f. exper. Path. 55, 89 (1906); 56, 92 (1906).

Figure 13. Neubauer (1928), p. 776. A possible source for the hypothesis that the amino-N does not go through the stage of ammonia.

us to give a coherent and rather straightforward interpretation of Krebs' experimental strategy at this stage – more coherent, we claim, than the alternative explanation provided by Holmes.

SUMMARY

Laboratory notebooks are excellent sources for the investigation of scientific research processes. Yet these sources attest only specific aspects of a scientific discovery process as the experimental setup and empirical results. Only rarely does a researcher note his research strategies or his hypothetical considerations. One cannot simply read laboratory notebooks hoping for a comprehensive report of thoughts that take place during a scientific discovery process. Laboratory notebooks document a sequence of *research activities* which are the consequence of hypothetically assumed thoughts. Such hypotheses need to be formulated. In the case of the discovery of the urea cycle we introduced a cognitive model of scientific discovery processes for that purpose. It is implemented on a computer that simulated both the evolution of hypotheses by Krebs and his experimental work and led to a proposal for the sequence of actions as found in the laboratory notebooks. It differs from the standard interpretation and retrospective reconsiderations by Hans Krebs himself. We presented ample historical evidence in favour of our interpretation. On that basis we could show a computer model of a scientific discovery that for the first time provided new historical insights into the mechanisms of a creative research process.

NOTES

[1] We are grateful for helpful comments by Kärin Nickelsen and Etan Kohlberg.

[2] (Schultzen and Nencki 1872).

[3] Both notebooks are published in facsimile and transcription in (Graßhoff and Nickelsen 2001a) and (Graßhoff and Nickelsen 2001b).

[4] (Kulkarni and Simon 1990).

[5] (Holmes 1980).

[6] (Graßhoff 1995), (Graßhoff and May 1995a), (Graßhoff and May 1995b).

[7] (Graßhoff 1995), (May 1999), (Graßhoff and May 2001). The case is extensively discussed as well in (Graßhoff et al. 2000).

[8] (Holmes 1991).

[9] (Mackie 1980).

[10] (Graßhoff and May 2001), (Graßhoff and May 1995b), (May 1999).

[11] Formally, the underlying structure is a directed and-or-graph, where nodes represent types of events that can be instantiated by event tokens. A cycle corresponds to a sequence that contains type identical, but spatio-temporal distinguishable event tokens.

[12] The principle of action was inspired by a widely discussed proposal of Churchland, [1970], p. 221. However, the introduction of complex and elementary actions leads to substantial differences to Churchland's proposal.

[13] (Graßhoff and May 1995b).

[14] They tested for example pyruvic acid, lactic acid, and glucose in an experiment run on 13.11.1932. Hence, the tests of those substances is based on different considerations than the degradation of ornithine.

[15] (Neubauer 1928).

[16] (Neubauer 1928), p. 839.

[17] Note that this pathway is linked to the other (Neubauer 1928), p. 836, which we have discussed before with Figure 10.

[18] (Krebs 1973), p. 20, (Holmes 1991), p. 287: "At the very first, he [i.e. Krebs] was skeptical that it was even a valid observation. [. . .] By adding urease to a sample of the ornithine, and to a sample of ornithine together with an extract of liver containing arginase, he quickly eliminated these potential sources of error and satisfied himself that the phenomenon was real."

[19] (Holmes 1991), p. 288: "Formamide ($HCONH_2$) represented a different approach, a continuation of the earlier strategy involving the pyrimidines, to see whether compounds that contained portions of the central structure of urea might give rise directly to that substance."

[20] When we discussed this topic with Larry Holmes, he put this as a challenge: to construct a path on which Krebs could have come to know the article by Halsey. Here the electronic environment briefly described above really proved its value, since it was a matter of a few hours to find several such pathways by which Krebs could have come across this article and a second one by Franz Hofmeister, which is cited by Halsey.

[21] (Hofmeister 1986).

[22] (Kossel and Dakin 1904).

[23] Volume XXV of *Hoppe-Seylers Zeitschrift für physiologische Chemie*.

[24] (Löffler 1918). It is cited in (Krebs and Henseleit 1932a), p. 45 and (Krebs and Henseleit 1932b), p. 61.

[25] (Holmes 1980), p. 280, according to the interview of Holmes with Krebs discussing the notebook entry of 26.10.31, when Krebs tested alanine with ammonia: "The reason for testing the two substances together was, as Krebs recalled in 1976, that he 'had some guess' that one of the nitrogens in urea "comes from ammonia and the other comes directly from amino acids."

[26] (Holmes 1980), p. 285: "I made the conjecture that the choice [i.e. to test ornithine] might have been connected with his "guess" that one of the urea nitrogens comes immediately from an amino acid. Such an idea could, in principle, have directed his attention to the two diamino acids, ornithine and lysine [. . .]."

[27] (Krebs 1973), p. 20: "Is ammonia an obligatory intermediate in the conversion of amino nitrogen to urea nitrogen? If it is not, the rate of urea formation from amino acids might be more rapid than the rates from ammonia".

[28] (Neubauer 1928), p. 776.

REFERENCES

Graßhoff (1995): Graßhoff, G., "Die Kunst wissenschaftlichen Entdeckens. Grundzüge einer Theorie Epistemischer Systeme," Habilitationsschrift (Universität Hamburg, 1995).

Graßhoff et al. (2000): Graßhoff, G., R. Casties, and K. Nickelsen, *Zur Theorie des Experiments. Untersuchungen am Beispiel der Entdeckung des Harnstoffzyklus* (Bern: Bern Studies in the History and Philosophy of Science, 2000).

Graßhoff and May (1995a): Graßhoff, G. and M. May, "From historical case studies to systematic methods of discovery," in *Papers from the AAAI Spring Symposium Series 'Systematic Methods of Discovery' Stanford University* (Menlo Park: AAAI Press, 1995).

Graßhoff and May (1995b): Graßhoff, G. and M. May, "Methodische Analyse wissenschaftlichen Entdeckens," *Kognitionswissenschaft* 5 (1995): 51–67.

Graßhoff and May (2001): Graßhoff, G. and M. May, "Causal Regularities," in *Current Issues in Causation*, W. Spohn, M. Ledwig, and M. Esfeld, eds. (Paderborn: Mentis, 2001): 85–114.

Graßhoff and Nickelsen (2001a): Graßhoff, G. and K. Nickelsen, *Dokumente zur Entdeckung des Harnstoffzyklus. Laborbuch Hans Krebs und Erstpublikationen*, vol. I (Bern: Bern Studies in the History and Philosophy of Science, 2001).

Graßhoff and Nickelsen (2001b): Graßhoff, G. and K. Nickelsen, *Dokumente zur Entdeckung des Harnstoffzyklus. Laborbuch Kurt Henseleit*, vol. II (Bern: Bern Studies in the History and Philosophy of Science, 2001).

Graßhoff (1995): Graßhoff, G., "The Methodological Function of Surprises," *Foundations of Science* 1.2 (1995): 204–208.

Halsey (1898): Halsey, J. T., "Über die Vorstufen des Harnstoffs," *Hoppe-Seyler's Zeitschrift für physiologische Chemie* 25 (1998): 325–336.

Hofmeister (1986): Hofmeister, F., "Ueber die Bildung des Harnstoffs durch Oxydation," *Archiv für experimentelle Pathodologie und Pharmakologie* (1986): 426–444.

Holmes (1980): Holmes, F. L., "Hans Krebs and the Discovery of the Ornithine Cycle," *Aspects of the History of Biochemistry, Federation Proceedings* 39 (1980): 216–224.

Holmes (1991): Holmes, F. L., *Hans Krebs: The Formation of a Scientific Life*, vol. 1, (Oxford: Oxford University Press, 1991).

Kossel and Dakin (1904): Kossel, A. and H. D. Dakin, "Über die Arginase," *Hoppe-Seyler's Zeitschrift für physiologische Chemie* 41 (1904): 321–331.

Krebs (1973): Krebs, H., "The Discovery of the Ornithine Cycle of Urea Synthesis," *Biochemical Education* 1 (1973): 19–23.

Krebs and Henseleit (1932a): Krebs, H. and K. Henseleit, "Untersuchungen über die Harnstoffbildung im Tierkörper II," *Klinische Wochenschrift* 11 (1932): 757–759.

Krebs and Henseleit (1932b): Krebs, H. and K. Henseleit, "Untersuchungen über die Harnstoff-bildung im Tierkörper," *Hoppe-Seyler's Zeitschrift für physiologische Chemie* 210 (1932): 33–66.

Kulkarni and Simon (1990): Kulkarni, D. and H. A. Simon, "Experimentation in Machine Discovery," in *Computational Models of Scientific Discovery and Theory Formation*, J. Shrager. and P. Langley, eds. (San Mateo/Cal.: Morgan Kaufmann, 1990): 139–175.

Löffler (1918): Löffler, W., "Desaminierung und Harnstoffbildung im Tierkörper," *Biochemische Zeitschrift* 85 (1918): 230–294.

Mackie (1980): Mackie, J. L., *The Cement of the Universe. A Study of Causation*, (Oxford: Clarendon Press, 2nd ed., 1980).

May (1999): May, M., *Kausales Schließen*, Berichte des Graduiertenkollegs Kognitionswissenschaft, vol. 64 (Universität Hamburg, 1999).

Neubauer (1928): Neubauer, O., *Intermediärer Eiweißstoffwechsel* (Berlin: Julius Springer, 1928): 797–825.

Schultzen and Nencki (1872): Schultzen, O. and M. Nencki, "Die Vorstufen des Harnstoffs im thierischen Organismus," *Zeitschrift für Biologie* 8 (1872): 124–146.

FREDERIC L. HOLMES*

LABORATORY NOTEBOOKS AND INVESTIGATIVE PATHWAYS

I

The papers presented in this volume illustrate the many forms of research records that have been kept by scientists, and the diverse purposes to which they are being put by historians. The nature of the records that have survived and the historical uses to which they can be put are mutually interdependent. Fragments of former records, or records kept in the first place only in fragmentary form can, nevertheless, reveal, as several of these papers ably demonstrate, critical information about the research process not visible in its published outcome. When we have available records that are more nearly continuous, the opportunities opened for historical reconstruction are, correspondingly more comprehensive. In my own experience with the laboratory notebooks of three experimental scientists working in three different centuries, I have encountered records complete enough to make it possible to reconstruct what I refer to as an "unbroken investigative pathway" extending over periods ranging from several years to the lifetime of a professional career. In a fourth case, for which conventional notebook records have not been available, I have found alternative modes of recording operations and data that allow similarly full treatment.

My interest in reconstructing such pathways emerged at first accidentally. Confronted with the early notebook records of Claude Bernard, I tried to scan through them looking for significant turning points, the earliest evidence for the incipient emergence of novel ideas or methods, and such matters. I found, however, that one could not easily make sense of the record of any particular experiment read alone. Aside from the difficulties of deciphering semi-legible handwriting, of elliptical descriptions of operations, and the common absence of statements about the reason for performing a given experiment in a particular way, the experiment seemed to have little meaning apart from its location within an ongoing sequence of experiments. I found, to my surprise, that each experiment did seem to acquire meaning when examined in the light of the experiments that had preceded it. I began reconstructing the investigative pathway, initially only as a way to extract such meanings. I started with the earliest experiments recorded in Bernard's notebooks on the subject of animal chemistry, in 1842, and just kept

* Yale University, New Haven, CT

Frederic L. Holmes, Jürgen Renn and Hans-Jörg Rheinberger (eds.), Reworking the Bench: Research Notebooks in the History of Science, 295–308 © *2003 Kluwer Academic Publishers. Printed in Great Britain*

going until eventually I reached his first major discoveries in this field – the special action of pancreatic juice on fats, and the formation of sugar in the liver – both of which took place in 1848. I then stopped, in part because I had arrived at a resolution of some of the central problems and issues with which Bernard had been grappling up until then, but also because the reconstructed trail had become almost too long to fit into one book or to hold the interest of a reader.

Along the way I began to justify the value of such a continuous reconstruction, even though it included long stretches of investigation that seemed to lead Bernard either into dead ends or to nowhere in particular, as a more realistic portrait of how scientists proceed, than would selective attention just to the climactic periods that culminated in landmark discoveries. The price to pay, in the form of lengthy narratives filled with dense technical detail, seemed worth the reward it yielded: that is, to document that science in the making is more complex, less certain, and less efficiently organized to lead directly to the outcomes retrospectively seen as inevitable, than it otherwise appears.

The satisfaction that this experience provided led me to seek out other opportunities to reconstruct investigative pathways, with the long-term aim of saying something general about the nature of the everyday activity within which, I believed, the creative impulses of science lie hidden. Hans Krebs had saved the laboratory notebooks documenting his experimental activity continuously, from the time he entered the laboratory of Otto Warburg in 1926, until beyond his momentous discovery of the citric acid cycle in 1937. The laboratory notebooks of Antoine Lavoisier, long preserved in the Archives of the Academie des Sciences in Paris, I found, contain nearly complete records of the experiments that Lavoisier carried out between the time, in 1773 when he began his sustained study of the processes that "fix or release air," until 1789, when most of the work on which he mounted a "revolution in physics and chemistry," was completed. In the case of Krebs, I reconstructed essentially the entire research pathway that he followed during the auspicious first decade of his scientific career. For Lavoisier, whose experimentation was itself sometimes interrupted by the multitudinous other activities he carried on, I focused more selectively on those strands of his broader program related to the chemistry of life; but I tried, nevertheless, to reconstruct that trail in a manner that did not leave major gaps along the way.

What is the reason for reconstructing such lengthy trails in such continuous fashion, when more selective or more synoptic accounts would allow shorter, more readily marketable books and easier reading? I would not necessarily prescribe my approach for other historians, and admit to disadvantages that I have not been able to overcome; but my persistence in this quest derives from a conviction that the "unbroken" research trail is somehow deeply connected both to the continuity of the broader scientific enterprise, and to the personal identities of the individual scientists who participate in that enterprise. In reconstructing such trails I have found them to be both unpredictable in their course, and rational in the relation of each step to those preceding it. I think that the reason for these patterns is deeply embedded in the unpredictability of life in general, and in the fact that at each step, whether in life in general, or in a scientific career, the individual is figuratively "located" at that point to which all her previous steps have led her. Although, as in ordinary life, a scientist can occasionally make, either for positive or negative motives, a major leap into a different landscape, the natural, and most often, most effective way to proceed into the unknown is to choose at each step a direction leading from where one has already come, toward somewhere else that appears, from that location, to be promising.

II

The papers in this volume amply document that research notebooks are not transparent accounts of the progress of an investigator along the historical trajectory leading to a discovery or other significant conclusion. Even when they have a narrative character, such records are intended merely as notes to refresh the mind of the investigator, either about matters to which to return at some later stage along the route, or about information needed when it comes time to write up an outcome for publication. For the historian, however, these same records can serve as the foundation on which to reconstruct that historical trajectory. What is written down can be likened to footprints along a trail, from which the historian attempts to recreate the activity of the human being who once left those tracks. The nature and difficulty of the reconstructions vary according to the diverse nature of the records kept, and the extent of their survival. The papers presented in this volume show, however, that there are common problems, and similar ways to resolve them, so that our collective experience may provide guidance as well as encouragement to other historians.

Investigators sometimes write down in laboratory notebooks the reasons for which they perform particular experiments. Sometimes they note surprising results and reflections on the implications of such results for what they should do next. Most of them do not make such comments regularly enough, however, to account for the intellectual motives behind each of the operational moves the historian will want to interpret. Systematically they record actions and results, not intentions and responses. Even the meticulous experimental diary that Michael Faraday kept tells, according to Friedrich Steinle, "what he did, not so much what he thought." How can we infer the investigator's original purposes, changes of intention, and revisions of direction from the traces of her actions that survive in the records we may have? Such intentions and the vicissitudes that modify them along the way can normally not be extracted from the accounts of the investigation presented afterward in publications. Typically the published reports exclude experiments that become retrospectively unnecessary to support the argument that the investigator now wishes to maintain. The original reasons for performing even the select group of results chosen to be included are liable to be displaced by the retrospective placement of each experiment within the framework of an argument.

Sometimes we can directly recover earlier stages in the thought of an investigator from diaries, correspondence, personal memoranda, or preliminary drafts of publications; but these rarely are sufficient to fill in all the significant stages in the mental pathway we seek to reconstruct. The published reports that cannot be trusted in isolation to reflect these stages accurately are, however, more useful when interpreted in combination with the research record. The original reasons for performing certain experiments are seldom so unrelated to the retrospective rationale for including them in a paper that the latter cannot in any way illuminate the former. Sometimes the very design of an experiment can reveal that it was performed with an idea in mind that has survived in the published account. Sometimes its design shows, on the other hand, that it must have been performed with a different idea in mind.

For such "triangulations" between the published and unpublished documents it can often become an important question to identify within a laboratory notebook the original record of an experiment that the author has afterward described as playing a key role along the way to a discovery or to some other public claim. A comparison of the location of such an experiment within the investigative pathway traceable through the notebook record and its location within the published account can reveal much about the ways in which the scientist has recast his actual research trail to transform a chronological sequence of experiments into a logical chain of evidence to support the outcome reached. Sometimes the scientist has later related a discovery story that we may wish to compare with the narrative that we can reconstruct from the original record. The descriptions of individual experiments in published form are, however, often themselves sufficiently reconstructed so that the identification of the corresponding experiment in the record becomes problematic enough to test the interpretative skills of the historian. To illustrate such problems, I have chosen four examples, one from each of my studies respectively of the work of Claude Bernard, Hans Krebs, Antoine Lavoisier, and the research pair of Matthew Meselson and Frank Stahl. In each case I was able to locate with some confidence the record of the experiment in question, but the means of identification and its certainty varied according to the nature of the record itself and of the supporting evidence available. In each case the identification of the experiment in the laboratory record had significant implications for the larger interpretation of the pathway to the outcome of the investigation.

Before one can reconstruct an unbroken research trail, or even interpret the meaning of any single experiment within it, it is necessary to decide whether the research notes that have been preserved contain a full, or only a partial record of the experiments performed. Generally there is no direct way to determine that the record is complete. Rather, one comes gradually to the conviction that few, or no, experiments are missing, if one does not encounter in the published papers experimental results that resist identification with any of those in the notebook records; and, if interpreting the steps along the research trail on the assumption that the record is complete does not lead repeatedly to gaps in reasoning or other discontinuities unexplainable by contextual factors. In the case of Hans Krebs, these conditions were met, and his own testimony that his notebooks included all the experiments he had performed, reinforced my confidence in the assumption of completeness. At certain crucial points, however, the absence of original notebooks kept by an assistant required interpolation to infer what might have filled some gaps between recorded experiments. In spite of the fact that Lavoisier's notebook record has been assumed to be incomplete, I seldom found the need to postulate that he had performed experiments that had disappeared from the records. For Claude Bernard, on the other hand, I came to believe that his preserved notebooks included most, but not all of the experiments reported in his publications. These respective conditions clearly impinge on the reasoning that the historian must apply when she attempts to identify in the record of any of these scientists, or of other scientists for whose records similar conditions may be found, an experiment retrospectively described as a key to a discovery he has made.

To illustrate through actual examples, the application of appropriate experimental methods to the study of the phenomena of life, Claude Bernard recounted, in 1865 in his *Introduction to the Study of Experimental Medicine,* how he had made each of his major discoveries. He cited himself, rather than others, he wrote, "for the sole reason that in matters of reasoning and intellectual processes, I am more certain of what I advance than in interpreting what might have passed through the minds of others." To exemplify the importance of making comparative (control) experiments, he described his efforts in 1848 to determine where the sugar ingested by a dog disappeared during its course through the circulation. He was guided by the "prevailing theory" that the source of sugar in animals is always in its food, that animals produce none themselves. Nevertheless, he routinely examined blood taken from a dog fed exclusively meat. "But my astonishment was great," Bernard recalled, "when I confirmed that the blood of an animal that has eaten no sugar equally contains it." The pedagogical lesson Bernard drew from this experience, beyond the necessity for comparative experiments, was that "when the fact that one encounters is in opposition to a reigning theory, one must accept the fact and abandon the theory."[1] Bernard's reward for following this precept had been that it led him to the discovery of the glycogenic function of the liver.

In 1968 Mirko Grmek located an experiment in Bernard's laboratory notebook, dated "August 1848," in which the physiologist had isolated sugar from the portal blood of a dog fed for eight days on raw meat. After describing his procedures in detail, Bernard commented in his notebook: "This experiment is exceedingly strange. From it one can comprehend nothing. Would sugar form in the portal vein, by what organ, by what mechanism?" Quite naturally, Grmek inferred that this was the experiment to which Bernard referred in the story he afterward told, and accepted Bernard's account of its immediate consequence: that "Claude Bernard was immediately led to see that the old theory about the nutritive origin of animal sugar was false, and that the liver produced sugar."[2]

In my interpretation of this experiment I differed somewhat from Grmek. Having found other evidence that Bernard had long been skeptical of theories derived from chemical reasoning alone, as was this "reigning" theory of nutrition, I argued that Bernard's surprise was not focused on the general presence of sugar in the blood of an animal that had eaten no sugar for eight days, but on its specific presence in the portal vein. What he could not understand was how it could be produced just there, where the ingested food had not yet passed through any organ capable of converting some other aliment to sugar.[3] The details of this difference of opinion are not necessary for my present subject, which is the question of how the historian can be certain that the experiment found in the laboratory notebooks is the one to which Bernard later attributed this dramatic shift in his point of view.

That no other experiment during the period leading up to Bernard's earliest publication based on his discovery can be found in his notebook record that fits his description as well as this one does is a suggestive, but not conclusive argument, because there are several other experiments reported in the same publication that cannot be found in that record. The question is, does this experiment fit so closely that it can be identified positively with Bernard's story? The question cannot be answered beyond a doubt, because Bernard gave few experimental details in his story. There is nothing in the experimental procedure itself which conflicts with the story, and the expression of incomprehension that Bernard recorded seems, at first, to confirm his recollected astonishment at the result. Grmek found that confirmation so persuasive that he accepted Bernard's account also of its significance for the "intellectual process" that it evoked in his mind. I, too, found the identification of the experimental procedure and its outcome with the experiment recounted in Bernard's story convincing, but for reasons that would take us too far afield to summarize here, I concluded that he afterward came to attribute to the immediate event a significance that it only some weeks later came to have for him as he wrote the paper in which he presented this and subsequent results to his peers. The case illustrates both the power and the limitations of laboratory notebook records for probing beneath the published surface to recapture the investigative trail that the investigator himself inevitably reconstructs in the retelling.

In the publication in which he described, in April, 1932, the ornithine cycle of urea formation, since regarded as one of his two most important discoveries, and as the opening of a new era in the history of metabolic biochemistry, Hans Krebs wrote that "we find an extraordinary, unexpected effect of ornithine on the formation of urea from ammonia. If one adds d-ornithine to an experimental solution containing ammonia, lactate and living tissue, the synthesis of urea increases powerfully through the addition of the ornithine."[4] In a retrospective account of the discovery 44 years later, Krebs transformed this statement of fact into a concise narrative episode. Under the heading "The first Crucial Finding," he wrote:

> So we measured the rate of urea synthesis under many different conditions, and these included the presence of mixtures of ammonium ions and amino acids. It was in the course of these experiments that we discovered the

exceptionally high rates of urea synthesis when both ornithine and ammonium ions were present. The interpretation of this finding was not at once obvious.[5]

To identify in his laboratory record the experiment in question naturally became one of my central concerns as I began to reconstruct Krebs' investigative pathway from the bound volume covering that period during which it must have been performed. To my initial disappointment, the notebook contained none that even loosely fit this description. Krebs suggested to me that the experiment might have been carried out, under his supervision, by a medical student, Kurt Henseleit, who was working in his laboratory at the time, and who became the co-author of the published paper. By a stroke of very good luck, Krebs visited Henseleit's widow a few months later, and learned that she still had in her possession the laboratory notebook that her husband had kept during the crucial months. When I examined this document, I found an experiment that seemed to fit perfectly the retrospective description. Dated November 15, 1932, it was included among a set of manometer runs made simultaneously on the effects of various substrates on the formation of urea by surviving liver tissue slices. Whereas each of the other substrates tested that day had little or no effect, the combination of ornithine and ammonia more than doubled the rate with ammonia alone.[6]

The experiment was clearly like the one that Krebs later described, but was it the first of its kind, or only a subsequent confirmation? For such a crucial and unexpected finding, it seemed odd that there was no comment whatever in the notebook about its outcome, nothing corresponding to the strong, recorded reaction of Claude Bernard to a surprise of similar magnitude. That was, however, typical of their respective experimental styles. Bernard's notebooks contain narratives of his experiments, in which descriptions of the operations carried out often merge with reflections on their outcomes, or suggestions to himself about what to do next. Krebs recorded his manometer experiments in a standard, spare format that he had learned in the laboratory of Warburg. Mostly they contained columns of manometer readings, with initial conditions such as the composition of the solutions used, and a brief title of the experiment. At the bottom Krebs sometimes listed the results achieved, but with rarely more than the briefest comment on the outcome. Henseleit followed the same format.

To identify this experiment as the "first crucial finding," then, we must fall back on the assumption of completeness: that is, to infer from the fact that no earlier experiments of its kind are recorded in either the notebooks of Krebs or of Henseleit that none were, in fact, performed before this one. The records are dense, documenting experimentation by either Krebs or Henseleit that is nearly continuous through the period. Moreover, my reconstruction itself brought to light no evidence that any experiments later reported in the several publications on the topic, or essential to the argument, were missing. Finally, I could gain stronger reassurance by the fact that I was able to talk with Krebs about such matters. He, himself, believed that the notebook records were complete, and he agreed that the experiment I had located in the Henseleit notebook did constitute that first crucial finding described in his later accounts. Memory is, of course, not faultless, and it is always possible that, more than forty years after the event, he merely identified the experiment because it was the only plausible one we could find in the notebooks. Nevertheless, I think that the case is as strong as is ordinarily possible in historical interpretation.

Although the identification of this crucial experiment in the notebook of his assistant became unproblematic, its location in the chronological sequence – in particular, the nature of the experiments preceding it – raised significant questions concerning Krebs' published account of the reasons for which he performed the experiment. He had stated that he "decided to measure systematically the rate of urea synthesis in the presence of a variety of precursors," with no preconceived ideas or hypothesis about the chemical mechanism.[7] In his conversations with me, he reiterated that he had "tested the effects of all sorts of substances," without specific ideas. The choice of ornithine seemed to me, however, unusual enough to invite further questions. It was not included at the time among the amino acids regarded as constituents of proteins, and was more expensive to procure than many of the more common amino acids that, according to the notebooks, he had not tested at that point. From a remark he made at another time in our conversations about a "vague idea" he had briefly entertained that one of the NH_2 groups of urea might be derived from ammonia, the other directly from an amino acid, I constructed a plausible hypothesis that he might have had in mind that ornithine could play a special role because it contains an NH_2 group in addition to the one common to other amino acids. When I proposed my interpretation, however, he maintained his position that he had tried ornithine, merely "because it was there." I had come to trust the reliability of Krebs's memory too much to override it without compelling documentary evidence, yet there was still no explanation for why this relatively rare substance happened to "be there." I concluded that, even with such rich experimental records and the testimony of the still-living scientist, this critical step in his investigative pathway could not be definitively explained.[8]

One of the most famous experiments in the history of chemistry is the analysis of the process of fermentation published by Antoine Lavoisier in 1789 in his *Traité élémentaire de chimie*. "This operation," he wrote, "is one of the most striking and most extraordinary of all those that chemistry presents us." To determine the composition of the substances that entered into and resulted from fermentation in the form of a balance in which the weight of each element was the same before and after the operation, Lavoisier regarded as the climax of his long effort to establish a new kind of chemistry. "Because the must of grapes gives carbonic acid and alcohol," he asserted, "I can say that *must of grapes* = *carbonic acid* + *alcohol*," thus expressing for the first time a chemical operation in the form of an algebraic equation.[9]

The complicated tables in which Lavoisier listed the weights of each of the elements – carbon, hydrogen, and oxygen – before and after the operation[10] have impressed historians with the apparent accuracy of the balances, but puzzled them about how he could have obtained them with the incorrect values he gave for the compositions of the component substances. Lavoisier gave no public account of the history of this experiment, and it became for me a compelling question whether its history could be reconstructed from his surviving laboratory records.

A difficulty arose immediately. In the second volume of the *Traité,* Lavoisier described and represented in a drawing, an apparatus designed especially for fermentation experiments. Beautifully crafted, the device was intended to allow him to measure the sugar consumed, the alcohol produced, and the carbonic acid given off as a gas.[11] He implied, although he did not explicitly state, that he had used this apparatus to perform the experiment whose results he published in the first volume of the same text. The notebook records include, however, only one failed attempt to carry out such an experiment with this apparatus prior to the publication of the *Traité.* If we were to assume, as some scholars have thought, that the notebook record is incomplete, we might be driven to infer that we cannot recover the investigation underlying the published result. From my reconstruction of other phases of Lavoisier's investigative pathway, however, I had gained confidence that nearly all the experiments he reported in his published memoirs up until that time could be found somewhere in his laboratory records. Sometimes he went back to experiments carried out years before, reworked the results, and used them to support new arguments. It was plausible, therefore, that one of the older fermentation experiments recorded in his notebooks might have served as the foundation for the description found in the *Traité.* The trouble with this supposition was that all of the earlier experiments in his notebooks were carried out with a simpler apparatus, affording no means to collect the carbonic acid, which merely escaped through an open tube. How then would he have attained the measured weight of carbonic acid included in his table of the products of the fermentation?

Quantitative experiments offer the special advantage that, by working through the data and the calculations made on them, one can sometimes determine more definitively than with qualitative ones, whether or not an experiment recorded in a notebook is the same one reported in a publication. In this case it was possible to establish an identity, because Lavoisier left also detailed records of the recalculations he afterward made on the original data. Recapitulating what he had done with the original numbers left little doubt that a specific experiment conducted in 1785 was the source of the description Lavoisier gave of such an experiment in 1789. The identification also allowed me to explain how Lavoisier was able to produce the balance sheet that has mystified historians. The quantity of carbonic acid shown in the table, he had obtained, not by direct measurement, but by subtraction of the quantities of carbon and oxygen contained in the other products from the total quantities of these elements in the starting materials. His overall balance was exact, therefore, not because his measurements exactly balanced, but because he assumed the balance of starting and end products in order to calculate what he could not measure.[12]

The identification of the original notebook record of the published experiment explains how Lavoisier arrived at his famous result, but raises the question whether he misled readers about what he had done. In my detailed treatment of this subject I argue against such a conclusion; but the matter is subtle, and instructive about the complexities of the processes through which scientists convert pieces of ongoing, open-ended investigative pathways, into discrete, bounded public assertions.

In 1958 two young scientists at Caltech, Matthew Meselson and Frank Stahl, published the results of an experiment designed to test predictions made by various "hypotheses for the mechanism of DNA replication ... concerning the distribution among progeny molecules of atoms derived from parental molecules." The most prominent such prediction derived from the double helix proposed by James Watson and Francis Crick in 1953. According to this model, the two strands of the helix would separate, and on each of them a complementary strand would form to produce a progeny molecule. The atoms from the parental molecules would be distributed "semiconservatively" between the two newly formed ones. Because of the difficulty some prominent members of the emerging field of molecular biology felt in imagining how the two strands wound many times around one another in the helix could come apart, however, alternative schemes were suggested. Max Delbrück proposed that the strands break apart at each turn of the helix. Others thought that the parental molecule might remain intact as a template on which daughter double helixes would form. In 1954, Meselson had the idea that he could test these hypotheses by growing organisms in a medium in which they would produce DNA denser than normal, then switching them to a normal medium. At intervals afterward he would extract the DNA from the organisms, and separate the generations according to their densities, by centrifuging them in an appropriate medium. Beginning the investigation together in September, 1956, Meselson and Stahl arrived by the fall of 1957 at a workable method, using the bacterium *E.coli,* and a medium containing ^{15}N to produce the dense DNA. Having found earlier that a solution of CsCl forms a density gradient in which the DNA collects in well-defined bands at the level at which its buoyant density equals that of the medium, they were now able to show that at the end of the first generation, a band forms that is exactly intermediate between those representing fully "heavy" and fully "light" DNA. Even before their paper reporting these results appeared in July 1958,13 colleagues informed of it by correspondence regarded the experiment as a beautiful confirmation of the Watson-Crick hypothesis of DNA replication. The "Meselson-Stahl experiment" soon became a classic in the early history of molecular biology.[13]

Like Bernard and Krebs, Meselson and Stahl themselves provided, for a volume of reminiscences about *Phage and the Origins of Molecular Biology,* which appeared in 1965, a retrospective story about how they had come to perform this landmark experiment. After summarizing their initial plan, briefly mentioning some of their detours along the way, and their decision to abandon their early efforts to incorporate 5-bromouracil substituted DNA into bacteriophage in favor of the system using bacteria and ^{15}N, they wrote that "the second and third experiments along these lines worked beautifully; so we renumbered them 1 and 2 and began to write the paper." A glance at the figure published in their original paper shows that the ultra-violet absorption bands which constitute the immediate data of the experiment are a composite of the results of experiments numbered "1" and "2."

When I began to reconstruct the history of the investigation leading to the Meselson-Stahl experiment, one of my central concerns was to identify, in whatever record of their experimental pathway remained, these two experiments. I was even more interested, however, in learning what I could about the "real" experiment number 1, which, by implication, must not have turned out so beautifully.

The first piece of information I learned about that initial experiment came from Meselson's response to my question, could he remember a moment when he first realized how the experiment would turn out? The question evoked a vivid recollection of a complicated experiment that he had conducted while Stahl was away, in which he had attempted simultaneously to grow some bacteria in heavy medium that he switched to normal medium, and others in normal medium that he switched to heavy medium. When he developed the films, he saw too many bands – there were, he recalled, three, where there should have been only two – but their placement told him immediately that the experiment would work, and he became greatly excited. This recollection, vivid as it was, was not detailed enough to reconstruct the procedure that he had followed, nor did his mistaken memory that it happened on New Years eve enable me to locate it chronologically.

Full laboratory records of the experiments Meselson and Stahl performed in this investigation have never turned up, and it is not even clear if they were ever kept. Fortunately, Meselson had in his possession the loose-leaf log book of the runs on the analytical centrifuge on which these experiments, as well as a long series of prior experiments, were performed. These sheets tell little about the experimental plan, or even the object of each experiment, but do list the contents placed in the centrifuge cells, give the RPM at which the machine was run, and sometimes some further details. In addition, he had kept the original films on which the absorption bands revealing the location of the DNA bands in the centrifuge cells appear: they constitute a complete record of the raw data from the experiments. Because the dates of the runs are also entered in the log, these documents enable one to reconstruct a narrative comparable to those that full laboratory notebook records made possible in the cases previously described. Extended conversations with Meselson and Stahl, in which we examined together the centrifuge logs and critical films, often revealed further information about the purpose and nature of individual experiments.

Here too, identification of the experiments recorded in the log and on the films depended in part on the supposition that these records included all the experiments performed. Both the dating of the runs and Meselson's recollection supported that assumption. The key to the identification of the "second" experiment – that is, the experiment labeled "1" in the published figure – was that I was able to match each of the films resulting from a set of consecutive runs on the centrifuge with one of the films shown in the figure. That identification left in turn only one possible set of runs that could be identified with the true first experiment of its type.

The films from this set of runs did fit with the recollection of the two investigators that the experiment was both flawed and that it revealed what the outcome of a more successful experiment would be. But it proved impossible to reconstruct exactly how the experiment had been planned and performed, because neither Meselson nor I could interpret the codes with which he had labeled the contents of the individual centrifuge cells. Moreover, none of the films showed the three bands that he remembered having seen. In my forthcoming study I discuss in considerable detail the limits in my ability to resolve the questions raised about this experiment, and to reconcile fully Meselson's memory of the event with the surviving documentary evidence.[14]

The four examples I have discussed represent one of the many kinds of interpretative tasks that come up in the reconstruction of investigative pathways from surviving records. In each case the basic problem was to identify an experiment later described by its author or authors as key to the outcome of an investigation with the original record of an experiment in a laboratory notebook or equivalent document. The procedures necessary to establish the identification differed in ways that reflect differences in the nature of the experiments, in the ways in which the investigators recorded them, in the degrees to which the original records have survived, and in the kinds of supplementary evidence that could be brought to bear. In two of the four cases I was able to go over these records with the investigator; in the other two, the passage of time made such an interaction impossible, and I could rely only on my own encounter with documents that had long survived the passing of their authors.

In some studies, such as the measurement of the mechanical equivalent of heat discussed by Otto Sibum in this volume, intense scrutiny of the original record of an experiment identified with the published account can be self-contained and rewarding in itself. In my studies, however, this identification was a step toward the integration of these noteworthy individual experiments into the pathways that led toward them, and from them toward a denouement, within a prolonged investigation. Placed within its temporal context, the experiment may be shown to have originally had a meaning different from that which it took on for the investigator by the time he was ready to write up his conclusions for publication. The experiment becomes a nodal point within an ongoing progression of events that make up the life of the experimentalist. It is that day-by-day encounter of the thoughts and intentions of scientists with the realities of laboratory experience, with operations that sometimes go as expected but often do not, with hopes sometimes realized but sometimes left unfulfilled, that I have tried in my studies to entice back to life from the repositories in which their traces remain.

NOTES

[1] Claude Bernard, *Introduction a l'étitude de la médecine expérimentale* (Paris: Bailliere, 1865).
[2] M. D. Grmek, "First Steps in Claude Bernard's Discovery of the Glycogenic Function of the Liver," *Journal of the History of Biology*, 1 (1968): 141–154, on pp. 148–153.
[3] Frederic Lawrence Holmes, *Claude Bernard and Animal Chemistry* (Cambridge, MA: Harvard University Press, 1974), pp. 422–425. See also, Mirko D. Grmek, *Le legs de Claude Bernard* (Paris: Fayard, 1997), pp.237–239; and Frederic L. Holmes, "The Legacy of Mirko Grmek's Historical Studies of Claude Bernard," *Medical History*, 43 (1999): 114–118.
[4] Hans Adolf Krebs and Kurt Henseleit, "Untersuchungen über die Harnstoffbildung im Tierkörper," *Klinische Wochenschrift* 11 (1932): 757–759, on p. 758.
[5] Hans A. Krebs, "The Discovery of the Ornithine Cycle," in *The Urea Cycle*, ed. Santiago Grisolia, Rafael Bágueria, and Federico Mayor (New York: John Wiley, 1976):. 1–12, on pp. 4–5.
[6] Frederic Lawrence Holmes, *Hans Krebs: the Formation of a Scientific Life, 1900–1933* (New York: University of Oxford Press, 1991), p. 283.
[7] Krebs, "Ornithine Cycle," p. 4.
[8] Holmes, *Krebs*, pp. 284–287.
[9] [Antoine-Laurent] Lavoisier, *Traité élémentaire de chimie* (Paris: Cuchet, 1789), 1: 139–152.
[10] *Ibid.*, 1: 144, 147, 148.
[11] *Ibid.*, 2: 461–464, and Plate X.
[12] Frederic Lawrence Holmes, *Lavoisier and the Chemistry of Life* (Madison: University of Wisconsin Press, 1985), pp. 340–352, 371–384, 388–402.
[13] Matthew Meselson and Franklin W. Stahl, "The Replication of DNA in *Escherichia Coli*," *Proceedings of the National Academy of Sciences* 44 (1958): 671–682.
[14] Frederic Lawrence Holmes, *Meselson, Stahl, and the Replication of DNA: a History of "the Most Beautiful Experiment in Biology"* (New Haven: Yale University Press, 2001).

JED Z. BUCHWALD*

THE SCHOLAR'S SEEING EYE

The proliferation during the 1980s and 1990s of the desktop computer, it was often predicted, would soon lead to the paperless office. Reports, brochures, advertisements, research papers would all appear on-screen; electronic mail would abolish letters, and the ubiquitous pink slips announcing missed telephone calls would evaporate into the ether. Email has perhaps obliterated what little remained of artful scribbling, yet in all other respects the paper world in the year 2001 is more populated than ever before. Indeed, fast printers consume reams of paper at rates undreamt of by typists, much less by toilers with pen and ink. Wastebaskets are fuller, and more frequently filled, because the ephemera generated by the products of such corporate titans as Hewlett-Packard, IBM and Epson are known by everyone to be just that – things that need not be kept because they were born and live on some computer somewhere. As a result, paperwork, though more ubiquitous (and colorful!) than ever has nowadays considerably less longevity than in centuries past. What records are kept tend to be digital, and, increasingly, even laboratory "notebooks" may begin and end their lives in electronic storage units of various kinds. And because it is so simple to overwrite previous remarks, it may very well be impossible to produce a collection one hundred years from now that will be anything like this one. Notebooks of the future may be altogether ephemeral, informative jottings having been neatly eliminated by the delete key or whatever will soon replace it (perhaps a verbal command to the obedient machine). Emails may document scientific interchanges, if they are kept and can be retrieved, but they won't have the odd hand-written note, added quickly just as the letter or postcard was put in the mail, that overwrites a previous remark and reveals a now-forgotten byway of discovery. Perhaps the computer itself can help recover such things – as Grasshoff and May would have it – but the historian steeped in the human reality of the written and drawn object must remain, if not altogether skeptical, then at least hopeful that the scholar's seeing eye will not be replaced by an algorithm.

Perhaps future historians of contemporary science will of necessity rely altogether on an electronic equivalent of the "printed" record – the formal, or quasi-formal, final product – because their subjects work in a world so very different from that of the past. Drafts of the sort that had been common in the pre-

* California Institute of Technology, Pasadena, CA

Frederic L. Holmes, Jürgen Renn and Hans-Jörg Rheinberger (eds.), Reworking the Bench: Research Notebooks in the History of Science, 309–325 © *2003 Kluwer Academic Publishers. Printed in Great Britain*

electronic era scarcely exist today, because they are often just deleted or overwritten with succeeding versions. And yet the production of work has vastly increased, in part because it is now so easy to generate words and images. What is retained may be the result that lies at the end of a long chain of evaporated variants, but there are vastly more results than ever before. Projects are today underway that aim to capture as much as possible of living science and technology just because it is so astonishingly voluminous and comes in so many different forms, many if not most of them electronic in nature. This proliferation of entities and extravagance of form may or may not be a bad thing, depending on one's views of the proper aims of science. But whether good or bad, it is increasingly the way things are; scientific practice has changed, and will continue to change as it always has, and it is today bound – also as it always has been – to the technical world.

When historians in the future look back to the analog world – to the world of paper – it may be difficult for them to spy the residue of human thought and action, because they will be used to seeing only a perfected digital record, which will rarely leave informative traces of roads not taken or beginnings forgotten. Paper retains its human stain; the digital record is spotless. Of course, digital images can be made of paper objects, thereby making the human stain virtually available, as it were. Nearly perfect facsimiles of books and notebooks can be and have been produced – the purpose being to give the buyer a sense of direct contact with the object's original authors. But records originally digital have no marks to copy, and their authors cannot be made present in this way. Several of the papers in the present volume bring their subjects powerfully to life precisely because they have worked with the messy paper world left behind. Renn and Damerow have even tortured Galileo's ink itself to elicit cries of origin. Others, like Steinle and Shapiro, have reconstructed lost chains of practice from the words on remnant paper.

There is little that one would wish to add to the colorfully diverse array of methods that the authors of this collection have brought to bear on the recorded debris of past science, or at least little by way of general argument. No "epistemological" points or "epistemic" claims, nor even any remarks about scientific method, follow here. But it would perhaps be useful briefly to examine three instances in which scraps of paper provide historical evidence that in the one case would not exist at all, and in the others would probably have been eliminated, had electronic methods been available at the time. All three provide critical evidence concerning the working practice of one of the greatest experimenters and theoreticians of 19th century physics, Heinrich Hertz.

JOURNAL ARTICLES

My first two objects are not laboratory notebooks but drafts of articles written for journal publication.[1] Nevertheless, both they and the lab book that provides my third instance have in common informative marks left by direct contact with the author's hands. The first of these two consists of twelve numbered sheets beginning with 7, the preceding six being absent. These sheets were placed in a folder marked 250, attached to the complete manuscript for an article on elasticity and hardness that was printed in 1882 in a journal for engineers. The cover list for the manuscripts (apparently produced in 1959) describes these sheets in the following words: "attached are pages 7–18 which have no connection with this paper and which appear to be an incomplete draft of a paper dealing with electrical co-ordinates." In fact, the attached pages have quite a lot to do with the completed paper and nothing to do with electricity. Written by two different hands, with many crossings-out and additions, these apparently orphaned pages tell a story of the young Heinrich Hertz's struggle for originality in the face of what he saw as pedantic rigor on the part of a powerful man to whom he was beholden.

Early in 1881 Hertz handed a paper on elasticity that he had just completed to the mathematician Leo Kronecker for inclusion in Borchardt's *Journal für die reine und angewandte mathematik*. He was at the time an assistant in his mentor Hermann von Helmholtz's laboratory in Berlin. Having heard nothing for nearly three months, Hertz visited Kronecker, who told him that the paper had been handed to Gustav Kirchhoff, who "had some criticisms of the form of the paper".[2] Hertz knew Kirchhoff personally since he had been one of Hertz's doctoral examiners the previous year, when Kirchhoff had agreed with the others in evaluating Hertz's performance at the (for them) unusually high level of *magna cum laude*.[3] Evidence from Hertz's letters, parts of which were published years later by his daughter Johanna, clearly indicates that Kirchhoff had been rough with the manuscript. On May 4 Hertz wrote his parents that Kronecker

> ... showed me how Kirchhoff had thoroughly annotated the paper and had rewritten three or four pages in another form in the margins. At first I was surprised and even flattered that Kirchhoff had gone over it so thoroughly, but apart from a wrong sign that I had indeed overlooked, his comments seemed only to say the same thing (and by no means better) that was in the paper. In part the points were expressed in a manner peculiar to Kirchhoff which I do not like at all, and which I should be very unwilling to have imposed on me.[4]

Hertz went to see Kirchhoff, who was "very friendly and praised the work greatly", but who "seemed to believe that I had reached the right results so to speak by accident". Annoyed, Hertz was reluctant to make any changes, but decided to "try to deal with it by substituting his formulation for mine ... although I do not believe the paper will be any better for it". He had sent the paper back by the end of June, having by then "found that Prof. Kirchhoff himself had made the main error with which he had reproached me (what I had written was merely not quite clear) and I have demonstrated it to him".[5]

The story might have ended there, in which case we might have concluded that Kirchhoff had at first found something wrong with the paper that Hertz thought

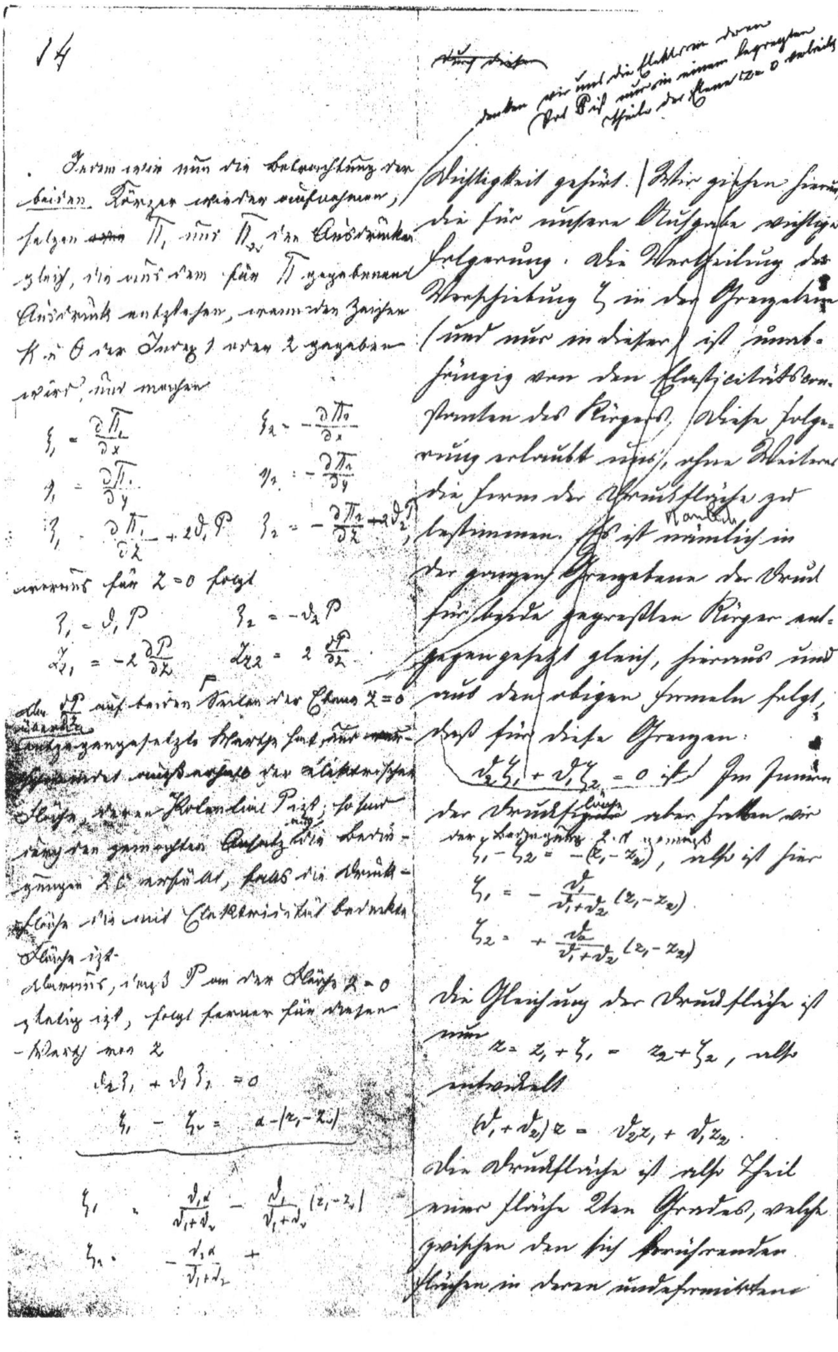

Figure 1. Kirchhoff's emendations.

not to be incorrect, that Hertz nevertheless initially decided to substitute Kirchhoff's formulation for his own, but that a month later he convinced Kirchhoff that he, Kirchhoff, had himself made whatever error it was that he had ascribed to Hertz. One might easily think that in the end Hertz won out, having shown the great man the error of his ways, and that the printed paper was essentially the one that Hertz had originally written, possibly emended a bit here and there in minor ways to keep the peace. In any case it would certainly seem from Hertz's letters that whatever Kirchhoff wanted done was not significant.

Return now to the orphaned twelve pages mentioned above and examine Figure 1, a sample drawn from among them. The hand on the right is Hertz's, but all the writing on the left is in a different hand; other changes occur throughout these twelve pages, including major sections marked for deletion. It is not hard to discover that these twelve pages are in fact remnant parts of the original manuscript that Hertz had given to Kronecker, and that the additional hand is none other than Kirchhoff's. Comparing these pages with the printed version provides important evidence for understanding what Hertz was willing to do at this stage of his career when faced with critical remarks from someone as powerful as Kirchhoff. In addition, we find evidence to show that what he, Hertz, thought to be a niggling point on the part of Kirchhoff in fact reflected the puzzling novelty of Hertz's formulations as seen by others – a novelty that was not apparent to Hertz himself.

One large passage that Kirchhoff crossed out was entirely rewritten by Hertz. Another deletion Hertz did not accept. Most significant of all, the quite considerable changes that Kirchhoff made to Hertz's mathematics were inserted virtually verbatim into the article as finally printed. Specifically, Kirchhoff asked Hertz to put in mathematics that he, Kirchhoff, insisted was missing. Hertz disagreed, though he acquiesced in the end, and in their disagreement we can see resistance to a novel, unfamiliar way of attacking a problem in an area, elasticity, in which Kirchhoff had long been expert.

The nub of Kirchhoff's objection was this. First, Hertz had introduced unfamiliar coordinate systems (of which more below). Second, his analysis had the following objectionable structure. Hertz began with a differential equation and a set of boundary conditions, after which he introduced a (to Kirchhoff, mathematically suspect) potential function, from which he moved directly to expressions for elastic displacements, arguing along the way that all relevant conditions of the problem were thereby satisfied. Kirchhoff rejected almost all of Hertz's original argument here and wrote out several pages of direct analytical demonstration, reaching in the end the very same results that Hertz himself had (just as the annoyed Hertz told his parents). Kirchhoff's changes were used by Hertz verbatim in the printed version, without any mention that they differed considerably from what he had originally submitted.

What can we conclude from this? First, that despite disagreement Hertz was willing – or felt forced – to give in to the eminent Kirchhoff's demands for what Hertz thought to be unnecessary changes. Position, power, influence – and perhaps even respect – were not to be taken lightly in the German academic world, or in any other for that matter. But, even more important, a close look at Kirchhoff's changes shows that he was uncomfortable with the very foundation of Hertz's analysis, because it was entirely unfamiliar to him. The problem that Hertz treated concerned the deformations of a pair of elastic objects that impact one another. To analyze the system Hertz had introduced an unorthodox set of moveable coordinate systems, one for each body, that very nicely encapsulated what he, Hertz, thought to be the physical essence of the problem – namely, the bipartite interaction of bodies in certain states (here of deformation). This way of thinking was characteristic of nearly all of Hertz's work during the 1880s (as discussed in Buchwald, 1994). Kirchhoff however thought in very different terms, since for him the typical elastic problem involved specifying a fixed constraint for a single deformable object, and not an equally-affected pair. Kirchhoff would no doubt have preferred Hertz to begin with an exact description of the problem (perhaps by means of a fixed coordinate system of some sort), and then to incorporate the conditions of the problem by approximation. Hertz had originally jumped right over all that, at once substituting his physical intuition for mathematical rigor.[6]

The orphaned pages, with the incontrovertible evidence of Kirchhoff's emendations, tell a story that even Hertz's letters home (themselves a private paper record) partially hid, and that is missing altogether from the printed article. From them we learn that Hertz's novel way of looking at problems in terms of bipartite interactions, and his mathematical methods (grounded in an easy use of suitable potential functions), were not inherently clear or acceptable even to sympathetic readers, for they elicited a negative reaction from Kirchhoff, who was otherwise happy to support such a promising protegé of Helmholtz. Here, visible in these pages, we glimpse something that, in the modern era, might never have persisted at all, since a latter-day Kirchhoff's changes might have been written into the original electronic document, with, perhaps, no record of the pristine MS having been kept.

THE LABOR OF INSCRIPTION

Our second instance concerns another important, decidedly pragmatic aspect of the paper world: whether written by hand, or even typed, lengthy scientific records were for the most part singular items. Full copies of long records – including notebooks, material prepared for publication, correspondence, and so on – were not often made because so much labor went into the physical act of inscribing them on paper. Though we lack at present a fully-fledged history of copy-production, nevertheless scholars who work with archives are well aware that the objects in their hands were for the most part never produced more than once. Parts of them were often used and re-used to fabricate other objects, but the original remains unique. Yet it is an original that, though unique, was nevertheless often worked on more than once, at more than one time and place, and whose physical structure may accordingly bear witness to the process of fabrication in ways that would today be removed by the computer's perfect erasure.

No paper trail apparently remains of Hertz's laboratory work before he began the investigations in the spring of 1886 that culminated with his discovery of electric waves. Nevertheless, in one sense we do have something like a notebook for work he did in 1882 on hygrometry and evaporation, because both the printed product and the manuscript for it, which we do have, contain many clues to the course of his research. As we will see, even though Hertz found it necessary to make some changes, he retained in the published article itself substantial vestiges of a now-lost original draft that was probably drawn up early on when he thought that his experiments were providing good evidence for a new effect. Hertz almost certainly did so, even if inadvertently, just because he didn't want to rewrite the rather long article.

We need not discuss the details of Hertz's investigations in 1882 to understand that both printed article and manuscript provide information about the course of his experiments.[7] The essential point is this. Hertz was trying to produce evidence for a new effect that, he felt, might occur when a liquid evaporates into a space that is filled only with the liquid and its associated vapor. Specifically, he thought that a liquid might have a maximum possible rate of evaporation, just as it has other fixed properties. At first Hertz thought that he had been able to elicit the effect. The article (and the MS) describe a first set of experiments in the same words – indeed, the article is simply a printed version of the MS – concluding with remarks on possible errors that might vitiate the conclusion. Both pressure and temperature measurements were troublesome, but Hertz remarked at one point that neither type of problem could have been significant: "The first-mentioned error {in pressure} could not have occurred; nor do I believe that the second {in temperature} could".[8] But the article does not end here. Not at all – for Hertz went on to make other experiments which eventually convinced him that the effect could *not* in fact be shown to exist with the apparatus and detection equipment he had available. After discussing these experiments, Hertz remarked – in this same article – that "the positive results obtained by the earlier method had their origin partly, if not entirely, in the errors made in measuring the temperature".[9]

Both statements (the first, that the temperature measures in the initial experiments were reliable, and the second, that they must not have been) appear in the same published article. Both cannot be correct, nor would Hertz likely have held both consciously and simultaneously. We can use this contradiction to understand how Hertz may have constructed his report. The article as printed divides into (but was not so divided by Hertz) five distinct parts: first, an introduction that raises the issue of a new effect and elliptically discusses the results of the experiments; second, initial experiments that immediately revealed flaws in his assumptions concerning the behavior of his apparatus; third, experiments that seemed to reveal the new effect; fourth, more elaborate experiments that contradicted the results of the third set; finally, "theoretical" considerations that justify the ultimately negative results without giving up belief in the effect's existence.

The manuscript (with, we shall momentarily see, one very important exception) has no physical marks of discontinuity between these several parts. And yet we have just seen that statements in parts three and four concerning temperature measurement manifestly contradict one another. It seems reasonable to conclude that the article as it was finally written contains in its second and third parts substantial vestiges of a first version in which Hertz had announced success. After, or perhaps even while, he was writing this first account, he decided that the force of the paper would be considerably strengthened if he could provide something more than evidence that the new effect simply exists – if he could, that is, pin it down quantitatively to something better than "a few degrees": "I could not conceal from myself", he wrote in explanation of what prompted him to undertake additional experiments, "that the results, from the quantitative point of view, were very uncertain".[10] He accordingly built the apparatus of part four to do so, and he then discovered, undoubtedly to his great surprise and consternation, that the two sets of experiments do not agree.

This much can be deduced from the printed article itself. Does the corresponding manuscript provide any physical evidence? Indeed it does, at least in respect to the repercussions of these disappointing (for the young and ambitious Hertz) events. The second paragraph of the manuscript (and of the printed article) informs the reader that Hertz had undertaken experiments "which have only partly achieved their aim", the aim being to arrive "at an experimental decision" concerning a limit to evaporative rates.[11] At that precise point in the manuscript (beginning at the start of the paragraph and ending with "have only partly achieved their aim") Hertz pasted over whatever he had first written (see Figure 2). He did not want to spend considerable time rewriting everything from the beginning, so he decided instead to paste in just an amended paragraph, which of course meant, as we have seen, that the original version, with its rather different tone, remained substantially in place. In writing the amended paragraph, moreover, Hertz likely had difficulty deciding how to phrase his unfortunate, new results (given their essentially negative character), and he tried with considerable effort to put things in as positive a way as possible. This is why his emendation asserts that the experiments "have partly achieved their aim"; but in part four – the altogether new section, deep within the article – Hertz wrote much less positively that "the net result of the experiments is a very modest one".[12] The partial success amounted to this: that, within the temperature and pressure ranges Hertz had examined, an effect which no one had previously thought to exist in fact doesn't seem to. He might with greater accuracy have written right at the beginning that his experiments provide no evidence whatsoever for the existence of an evaporative limit – but then there would have been no reason to publish the article at all! Hertz's specific phrasing in the pasted-on emendation near the article's beginning was, it seems likely, a rather artful attempt to pull something out of the ashes that would justify going into print.

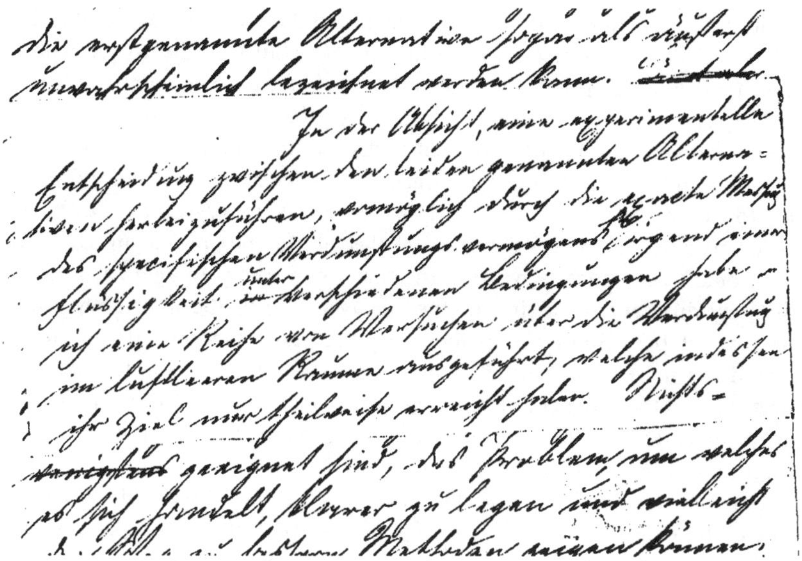

Figure 2. The pasted paragraph.

What can we conclude from this? Two points at least seem pertinent. First, that the printed article is itself an *artefact*, the result of a complex process during which decisions concerning the amount of labor to be devoted to task *x*, as well as specific issues of content, are wrangled over by the artefact's maker, here a single investigator. Limitations of time, energy, and available assistance can have palpable effects on whether something is completely rewritten, or not, and so on what eventually appears in print. Close attention to the logic of a printed argument may reveal these effects and their significance. And this not only because of practical considerations, but because scientists are not all Galileos, masters in the art of rhetoric, who succeed in erasing from the public record most evidence of the tortured paths they had originally followed. Second, in part because of these same practical issues, the manuscript prepared for printing may on occasion betray evidence of changes that themselves bear witness to an otherwise hidden course of events. Laboratory notebooks are not the only paper trails that can be informatively followed.

A LABORATORY NOTEBOOK

Our third instance does involve a laboratory notebook, the only one of Hertz's, strictly speaking, that has been found. It is however the most interesting one of all, for it contains Hertz's notes on the very set of experiments that eventually led him to conclude that electric waves exist.[13] The "notebook" actually consists of loose sheets concerning a sequence of experiments undertaken between September 7 and December 30, 1887.[14] We need not go into great detail in order to see what the notebook (and another private source) provide that would otherwise have been lost.[15] The experiments, as Hertz understood them, consisted in producing interference between a very high frequency electric wave in a wire (itself an effect that he had been the first stably to produce and to detect the year before) and the electric "force" in air. If the latter acted instantaneously everywhere – if, that is, it did not propagate – then the character of the interference picked up by Hertz's detector would track the wire wave; if the force in air did propagate at a rate not unreasonably different from that of the wire wave, then the interference would vary in a calculable fashion from place to place along the wire; if the air force propagated at a rate close to that of the wire wave, then the interference would have pretty much the same character everywhere. Hertz could deduce the speed of the air propagation from the sequence of interferences, combined with the wavelength and frequency of the wire wave.

Hertz's published article carefully lays out the two kinds of interference experiments that he undertook, the difference between the two types consisting essentially in the position of his detector. In the one, the detector was held vertically, in the other horizontally.[16] The vertical detector would respond to both kinds of force – electrostatic (from charge) and electrodynamic (from rapidly-changing current) – the horizontal detector only to the electrodynamic.[17] In the printed article Hertz first describes the vertical-orientation experiments, providing a table of his results, and noting that from the table it would seem that the interference tracks the wire wave. He does not emphasize the apparently inevitable conclusion that the force in air would not accordingly propagate at a finite rate. Instead, he turns immediately to the results of his horizontal-detector experiment. Here the interference decidedly did not track the wire wave, yielding clear evidence for finite speed in air (albeit at a rate different from that of the wire wave, a conclusion that Hertz very much insisted on at the time[18]).

If all we had were the printed article, then it would be difficult to conclude very much about what went on in Hertz's laboratory. We would know little about the specific difficulties of observation that he endured, neither would we know precisely how and when he first became convinced that electrodynamic force in air propagates. Compare the following two figures. Both are the same table, but Figure 3 comes from the published article, Figure 4 from Hertz's laboratory notebook:

The table marks points where the interference between wire wave and electric force in air has a certain character (denoted by +, –, or 0). The marks in the printed table are firmly unambiguous, but some marks, and four oblique lines, in the notebook tell a somewhat different story. In the notebook we see telltale signs of uncertainty, so that when Hertz tried to draw lines linking together sets of "0" marks he wavered, producing differences that were in fact quite significant. And several of the marks were themselves changed or overwritten, indicating some doubt concerning their true character. Yet there is no hint of *observational*

	0		1		2		3		4		5		6		7		8
100	+	+	O	—	—	—	—	—	O	O	O	O	O	+	+	+	+
150	+	O	—	—	—	—	O	O	O	O	O	+	+	+	+	+	O
200	O	—	—	—	—	—	O	+	+	+	+	+	O	O	O	O	O
250	O	—	—	—	—	O	O	+	+	+	+	O	O	O	O	O	O
300	—	—	—	—	O	+	+	+	+	+	O	O	O	O	—	—	—
350	—	—	O	+	+	+	+	+	+	O	O	O	—	—	—	—	—
400	—	—	O	+	+	+	+	O	O	O	O	—	—	—	—	—	—
450	—	O	+	+	+	+	+	O	O	O	—	—	—	—	—	—	O
500	—	O	+	+	+	+	O	—	—	—	—	—	O	O	O	O	+
550	O	+	+	+	+	O	O	—	—	—	—	—	O	O	O	O	+
600	+	+	+	+	O	O	—	—	—	—	—	O	O	+	+	+	+

Figure 3. From the published article.

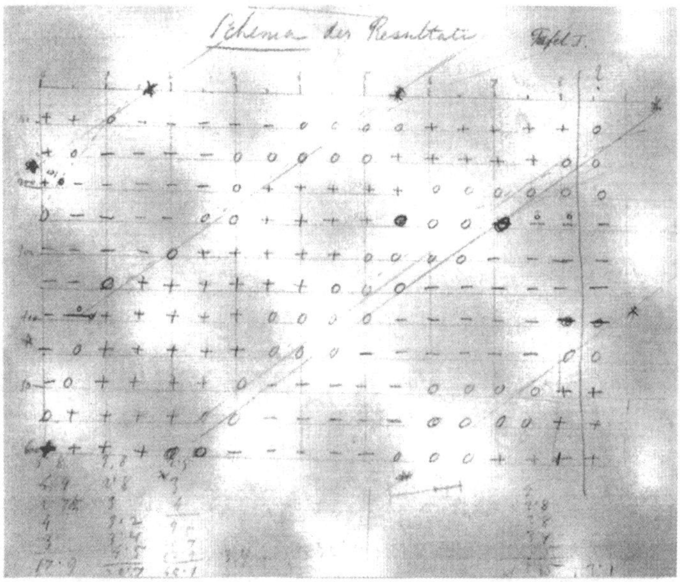

Figure 4. From the laboratory notebook.

uncertainty in the accompanying printed text, where we find Hertz treating the marks as though they were quite accurately placed. In print he offered reasons for not concluding from these apparently unambiguous marks what they seem to imply – namely, that the interference tracks the wire wave, with the result that the force in air does not propagate at all. The primary explanation Hertz offered his readers relied upon a possibility[19] that could only have significance in the event that other observations provided positive evidence for a finite rate of propagation of electrodynamic force. Yet when Hertz produced this same table in his notebook he had no such evidence! Quite the contrary: until sometime on December 24

Hertz was convinced that propagation in air does *not* occur, just as the table, even with its occasionally-uncertain marks and lines, seems to indicate.

The laboratory notebook alone does not directly tell us that Hertz initially thought he had produced evidence against propagation, because there are no remarks near his mute plus, minus and zero signs. We can however be quite certain that he had drawn this conclusion, because on December 23 he says so in a letter to his parents, where we read "...what is the unexpected and to me displeasing result of my endeavors? The velocity is not that of light, but certainly much greater, perhaps infinitely great, at all events not measurable".[20] The printed article continues with results from experiments in which the detector is held horizontal. And here we are told of success: "This experiment, repeated often and never with an ambiguous result, is sufficient to prove the finite rate of propagation of the electromagnetic action".[21]

A careful reading of the published article alone might lead someone attuned to the sequence of Hertz's research to suspect that he had initially been troubled by negative results but, certain that propagation must occur, had forged ahead with other, successful experiments, having known all along just why the first results might be misleading. The notebook, combined with the letter to his parents, tells an altogether different story. We can see from the notebook that experiments with a vertical detector were not untroubled (those wavering lines and drawn-over signs), but that Hertz was unable to manipulate the results into a form that might yield propagation. "There is", he accordingly wrote his parents, "no arguing with nature" because "the experiments seem all too clear to me". At this point – some time on Dec. 23 – Hertz had not as yet thought to use a possible distinction between the speeds of propagation of electrostatic and electrodynamic force, which he relied on in the published article, to explain away the negative results.

But Hertz was ever the careful experimenter, particularly when he was going to present results (here the absence of propagation) that were bound to generate much discussion. In Hertz's work on evaporation he went ahead with experiments that in the end disappointingly contradicted his first outcome. Here too Hertz went ahead with experiments to test his – here displeasing, though important – negative results in a different way. Which was why he carried out experiments with a horizontal detector. Now, we may ask, did Hertz, in carrying out these additional experiments, do so because he had good reason to suspect that his first results (with the vertical detector) might have been misleading, due to a theoretically-permissible difference in propagatory speed between electrostatic and electrodyamic force? One might easily draw that conclusion, given only the printed article.

The notebook makes no mention of the possibility before the additional experiments. And when we examine Hertz's immediately preceding work, in which he had used his oscillator and resonator to detect the inductive action of polarization currents in dielectrics, we find – not theory, but pragmatics. Hertz had indeed previously distinguished between static and dynamic force, but not because of issues concerning propagation. Rather, the difference was important for understanding how the resonator was affected by the oscillator, which was not merely a piece of straight wire with changing current – the oscillator was terminated by large metal plates, with great capacitances, on which huge charges accumulated, exerting static forces that powerfully affected the resonator. Because static and dynamic induction have different strengths, and because of the specific geometry of Hertz's apparatus, it was pragmatically important to distinguish between their effects. Hertz had carefully done so in his earlier experiments, specifying configurations that could be used to probe how the two kinds of force combine to activate sparking in his detector.[22] We may reasonably conclude, given the preponderance of the evidence, that Hertz pushed ahead with additional experiments when trying to detect propagation just because he knew well the vagaries of his apparatus – and *not* because he initially suspected a difference in speed between the static and dynamic force. There most certainly was a difference between the activities of the forces, but it had rather to do with the pragmatics of experimental design than with issues of deep theory.

What then happened, on this understanding of Hertz's state of mind, when he turned his detector into the horizontal? The published article certainly tells us that this was the very configuration that gave Hertz evidence for a finite speed of propagation for dynamic force in air, provided that he distinguished between it and static force. But when did it occur, and how rapidly did Hertz become convinced? A second letter to his parents tells us precisely. On December 26 he wrote that he "received a great gift in my work on the night before Christmas, at least that's how it seems; nature does seem to be somewhat more kindly disposed towards me than she previously appeared to be. We shall see!".[23] The notebook does not place the event so precisely as this, because there are no dates between Dec. 23 and Dec. 26. But the notebook provides something even better than an exact moment: it permits us to be virtually present with Hertz when he spied a positive result.

Figure 5 reproduces a part of the notebook between Dec. 23 and Dec. 26. Hertz's procedure for finding the character of the interference between the wire wave and the force in air required, in these experiments, flipping the wire to the opposite side of the detector after an initial observation. If the force in air did not propagate, then the effect of flipping should be completely different between points on the wire separated by half a (wire) wavelength: if flipping increases sparking at point x, then it should decrease sparking a half wavelength past x, because only the induction due to the wire would change character. Experiment 51 begins "The same except at 3 meters from the null-point." That is, we are now 3 meters down the wire – which is just about half a wavelength. Something different should happen at this point if the force in air travels at infinite speed, because the wire wave reverses its effect, whereas the force in air does not. Hertz continues: "Passing the wave by the side away from P strengthens sparking", which is just what happens near the origin – and just what *shouldn't* occur if the force in air does not propagate. It's worth quoting Hertz's original at this point, for he continues

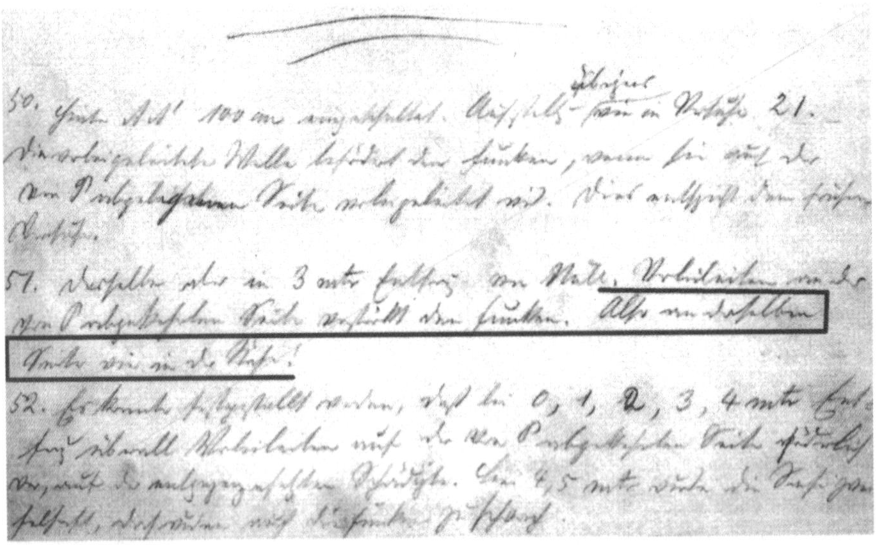

Figure 5. The moment of discovery (outline added).

emphatically: "Also von derselben Seite wie in der Nähe!" ("Therefore along the same side as at closer distances!").

The final sentence in 51, with its telling exclamation point, puts us in the room with Hertz at the very moment when he – unexpectedly! – obtained positive evidence for a finite speed of propagation in air. Hertz followed the result with more measurements, confirming it at many distances. It was almost certainly near this moment of discovery that he began to consider his earlier results as having provided evidence that static force does not propagate, just because the new experiments indicate that dynamic force does. As Hertz now saw it, the infinitely-fast static force simply overwhelmed the weaker dynamic force in the experiments with a vertical detector. Here, then, we see that on Christmas Eve, 1897 Heinrich Hertz had made two, not one discoveries. The second – the near-infinite speed of static force, which we no longer consider to be meaningful, and which he soon abandoned[24] – was conceptually possible just because Helmholtz's theory, which Hertz knew well, allowed for the distinction. But Helmholtz's permissible distinctions did not guide Hertz's experiments – not at all, for they were used as a resource on which Hertz drew only *after* he needed it in order to explain what went wrong with his first measurements.

Subsequent events, not least Hertz's own binding of his oscillator to equations for field theory that he himself produced, have obscured these early conclusions, though the distinction between static and dynamic force was clear enough in Hertz's own publications at the time (and therefore required him later to explain why it was not so compelling after all).[25] But without the laboratory notebook, and without Hertz's letters to his parents, we would not have seen the precise sequence of events. Nor would we have understood the pragmatic character of Hertz's playing with configurations in which static and dynamic force worked with and against one another. We might instead have thought that Hertz had always had in mind the possibility of using a distinction between these two kinds of force, in which case we would – reading only his published account – have concluded that he moved unhesitatingly through a series of experiments designed to separate the forces *theoretically* and not just *pragmatically*.

TECHNOLOGIES OF INSCRIPTION

Technologies of inscription – the means whereby informative marks are made, modified, preserved, retrieved and circulated – have been central elements, perhaps the most central of all, in the broad development of culture since their first production in Egypt and Mesopotamia during the late fourth millennium BCE. Cuneiform wedges were adapted to the material realities of pointed stick and clay, hieroglyphs and alphabets to ink tablet, papyrus and, eventually, vellum, parchment and paper (what one might call flexible media). Clay tablets, once baked, could not be modified, whereas flexible media could be overwritten and even scraped off – producing in the latter case the invaluable palimpsest, whose underscript may hold otherwise lost texts and even images. Since the 1980s electronic means of inscription have increasingly obliterated the ordinary use of paper, pen and typewriter for common tasks. Personal notes for culturally important events are still occasionally written by hand, albeit with something of the same spirit that millennia ago led Egyptian priests to emphasize rarity and importance by writing in the arcane hieratic rather than in the more accessible demotic script.

Stored inscriptions in contemporary electronic machines have a great deal more in common with Mesopotamian clay tablets than with their immediate predecessors, the typed or written page. Barring physical annihilation, baked clay tables are as nearly permanent as any human product can be: they do not decay, neither do they crumble with time. Paper and papyrus in even moderately damp climates do. Which is why there are vast records of Mesopotamian affairs, and why one of the most significant collections of documents for Egypt, found at Amarna (ancient Akhetaten) and dating of course to the reign of Akhnaten (Amenophis IV, 1366 to 1354 BCE at Akhetaten), also consists of clay tablets inscribed with cuneiform in Akkadian, the *lingua franca* of the period. Although electronic documents reside on less permanent materials (as of *circa* 2000), nevertheless they are so simple to copy that they proliferate and may thereby endure as one digital format succeeds to the next and incorporates its ancestor by electronic translation. Electronic materials will show little trace of the human work that went into producing them, unless authors make concerted efforts to do what scarcely anyone does – namely, to store away distinct stages of composition.

Usually a document is simply overwritten by its edited replacement, leaving only the latest and eventually the final – in effect, the "published" – form.

There is much to be said for careful examination of such a final product, whether clay tablet, book or journal article, even though scholars have a tendency to deprecate the informative character of book and article, and that for at least two reasons. First, the final product, destined as it is for public examination, has inevitably been forged with that very end in mind. If its author is a master of rhetorical technique, then one can reasonably conclude that the final work will probably hide many of the author's deepest aims and motivations. But, as we saw above, scientific authors are not inevitably endowed with rhetorical genius; neither did they in years past always have the assistance to produce entirely new versions of their work as they revised. Many clues may accordingly remain even in the finished product. The second reason for scholarly deprecation of works the author intended to be made public pertains rather to the social world of scholarship than to the world of the scholar's subject. Unpublished material requires – or at least has required – considerable physical effort to find, to organize, and even to read. The scholar who undertakes the task has an understandable pride in the labor expended, and perhaps also in being, as it were, an initiate into the *arcana* of manuscript, palimpsest, codex, or parchment. No doubt pride in mastery of the *arcane* will diminish as the tide of originally-private materials swells across the electronic world, making them equally available to any armchair scholar with access to the Internet.

There are limits to the interpretation of less-than-perfect rhetoric, and only materials not intended for the public eye when first produced can carry us further. Some of these – like the drafts for Hertz's articles – are directly connected to public materials, but they can tell us much that the printed document conceals. The authorial hand physically marks things in ways that the electronic world, as we know it today, altogether erases. Of course, the most informative materials of all are notes, remarks, notebooks and other such things that were never intended for publication at all, for they can tell us much about their author's quotidian practice, both physical and conceptual. These things show every sign of vanishing altogether in our digital world, and this will inevitably bring into being a new kind of history. We may never know just when and how, say, *Dr. X* decided to measure more closely that little bump in a spectrum which opened a path to a new understanding of galactic evolution. Whereas we do know just when, how, and even why Heinrich Hertz detected electric waves.

NOTES

[1] These, as well as the manuscripts for many of Hertz's other published articles through 1889, are today held in London's Science Museum, to which they were loaned by Hertz's daughter, Dr. Mathilde Hertz in 1938 and bequeathed in her will in 1975.

[2] M. Hertz, *Heinrich Hertz: Memoirs, Letters, Diaries* (San Francisco: San Francisco Press, 1977), p. 147. Full details of Hertz's work on this subject can be found in J. Z. Buchwald, *The Creation of*

Scientific Effects: Heinrich Hertz and Electric Waves (Chicago: The University of Chicago Press, 1994b), chap. 8.

[3] The other examiners were Helmholtz himself, Ernst Eduard Kummer in mathematics, and Eduard Zeller in philosophy. Zeller was a famed scholar of Greek philosophy who retired in 1895 at the age of 81. Hertz remarked in a letter to his parents that doctorates awarded *magna cum laude* at Berlin were rare, particularly when Helmholtz and Kirchhoff were involved.

[4] Hertz, *Memoirs, Letters,* p. 147.

[5] Hertz, *Memoirs, Letters,* p. 149.

[6] This claim is not speculative since the paragraph that Hertz rewrote at Kirchhoff's behest had originally treated the new coordinate systems as conveniences for calculation without explaining anything about them. The new one was much more careful.

[7] See Buchwald, *Creation*, chap. 9 for details.

[8] H. Hertz, *Miscellaneous Papers*, translated by D. E. Jones (London and New York: Macmillan and Co., 1896). p. 192.

[9] Hertz, *Papers,* p. 194.

[10] Hertz, *Papers*, p. 192.

[11] Hertz, *Papers*, p. 187.

[12] Hertz, *Papers*, p. 194.

[13] A transcription and facsimile of the notebook, with discussion, are printed in H. G. Hertz and M. G. Doncel, "Heinrich Hertz's laboratory notes of 1887," *Archive for History of Exact Sciences* 49 (1995): 197–270

[14] They include as well notes concerning the photoelectric effect from the previous May and July.

[15] For discussion of the experiments see [Buchwald, 1994b], chap. 16.

[16] Hertz's article was printed in two journals, the most widely-read account being "Ueber die Ausbreitungsgeschwindigkeit der elektrodynamischen Wirkungen," *Annalen der Physik und Chemie* 34 (1888): 551&ff. It was reprinted in both German editions of his *Untersuchungen über die Ausbreitung der Elektrischen Kraft* (Leipzig: Johann Ambrosius Barth, 1894). See Hertz, *Electric Waves* for the English translation.

[17] Hertz at this stage distinguished between electrostatic (due solely to charge) and electrodynamic (due solely to changing currents) force, a distinction that was permitted by Helmholtz's system, though not by field theory, according to which all current-generating actions derived entirely from a unitary electric field.

[18] See Buchwald, *Creation*, chap. 17.5.

[19] Namely (see below), that electrostatic and electrodynamic force might propagate at different rates.

[20] Hertz, *Memoirs, Letters*, p. 241.

[21] H. Hertz, *Electric Waves, Being Researches on the Propagation of Electric Action with Finite Velocity Through Space* (New York: Dover Publications, 1962), p. 119.

[22] See Buchwald, *Creation*, chap. 15 for discussion of Hertz's effort to understand how the resonator behaves.

[23] Hertz, *Memoirs, Letters*, p. 241.

[24] The difference between static and dynamic force, which is built into Helmholtz's original system (as well as Wilhelm Weber's for that matter), is not meaningful in field theory since, in it, an electric field is unitary. Field theory does however distinguish pragmatically between the near-field (close to the oscillator, where quasi-static conditions dominate, making the oscillator appear to behave like a statically-charged dipole) and the far-field (at a distance, where time- and distance-dependence are clearly evident). Hertz realized this when, the following summer, he applied field theory directly to his apparatus. Indeed, he created the distinction.

[25] See Buchwald, "How Hertz fabricated Helmholtzian forces in his Karlsruhe laboratory or why he did not discover electric waves in 1887," in *Universalgenie Helmholtz*, ed. L. Krüger (Berlin: Akademie Verlag, 1994), pp. 43–65.